Autorenkollektiv

Synthetic Organic Chemistry

Steric and Electronic Substituent Effects on the Carbon-Carbon Bond

Ch. Rüchardt and H.-D. Beckhaus

Selective Hydroboration and Synthetic Utility of Organoboranes Thus Obtained

A. Suzuki and R. S. Dhillon

Synthesis of Ynamines

J. Collard-Motte and Z. Janousek

Electrochemically Reduced Photoreversible Products of Pyrimidine and Purine Analogues

B. Czochralska, M. Wrona and D. Shugar

High-Pressure Synthesis of Cryptands and Complexing Behaviour of Chiral Cryptands

J. Jurczak and M. Pietraszkiewicz

AKADEMIE-VERLAG BERLIN

Autorenkollektiv

Synthetic Organic Chemistry

Autorenkollektiv

Synthetic Organic Chemistry

With 15 Figures and 23 Tables

AKADEMIE-VERLAG BERLIN
1986

Die Originalausgabe erscheint im Springer-Verlag Berlin · Heidelberg · New York · Tokyo
in der Schriftenreihe „Topics in Current Chemistry",
Volume 130

Vertrieb ausschließlich für die DDR und die sozialistischen Länder

ISBN 3-540-15810-3 Springer-Verlag Berlin · Heidelberg · New York · Tokyo
ISBN 0-387-15810-3 Springer-Verlag New York · Heidelberg · Berlin · Tokyo

ISBN 3-05-500195-8

Erschienen im Akademie-Verlag Berlin, DDR-1086 Berlin, Leipziger Straße 3—4
Lizenznummer: 202 · 100/509/86
Printed in the German Democratic Republic
Gesamtherstellung: VEB Druckerei „Thomas Müntzer", 5820 Bad Langensalza
Umschlaggestaltung: Karl Salzbrunn
LSV 1275
Bestellnummer: 763 602 2 (3072/130)
13800

Table of Contents

Steric and Electronic Substituent Effects on the Carbon-Carbon Bond

Christoph Rüchardt and Hans-Dieter Beckhaus

Institut für Organische Chemie und Biochemie
der Universität Freiburg, D-7800 Freiburg, FRG

Table of Contents

The factors influencing dissociation energies of C—C bonds have been investigated by thermochemical (ΔH_c°, ΔH_v, ΔH_{sub}) and kinetic methods and by molecular mechanics (MM2 force field). Quantitative analysis of the influence of strain H_s in *1* and in *2* and of the resonance energies H_r of the substituents X in *2* (X = C_6H_5, CN, OCH_3, COR, $COOCH_3$) has been successfully achieved.

$$X - CR^1R^2 - CR^1R^2 - X \quad \mathit{1} \rightarrow 2\,X - \dot{C}R^1R^2 \quad \mathit{2}$$

Enthalpy and entropy effects and their interrelationships are discussed. Resonance stabilization of radicals by more than one substituent (including capto-dative substitution) is frequently additive and in no example higher than this. A linear correlation is found between the central C—C-bond lengths in *1* and the strain enthalpies H_s, quite independent of the substituents X and their resonance contribution H_r.

1 Introduction

The Carbon—Carbon Bond is the backbone of Organic Chemistry. For a covalent bond between two like atoms its strength is exceptional, a phenomenon which is pointed out in most beginners' text books of Organic Chemistry. Due to this exceptional bond energy and due to their chemical inertness C—C bonds in carbon structures have been ideally suited for the storage of solar energy of past times in primary fossil fuels as well as in renewable feedstocks such as cellulose, starch and fat.

From information related to linear saturated hydrocarbon structures bond strengths of about 80 kcal · mol^{-1}, bond lengths of about 154 pm and bond angles of about 109° are quoted as standard reference values for C—C bonds.

However, with the exception of small ring compounds, much too little is known about the range of these dimensions in different carbon structures and even less about the factors responsible for observed variations. The dimension of this question is recognized immediately when the C—C bond strength in ethane (88.2 kcal · mol^{-1})[1] is compared with that of the central bond in the Gomberg dimer *1* (12 kcal · mol^{-1})[2,3].

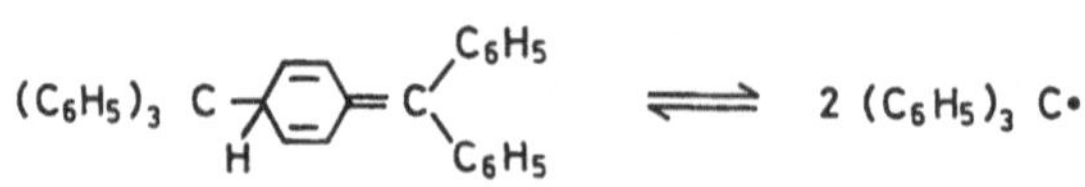

1 *2*

This difference of 76 kcal · mol^{-1} in bond strength is translated into a rate factor of $1:10^{30}$ (at 300 °C) for the thermal cleavage of ethane into methyl radicals (at ~700 °C) or *1* into trityl radicals *2*[2] (at ~25 °C). This is an incredible factor[4] for such a simple and basic phenomenon as the substituent effect on the C—C bond strength. Therefore it is even more astonishing that the traditional hyphen between two C's is considered to be a satisfactory symbol for this bond.

The enthalpy required for the thermal cleavage of a C—C bond into two carbon radicals is the defining reaction for the bond dissociation enthalpy H_D [1, 5, 6]. The reaction coordinate of this process on the enthalpy scale (Fig. 1) generally has no separate transition state (enthalpy maximum), because it is known that the rate of the back-reaction, the dimerization of simple alkyl radicals, is a non-activated process controlled by diffusion (see later).

Therefore the bond enthalpy H_D and the activation enthalpy $\Delta H^{\neq}$ for the dissociation process are generally identical[6a]. Consequently, bond enthalpies H_D can be deduced from the temperature dependence of the rate constants k of thermal bond cleavage reactions with the aid of the Eyring equation

$$\ln k = \ln \frac{k_B T}{h} - \frac{\Delta H^{\neq}}{RT} + \frac{\Delta S^{\neq}}{R}$$

The question as to which factors determine the dramatic substituent effect on the C—C bond strength mentioned above has been discussed since Gomberg's days. A particularly important contribution was made by Karl Ziegler in his pioneering work of the fourties[7]. In this early demonstration of the power of kinetics for the investi-

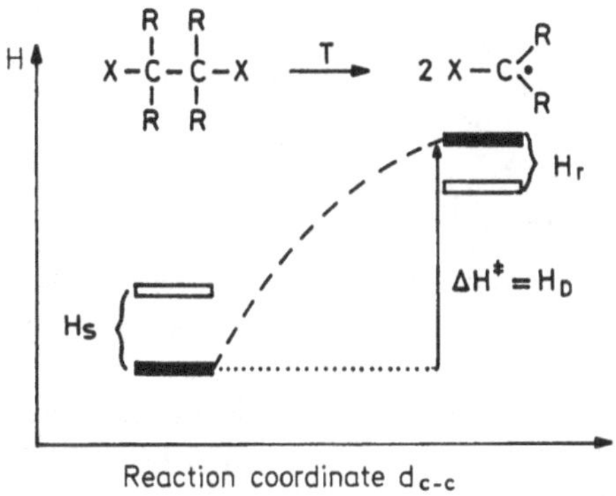

Fig. 1. Reaction coordinate of the bond dissociation process with non activated radical recombination; the influence of ground state strain H_s and of resonance stabilization H_r of the radical centers by the substituent X on the bond strength H_D is qualitatively indicated

gation of reaction mechanisms it was shown that the rate of C—C bond dissociation (see Fig. 1) is increased by bulky substituents R and by X-groups such as phenyl which can conjugatively stabilize the radical being generated. In qualitative terms Ziegler recognized ground state strains H_s [8] and resonance stabilization of the carbon radicals H_r as the two factors contributing to the modification of C—C bond strengths.

Work towards a quantitative analysis of these effects was initiated in our laboratory in the early seventies [9] when an unusually large ring size effects on the thermal cleavage reaction of 1,1′-diphenylbicycloalkyls *3* into 1-phenylcycloalkyl radicals *4* was observed; this was inexplicable by i-strain effects alone and pointed to the importance of f-strain in this phenomenon [10].

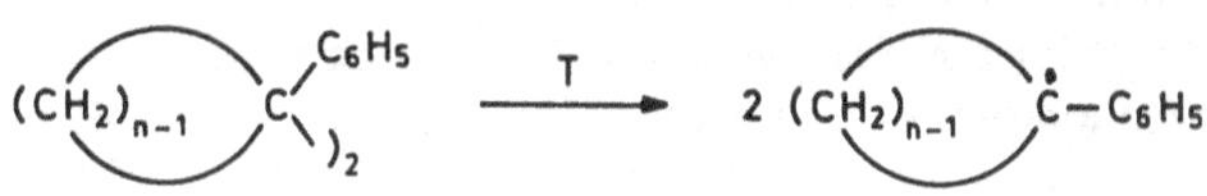

3 *4*

As a consequence, the aim of our own work was to attempt a quantitative analysis of the relationships between thermal stability, ground state strain and resonance effects of substituents and of the influence of strain on structure [9b]. The results were expected to be valuable not only as fundamental knowledge but were expected to contribute to fields of applied chemistry such as carbon initiators [11], coal pyrolysis [12] or thermal stabilities of polymers [13].

Models

With this aim in mind we began the investigation of the thermal stabilities, the thermochemistry and the structures of several series of compounds *5*. The substituents X and their characteristic resonance effects on the bond strengths are typical for each series (transition state effect). Within each series the bulk of the alkyl side chains R is changed in order to analyse the steric ground state effect.

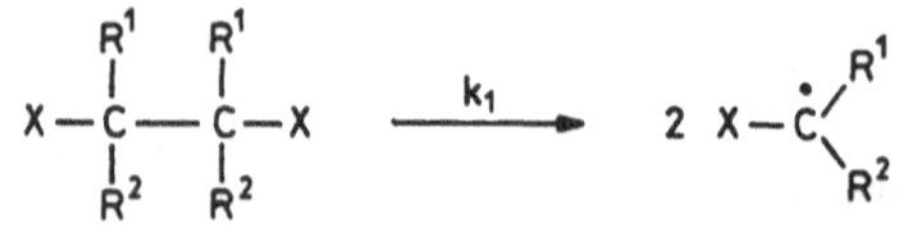

5 6

Additionally, results for some unsymmetrical compounds and the question of radical stabilization by more than one substituent will be discussed. Finally, some of the consequences of substitution and strain on structural parameters will be briefly addressed [9b)]. The syntheses of all compounds referred to in this article and the determination of their structures and their configurations have been published or will be reported elsewhere [9b)]. All compounds were obtained on at least a 100 mg scale and their purity was confirmed by standard analytical procedures.

2 Methods

The experimental approach to uncover relationships between bond strength, steric and resonance effects of substitutents and structure required the application of the broad methodology of Physical Organic Chemistry which can be outlined only briefly in this article [9a)]. The thermolysis reactions were conducted in ampules with strict exclusion of oxygen but with the addition of good radical scavengers, such as mercaptans, tetralin or mesitylene [14–19)]. At least 80–90% of the products were derived from the radicals, e.g. *6* generated by homolysis of the weakest C—C bond in *5*. Qualitative and quantitative analyses were performed by gc, capillary gc and coupled gc-MS-experiments. The products, generally obtained in high yield under proper conditions [14–18)], are convincing evidence that induced decomposition is not a serious disturbing factor in the systems investigated.

The high selectivity for the cleavage of one C—C bond in preference to all others, which was observed almost without exception [20)], is understandable in view of the enormous overall spread of the rate data mentioned above.

The kinetics of these pyrolysis reactions were followed by several complementary methods under conditions as close to the product studies as possible. The most frequently-used ampule technique [14–17)] with gc analysis of *5* and the scavenger technique, with chloranil or Koelsch radical as scavenger [18)], for very labile compounds *5* were complemented by the DSC method, in which the heat flow under conditions of linear temperature increase is analysed. It proved to be a particularly convenient and reliable technique [18, 21)]. Rates were followed over a temperature span of at least 40 °C with temperature control of ± 0.1–0.2 °C. All rate data and activation parameters were subjected to a thorough statistical analysis including statistical weights of errors. The maximum statistical errors in k were $\pm 3\%$, in $\Delta H^{\neq} \leqq 1$ kcal · mol^{-1} in $\Delta S^{\neq} \leqq 3$ e.u. and in $\Delta G^{\neq}$ (at the temperature of measurement) $\leqq 0.5$ kcal · mol^{-1}.

The question of cage recombination [22)] merits special consideration in these systems. The most sensitive way to check for it was to test if meso/DL equilibrations occurred in the course of the thermolysis reaction of a pure diastereomer [18, 20)]. Additional evidence for the unimportance of cage dimerizations are the high disproportionation-recombination ratios found for most of the radicals involved [9, 23)] and the high fluidities[23] of the medium at the high temperatures which were required for most reactions[9].

In a few cases the existence of intermediate radicals and their equilibrium constants with the dimers were established by esr. In a few instances rates of radical recombinations were measured by product-resolved kinetic esr experiments[24].

Thermochemical data were required for the estimation of ground state strain. Heats of formation (± 0.5 kcal $\cdot$ mol^{-1}) were obtained by the experimental determination of heats of combustion[25-27] using either a stirred liquid calorimeter[25] or an aneroid microcalorimeter[26]; heats of fusion and heat capacities were measured by differential scanning calorimetry (DSC), heats of vaporization[21, 25, 27] by several transport methods, or they were calculated from increments[28]. For the definition of the strain enthalpies Schleyer's single conformation increments[29] were used and complemented by increments for other groups containing phenyl[30] and cyano substituents.

Experimental thermochemical results were mainly required to extend the parametrization of the current force fields to highly strained compounds. Heats of formation calculated with Allinger's MM2 force field for alkanes[32] and its extension to alkylbenzenes[30] proved to be in by far the best agreement with the experimental results[27]. A few examples which demonstrate the quality of this agreement are shown in Table 1.

Table 1. A comparison of experimental and calculated strain enthalpies, [kcal $\cdot$ mol^{-1}][a] of some highly crowed hydrocarbons

	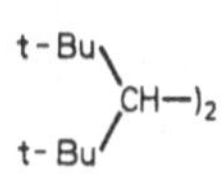	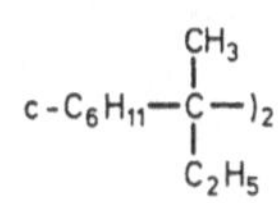 (meso)	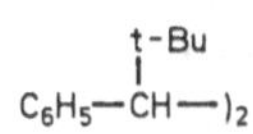 (meso)	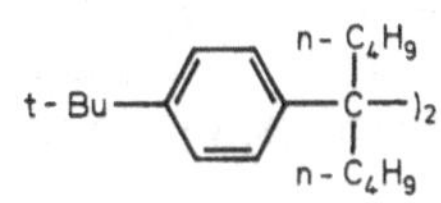
	7	*8*	*9*	*10*
H_s (exp.)	66.3 ± 0.7[b]	35.0 ± 0.7[b]	17.8[c]	22.4[c, d]
H_s (MM2)[e]	57.7	34.8	15.1	22.3
Ref.	33)	25)	15)	27)

[a] $H_s = \Delta H_f^\circ(g) - \Delta H_f^N$; the strain free reference value ΔH_f^N is defined by group increments[29, 30].
[b] heat of sublimation determined experimentally.
[c] heat of vaporization calculated from increments[28].
[d] corrected for strain introduced by the p-t-butyl substituent.
[e] calculation of $\Delta H_f^\circ(g)$ using the MM2 force field, see Ref.[32c]

Only in the most extremely strained compound, tetra-t-butyl-ethane *7*[33], is an appreciable difference found between experiment and calculation. Even this can probably be overcome by a slight increase in the force constants for the van der Waals repulsion in the MM2 force field[34].

The quality of the MM2 force field was in addition tested for its ability to predict structural parameters. Comparison with X-ray data for many compounds[9, 31, 35], a few of which are shown in the last section of this article, indicated that the agreement was in general excellent.

An additional advantage of the force field method [32] is its power to predict the energy levels of conformations which are not populated and even complete rotational potentials of bonds. Again, statisfying agreement with results from dynamic nmr-measurements for a series of crowed hydrocarbons was found [36]. Knowledge of the shape of rotational potentials proved to be helpful for the interpretation of entropy effects in these series of compounds and in their thermolysis reactions.

3 The Steric Effect

To test the relationships between ground state strain and thermal stability independently of substituent effects several series of unsubstituted aliphatic model compounds are used.

In Fig. 2a the free enthalpies of activation $\Delta G^{\neq}$ (300 °C) of the thermolysis reactions of symmetrical hexaalkylethanes *11* (C_q—C_q series) — the weakest bond connects two quaternary carbons — are plotted against their ground state strain H_s as obtained from MM2 calculations [14,32,37]. The large range of stability differences encompassed by this series is easily judged from the scale on the right side of Fig. 2 in which is given for each compound the temperature at which the half-life is 1 h.

The high quality ($r = -0.987$) of the linear correlation in Fig. 2a, for which Eq. (1) is given in the caption, is quite surprising for several reasons. In particular, because free enthalpies of activation $\Delta G^{\neq}$ are correlated with strain enthalpies H_s despite the fact that there is neither an isoentropic ($\Delta S^{\neq}$ = const.) [41] nor an isokinetic relationship ($\Delta H^{\neq} \alpha \Delta S^{\neq}$) [41] within this series. Indeed $\Delta S^{\neq}$ varies from 13 to 26 entropy units [14]. In a kind of Exner test [42] it was shown, however, that the order of decreasing $\Delta G^{\neq}(T)$ values is independent of temperature and therefore significant for structural interpretation [14].

From the axis intercept in Fig. 2a, $\Delta G^{\neq}$ (300 °C) = 62.1 kcal · mol^{-1}, and from the mean entropy of activation, $\Delta S^{\neq}$ = 15 e.u. [14], an activation enthalpy $\Delta H^{\neq}$ = 71 kcal · mol^{-1} is calculated for a hypothetical unstrained compound. This is in close agreement with the value for the bond dissociation energy expected from the literature values for this type of bond [1]. When the seemingly more proper $\Delta H^{\neq}$ values are plotted against H_s for this reaction a distinctly poorer correlation is found. It has, however, almost the same slope (—0.62) and an intercept corresponding to $\Delta H^{\neq}$ = 72 kcal · mol^{-1} as in Fig. 2a [14].

The clue to an understanding of these unexpected phenomena is found in the "compensation effect" discussed by Benson [43]. $\Delta H^{\neq}$ is measured at much higher temperature — and for each member of the series in an individual temperature range — than the standard temperature 25 °C to which the strain enthalpy H_s corresponds. For a precise comparison $\Delta H^{\neq}$ values should be extrapolated over large temperature ranges down to 25 °C. This is not possible because $\Delta C_p^{\neq}$ values, the differences in heat capacity between ground and transition states, are not available. Benson points out [43] that the main factors determining $\Delta C_p^{\neq}$ are the changes in the degrees of freedom of translation, of internal and external rotation and of vibration. Just the same factors determine $\Delta S^{\neq}$. Therefore the temperature effects on $\Delta H^{\neq}(\Delta C_p^{\neq}\, dt)$ and on $T\Delta S^{\neq}$ are very similar. Due to the opposed signs of these two contributions to $\Delta G^{\neq}$ the temperature effect is largely compensated in $\Delta G^{\neq}$ [44], which is a term whose

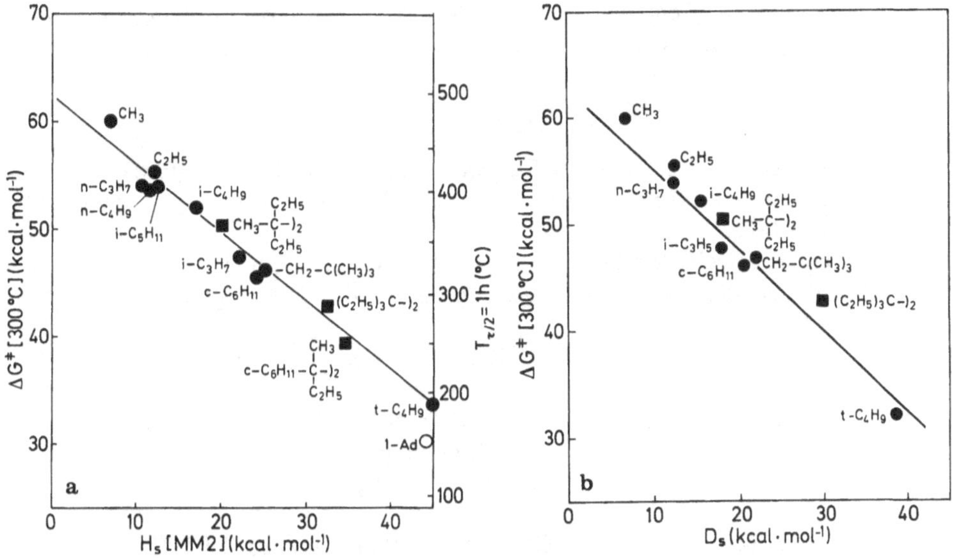

Fig. 2a and b. Relationships between ΔG* (300 °C) of thermal decomposition of sym-hexyalkyl-ethanes and
a ground state strain H_s (MM2 values) [32d]; **b** change in strain enthalpy during dissociation D_s (MM2 values) [32d]

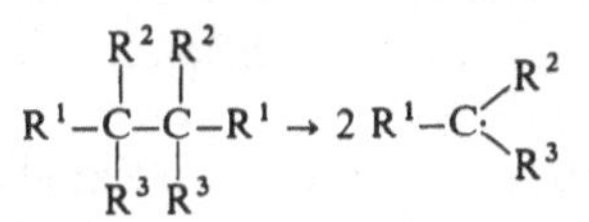
C_q–C_q series

11 *12*

● R^1 indicated, $R^2 = R^3 = CH_3$
■ $R^1R^2R^3C$ indicated
○ R^1 indicated, $R^2 = R^3 = CH_3$; H_s (MM2 value, 57.9 kcal · mol^{-1}) [39] corrected for inherent strain of adamantyl groups (7.9 kcal · mol^{-1}) [40].

Correlation equations:

Eq. 1. $\Delta G^{\neq}$ (300 °C) = 62.1 (± 0.7) —0.63 (± 0.03) H_s [kcal · mol^{-1}] [14b, 37]
r = —0.987; n = 13
Eq. 2. $\Delta G^{\neq}$ (300 °C) = 62.2 (± 1.1) —0.72 (± 0.05) D_s [kcal · mol^{-1}]
r = —0.981; n = 10

temperature dependence does not vary significantly within a reaction series. The correlations between $\Delta G^{\neq}$ and H_s are therefore better than those between $\Delta H^{\neq}$ and H_s and are more or less independent of temperature. We will take advantage of this in the following discussion.

An additional point of discussion is the slope —0.63 of the correlation in Fig. 2a. It suggests that ~37% of the ground state strain is still present in the radicals *12* being formed, as long as the reaction coordinate of Fig. 1 is valid. Figure 1 was based on the assumption of non-activated radical dimerization, which may no longer be the case for bulky radicals [45]. In order to get deeper insight into the situation the strain enthalpies of the radicals *12* involved were calculated by the MM2 force field which was extended to radicals for this purpose [46]. In Fig. 2b $\Delta G^{\neq}$ (300 °C) for the same series of compounds is therefore plotted against the change in strain D_s accompanying the dissociation process:

$$D_s = H_s\ (\text{dimer}) - 2\ H_s\ (\text{radical})$$

A correlation of similar quality as in Fig. 2a is obtained Eq. (2) ($r = -0.981$, $n = 10$) with almost unchanged intercept (62.2 ± 1.1 kcal · mol^{-1}). The slope is increased however to -0.72.

Very similar correlations Eq. (3 and 4) were found for a series of unsymmetrical $C_q \cdots C_q$ compounds *13* which dissociate into t-alkyl *14* and t-butyl radicals *15*

$$R^1R^2R^3C{-}C(CH_3)_3 \rightarrow R^1R^2R^3C\cdot + \cdot C(CH_3)_3 \qquad (n = 9)$$

eq. (3) $\quad \Delta G^{\neq}\ (300\ °C) = 64.4\ (\pm 0.4) - 0.67\ (\pm 0.02)\ H_s$ [kcal · mol^{-1}] $r = -0.997$ [17)]

eq. (4) $\quad \Delta G^{\neq}\ (300\ °C) = 65.7\ (\pm 0.3) - 0.91\ (\pm 0.02)\ D_s^*$ [kcal · mol^{-1}] $r = -0.998$ [17)]

The change in strain is designated D_s^* in this case because the strain enthalpies of the radicals were estimated from the strain enthalpies of the corresponding hydrocarbons $R^1R^2R^3CH$ [47)].

Summarizing, one can say that 75–90% of the strain enthalpy released in the dissociation process is found as a reduction in $\Delta G^{\neq}$ or $\Delta H^{\neq}$. The missing 10–25% is either a tribute to inadequacies of the compensation effect discussed before or also the recombinations of alkyl radicals pass, in contradiction to Fig. 1, a small activation barrier which is equal to 10–25% of the strain enthalpy of the dimer [9)].

Barriers to recombination have been observed for bulky persistent alkyl radicals like di-t-butylmethyl *16* or tri-isopropylmethyl *17* [45)]. While *16* dimerized slowly but quantitatively [39,48–50a)] to *7*, *17* decomposes in a unimoleculat process and its dimer remains unknown [45)].

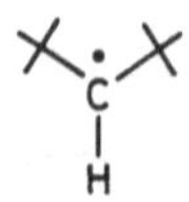

16

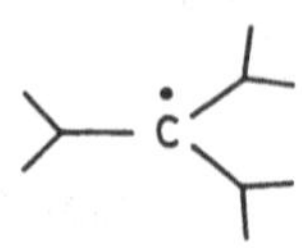

17

To obtain quantitative results the recombination and disproportionation rates of triethylmethyl radicals *19* were measured by kinetic esr spectroscopy [50)]. The radicals *19* were generated by photolysis of *18*.

Despite the building up of 32.6 kcal · mol^{-1} of strain in the dimerization reaction its rate was independent of temperature. It is therefore a non-activated reaction. The order of magnitude of this rate is typical for a diffusion-controlled process. The selectivity for disproportionation to *21* and *22* as against dimerization to *20* and the small rate retardation in comparison to t-butyl *15* [50b)] are therefore due to entropy control. This is probably quite generally the case for radical-radical reactions [17, 20)].

As suggested by Ingold [45,49,] and confirmed by independent rate data from our laboratory [50a)] only extremely crowded radicals like di-t-butylmethyl *16* have to pass an enthalpy barrier of recombination.

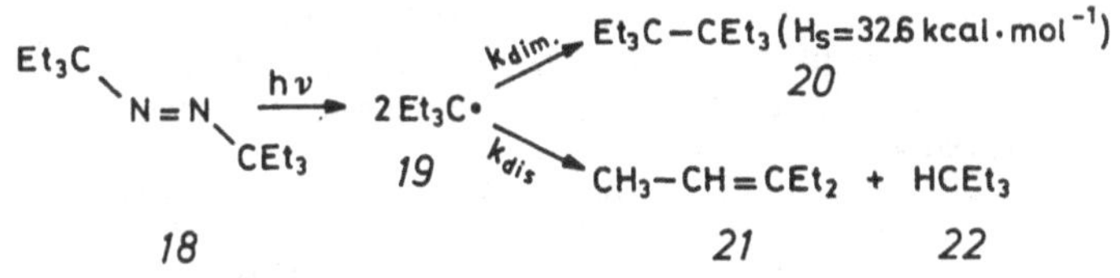

$k_{dim.}$ (290–400 k) ~ $1.0 \cdot 10^{8}$ $l \cdot mol^{-1} \cdot s^{-1}$

$k_{dis.}$ (290–400 k) ~ $17 \cdot 10^{8}$ $l \cdot mol^{-1} \cdot s^{-1}$

Very similar relationships were observed for other series of alkanes in which the weakest bond is that between two tertiary carbons (*23*, C_t—C_t series) Eq. (5 and 6) [16)] or a tertiary and a quaternary carbon *25*, C_t—C_q series) Eq. (7 and 8):

$R^1R^2CH—CHR^1R^2 \rightarrow 2\ R^1R^2CH\cdot$ C_t—C_t series [16)]
23 *24*

eq. (5) $\Delta G^{\neq}$ (300 °C) = 66.9 (± 1.0) — 0.65 (± 0.04) H_s
[kcal · mol^{-1}]
t = —0.975, n = 16

eq. (6) $\Delta G^{\neq}$ (300 °C) = 66.2 (± 1.5) — 0.79 (± 0.07) D_s
[kcal · mol^{-1}]
r = —0.97, n = 8

$R^1R^2CH—C(CH_3)_3 \rightarrow R^1R^2CH\cdot + \cdot C(CH_3)_3$ C_q—C_t series [17)]
25 *26* *15*

eq. (7) $\Delta G^{\neq}$ (300 °C) = 65.4 (± 1.1) — 0.70 (± 0.07) H_s
[kcal · mol^{-1}]
r = 0.971, n = 9

eq. (8) $\Delta G^{\neq}$ (300 °C) = 64.8 (± 1.5) — 0.82 (± 0.09) D_s^*
[kcal · mol^{-1}]
r = —0.97, n = 9

The axis intercept increases from the C_q—C_q series (62.1 kcal · mol^{-1}) to the C_q—C_t series (65.4 kcal · mol^{-1}) and the C_t—C_t series (66.9 kcal · mol^{-1}), reflecting the known fact that the bond dissociation energies of carbon bonds decrease with increasing alkylation. This is frequently attributed to radical stabilization by hyperconjugation. This is not conclusive, however, and there is good evidence for an alternative interpretation of this difference in bond strength as a ground state phenomenon due to differences in the quality of overlap in these systems [16, 51, 52].

In the context of this work we investigated several pairs of $C_t—C_t$ diastereomers *23* which differed in their thermal stability.

Table 2. Differences in ΔH^{*} [kcal · mol^{-1}] and in ΔS^{*} [e.u.] of thermal cleavage for *D,L* and meso diastereomers $R^1R^2CH—CHR^2R^1$ *23* and their comparison with corresponding differences[a] of ΔH^0 and ΔS^0 for the ground states

R^1	CH_3 [16]	CH_3 [16]	$c—C_6H_{11}$ [55]	C_6H_5 [54, 20]
R^2	$t—C_4H_9$	1-adamantyl	$t—C_4H_9$	$t—C_4H_9$
$\Delta\Delta H^{*}$ (D,L-meso)	6.3 (± 1.1)	9.2 (± 1.8)	4.3 (± 1.0)	−1.5 (± 1.2)
ΔH_S (meso-D,L)[a]	7.1	4.1	6.4	−3.2
$\Delta\Delta S^{*}$ (D,L-meso)	4.0 (± 1.7)	13.0 (± 2.3)	−0.4 (± 1.6)	−7.6 (± 2.1)
ΔS^0 (meso-D,L)[a, b]	2.9	3.2	0.3	−3.9

[a] force field calculations using the MM2 force field [30, 32b].
[b] the entropies were calculated [53a)] by the program DELFI [53b)], which calculates the full matrix of the second derivative of the energy

Because both diastereomers lead to the same radicals $R'R^2CH\cdot$ *24* on thermolysis this difference has to be due to differences in the ground state stability. This has been confirmed by EFF calculations and can be understood easily on conformational grounds [16, 56]. The minimum energy conformations of all members of the alkane series are gauche.

meso–*25* D,L–*25*

In the D,L-diastereomer D,L-*25* both bulkier R^1 groups can occupy the less hindered position opposite to hydrogen while in meso-*25* one R^1 group is in the less favorable position staggered with respect to two R groups; consequently D,L is more stable than meso. The conformational behaviour of the 1,2-diphenyl-1,2-dialkylethanes [15, 20] on the other hand is more complex due to the shape of the phenyl rings [9b, 54]. The diastereomers of di-t-butyldiphenylethane (see Table 2) show the reversed order in stability, because the meso isomer escapes strain by adopting the anti conformation.

This conformational situation is also responsible for entropy effects. It has been shown [16] that the entropy differences between two diastereomers in this series is mainly dependent on the shapes of the rotational profiles about the central bonds. The observed differences in $\Delta S^{\neq}$ (D,L-meso) for the thermolyses can be reproduced semiquantitatively by differences in ground state entropy (see Table 2) which were calculated by the force field method [53].

In summary, the relationship between ground state strain H_s and thermal stability of hydrocarbons which was suggested in a qualitative manner by Ziegler [7], has now been successfully developed into a quantitative one. It is particularly satisfying that the slopes of the $\Delta G^{\neq}/H_s$ correlations of several series of hydrocarbons are very similar. This supports the assumption that the steric effect is acting in a quantitatively analogous manner in these series.

4 The Resonance Effect

As a next step in this analysis we investigated [18] a series of 1,2-diphenyl tetraalkylethanes *27* which generate resonance stabilized tertiary benzyl radicals *28* at elevated temperatures (Fig. 3). Having worked out a method for analysis of the steric effect we hoped to succeed also in quantitatively separating it from the resonance effect of substituents. It is immediately recognized from Fig. 3 and the related correlation Eq. (9 and 10) that thermolysis occurs at much lower temperatures (100°–200 °C) and with much lower activation enthalpies than in the aliphatic C_q—C_q series *11*.

Again good linear correlations of $\Delta G^{\neq}$ (300 °C) and H_s [30, 32] Eq. (9) or D_s Eq. (10) are observed. The slopes of these correlations are very similar to those found for the aliphatic C_q—C_q series *11*, supporting the assumption that the steric effect is the same in both cases. Therefore the difference in the axis intercepts (Fig. 3) of the correlations of the aliphatic C_q—C_q series *11* and the C_q—C_q phenyl series *27* must be ascribed to the action of the resonance substituent effect in *28*. If the difference in mean entropy of activation in the two series ($\Delta S^{\neq}$ = 16 [14b] and 20 e.u. [18] respectively) is taken into account it is calculated that the resonance energy H_r is 8.4 (± 1) kcal · mol^{-1} for each tertiary benzyl radical *28*. This value corresponds numerically to the difference in bond dissociation energy of the tertiary C—H bonds in 2-methylpropane (93.2 ± 0.2 kcal · mol^{-1}) [57] and in cumene (84.4 ± 1.5 [57] or 86.1 [58] kcal · mol^{-1}). Values for the benzyl resonance energy [59] quoted in the literature and obtained by other methods are in qualitative agreement but are quite scattered. The method we used has the unique advantage that H_r is evaluated from a whole series of compounds or reactions and that possible steric accelerations by phenyl are explicitly separated in this analysis.

A very similar result is obtained for sec-benzyl radicals *30* by the analysis of the C_t—C_t phenyl series [15) 20)].

$$C_6H_5\text{—CHR—CHR—}C_6H_5 \longrightarrow 2\ C_6H_5\text{—}\dot{C}HR \quad (C_t\text{—}C_t \text{ phenyl series})$$

29 *30*

Eq. (11) $$\Delta G^{\neq}\ (300\ °C) = 51.4\ (\pm 1.4) - 0.48\ (\pm 0.11)\ H_s \quad [\text{kcal} \cdot \text{mol}^{-1}] \quad r = -0.86,\ n = 8$$

In the manner discussed above, a resonance energy [59,60] $H_r = 7.8$ (± 1.5) kcal · mol^{-1} per secondary benzyl radical *30* is calculated by comparison of the axis intercept with that of the aliphatic C_t—C_t series *23* [20]. It is remarkable that in both

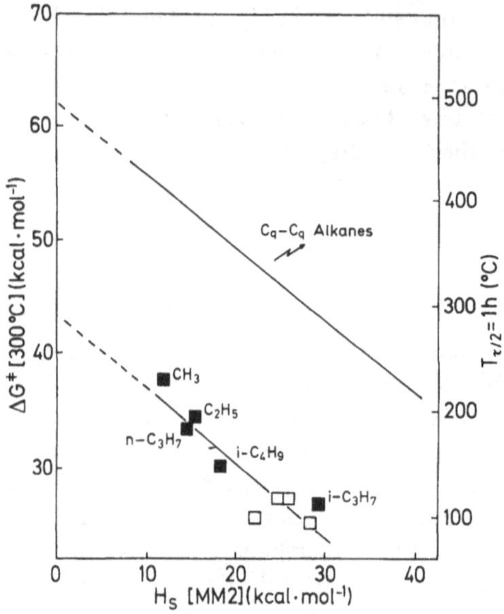

Fig. 3. Relationship between ΔG^* (300 °C) of thermolysis and ground state strain H_s for a series of 1,2-diphenyltetraalkylethanes (for comparison the correlation line for the thermolysis of *11* in fig. 2a is included)

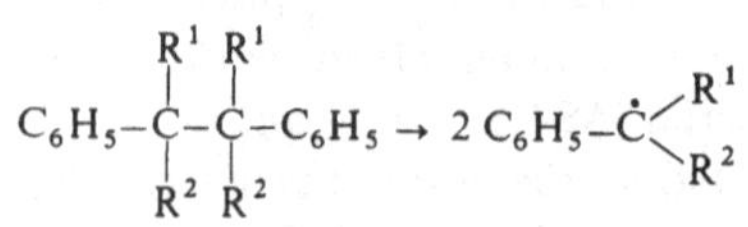
(C_q–C_q phenyl series)[18]

27 *28*

■ R^1 indicated, $R^2 = CH_3$
□ R^1, R^2 = n-alkyl
Eq. **9**. ΔG^* (300 °C) = 43.3 (± 2.3) —0.64 (± 0.11) H_s [kcal · mol⁻¹]
r = —0.92, n = 9
Eq. **10**. ΔG^* (300 °C) = 44.2 (± 2.1) —0.77 (± 0.11) D_s [kcal · mol⁻¹]
r = —0.93, n = 8

series meso and D,L diastereomers of quite different thermal stability (see Table 2 and Ref. [20]) were included.

The correlations for the two phenyl-substituted series are, as seen from Fig. 3 and from the correlation coefficients of Eq. (9–11), of sowhat lower quality than those of the unsubstituted alkanes. This is probably due to the greater variations in $\Delta S^{\neq}$ in the two phenyl series. This variation has been ascribed mainly to two factors [18, 20, 61]. When frozen rotations around bonds in the ground state are set free on dissociation in the transition state an increase in $\Delta S^{\neq}$ results. This effect does not necessarily run parallel with the strain H_s, because rotational barriers of highly strained compounds are sometimes flatter than those of less strained ones [31, 35, 36]. The decisive question for estimating $\Delta S^{\neq}$ is the following: is there a rotamer of particular low energy, i.e. a steep minimum available? Due to the flat geometry of phenyl substituents this is the case in the phenyl series [35]. On the other hand, resonance stabilized benzyl radicals are more restricted in their freedom for internal rotation than alkyl radicals [18]. A second effect influencing $\Delta S^{\neq}$ seems to be caused by very tight pairing of the radicals in the activated complex of bond dissociation [18]. The attractive forces responsible for this phenomenon could but need not lead to the formation of real tight radical pair intermediates. There is good evidence that benzyl type radicals form particularly tight radical pair complexes in which several of the internal rotations of the side chains cannot be set free [20, 62]. This has been discussed as the reason why the members with the ethyl side-chains in several of the investigated series have higher activation entropies $\Delta S^{\neq}$ than those with bulkier or longer side chains [18]. Whereas the small ethyl side-chains gains all its possible freedom of rotation, even in a tight radical pair, this is not possible for larger groups.

At this point we conclude that our studies have led to a deeper understanding and a quantitative seperation of steric and resonance effects on C—C bond strengths, a goal which has been discussed in the context of the hexaphenylethane story [2] for some decades.

Table 3. Resonance energies H_r[59] of substituted alkyl radicals $X-\dot{C}R^1R^2$ *6* ($\pm$ 1 kcal · mol^{-1})

X in *6*	n[a]	H_r [kcal · mol^{-1}]	Ref.
CH_3	13	$\equiv 0$	14)
C_6H_5	17	8.4	15, 18, 20)
$R^3-C(=O)-$	3	6.5	63)
$N\equiv C-$	7	5.5	61)
CH_3OOC-	6	3.5	64)
CH_3O-	7	1.3	65)

[a] number of compounds investigated in this series

A similar analysis, but not based on as extensive data in all examples, was consequently performed for other substituents. The results are reported in Table 3. The stabilizing effects of carbonyl, cyano and ester groups are in agreement with the most reliable literature references [1, 57, 61, 63–64]. The stabilization energies of Table 3 should be particularly reliable, because they were obtained from reaction series and not from single experiments. The high stabilization by the keto function was somewhat surprising. The very small stabilizing effect of methoxy was not unexpected [65, 66], even though alkoxy groups have frequently a large rate accelerating effect in radical chemistry, particularly when radical centers are generated [67] α to alkoxy groups, e.g. in autoxidations. These rates are usually, however, controlled by FMO-interactions [68] and a large "α-methoxy effect" is a safe indicator of an early transition state in a reaction generating a radical center and of SOMO-HOMO control [67].

5 The Question of Additivity of Substituent Effects

This question has become particularly popular since Viehe [69] postulated that "capto-dative substitution", i.e. interaction of a radical center with a donor and an acceptor substituent, leads to stabilizing effects clearly exceeding additivity. In order to get deeper insight into this question, stabilizing effects of more than one substituent at the same time were determined for the series of radicals shown in Table 4. Their accuracy is lower than that of the data in Table 3, because several of these resonance effects H_r were obtained from the thermolysis data for a single compound.

Table 4. Resonance energies H_r [59] [kcal · mol^{-1}] of doubly and triply substituted radicals

Radical		n	H_r (exp)	H_r (calc)[a]	Ref.
$(CH_3)_3C\cdot$			≡0	≡0	
$(C_6H_5)_2\dot{C}$–R	*31*	3	12[c]	16.8	70)
$(C_6H_5)_3C\cdot$[b]	*2*	2	19[c]	25.2	70)
R–$\dot{C}(CN)_2$	*32*	5	8.5	11	71)
C_6H_5–$\dot{C}(CH_3)(CN)$	*33*	1	15	13.9	61)
C_6H_5–$\dot{C}(CH_3)(OCH_3)$	*34*	5	9.4	9.7	66)
	*35*	1	3.2	—	21)
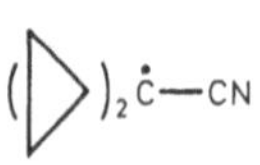	*36*	1	8.5	7.7	72)
C_6H_5–$\dot{C}(CN)(OCH_3)$	*37*	1	14	15.2	62)

[a] calculated by assuming additive stabilization by the susbstituents of Table 3.
[b] determined from thermolysis data for pentaphenylethane and 2,2-dimethyl-3,3,3-triphenylpropane [70].
[c] steric inhibition of resonance decreases H_r

The stabilization of benzhydryl *31* and triphenylmethyl *2* is less than additive, as expected for the non-planar propeller-like structures of these radicals, which do not allow the development of full conjugation. The angle of twist is probably very similar in benzhydryl and trityl radicals [73] and one is tempted to attribute to each twisted phenyl an additive stabilization of 6 kcal · mol^{-1}. On the other hand two cyano groups in *32* likewise stabilize a radical less than additively. Phenyl and cyano (*33*) and phenyl and methoxy (*34*) show additive stabilization. For one cyclopropyl group in *35* a little more than 1 kcal · mol^{-1} stabilization can be counted and additivity follows consequently for *36*. The captodative radical *37* is stabilized according to additivity and not more. A similar, but still somewhat preliminary result has been found for the α-cyano-α-methoxy-neopentyl radical *38*. Therefore a leveling effect by the phenyl substituent is not responsible for the failure to observe a larger capto-dative effect in *37*.

The data in Table 4 are not yet sufficient to give a final answer as to when additivity is to be expected and when not. They clearly show however that additivity of substituent effects definitely is not the exception.

The 1-cyano-1-methoxybenzyl radical *37* and its dimers *39* were investigated with particular care[62] in order to test the reliability and the persuasive power of the methods used and the results discussed so far.

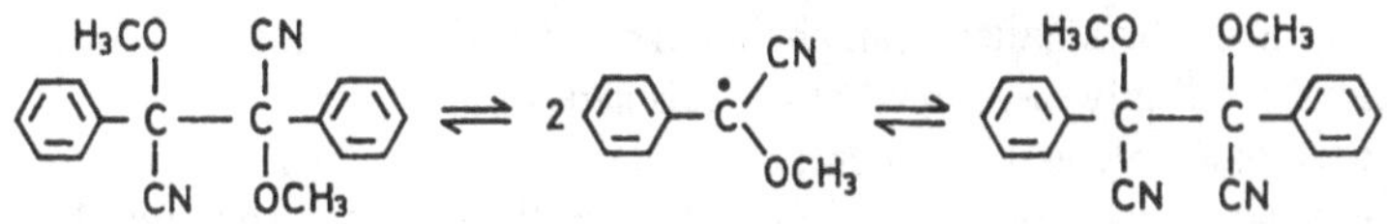

D,L-*39*   *37*   meso-*39*

Meso- and D,L-2,3-dimethoxy-2,3-diphenylsuccinonitrile *39* were obtained in a 1:1 ratio by a preparative dimerization procedure. Their configurations were assigned by crystal structure analysis [74]. According to nmr both diastereomers have a configuration with the phenyl groups in the anti position. From a conformational analysis by the force field method it was concluded that these conformations remain the preferred ones also for the isolated molecules; meso-*39* was calculated to be 1.2 kcal · mol^{-1}

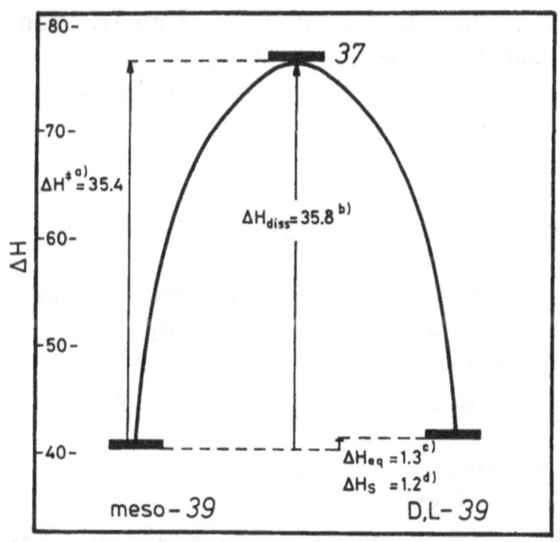

Fig. 4. Enthalpy diagram for the reversible dissociation of meso- and *D,L-39* [kcal · mol^{-1}].
The numerical values in the Fig. are from the following sources:
a) from the kinetics of the thermal equilibration of meso-*39* and D,L-*39* (nmr method) [75]
b) from measurements of the equilibrium constant K_{diss} between *39* and *37* by esr [62];
c) from measurements of the equilibirum constant K_{eq} between meso-*39* and D,L-*39* by gc and by nmr [62, 75];
d) from EFF calculations [62]

more stable than D,L-*39*. Figure 4 shows the complete reaction coordinate of the system which was constructed by independent determination of

1) the equilibirium constant K_{eq} over a temperature range of 40 °C by gc,
2) the equilibirum constant K_{diss} over a temperature range of 60 °C by esr [62] and
3) the activation enthalpies $\Delta H^{\neq}$ for the thermal dissociation of the two diastereomers independently [62, 75].

Figure 4 shows that the enthalpy difference between the two stereoisomers ΔH_{eq} is in excellent agreement with the prediction of the force field calculations.

The close agreement between ΔH_{diss} and $\Delta H^{\neq}$ in Fig. 4 allows the conclusion that the dimerization of the radicals *37* is a non-activated process. It has been confirmed independently by direct kinetic experiments [76] that $\Delta H^{\neq}$ of recombination for radicals *37* is similar to or smaller than their barrier for diffusion. Capto-dative substituted radicals accordingly have no kinetic stabilization.

Quite remarkable is the low entropy of activation $\Delta S^{\neq}$ of 10.9 e.u. for the dissociation of meso-*39* in contrast to ΔS_{diss} (31.1 ± 0.8 e.u.) for the complete dissociation into free radicals. In the activated complex of the dissociation of *39* the central C—C bond is no doubt almost completely broken but the system has gained only little additional freedom of mobility as compared to the ground state dimer. This strongly supports the formation of a sandwich-like arrangement of the two radicals in the activated complex as discussed above [20, 77]. According to the principle of microscopic reversibility, the dimerization reactions of radicals of this type should proceed via the same sandwich-like alignement and, therefore, may have to pass a barrier in $\Delta G^{\neq}$ due to the loss of entropy complex formation. Among all possible encounter pairs those which are sandwich like oriented should be favoured by secondary valence interactions [50a, 77]. They are therefore more populated than differently oriented pairs e.g. those leading to disproportionation. This could explain the high dimerization/disproportionation ratios which are found for benzyl type radicals [20] as well as for cato-dative radicals [69]. Hence simple steric effects may be responsible for the observed high regiospecificities of the termination of benzyl type radicals.

6 Strain, Structure and Bonding

The relationships between structure and strain constitute a very complex story which cannot be dealt with comprehensively in this context [9b]. For this reason only a few selected points of interest will be addressed briefly here:

1) the reliability of the force field method to predict structures of highly strained compounds and
2) the relationships between bond length, bond strength and strain.

In Fig. 5 the X-ray structure of *40*, the most highly strained member of the C_q—C_q alkane series is shown [39, 74] and in parentheses the predictions of MM2 force field calculations are given [39].

Despite the fact that in this compound the three central carbon bonds are 164 pm or longer and, therefore, are among the longest known bonds of this kind the predictions by the force field method are exceedingly precise. The very good agreement between experimental and calculated bond angles and torsional angles is even more impressive since these are controlled by much weaker force constants. A similar

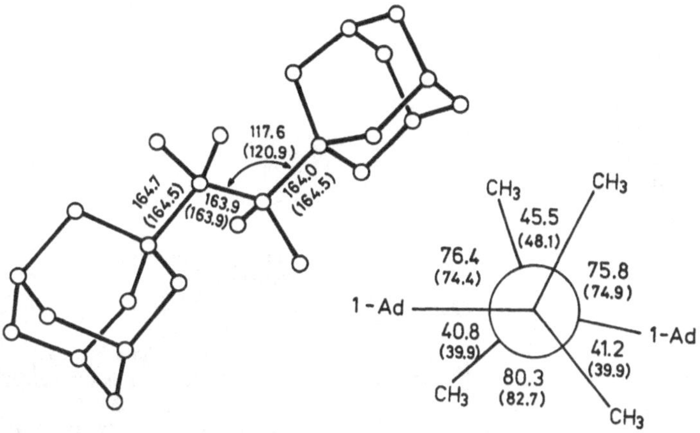

Fig. 5. Crystal structure of 3.4-di-(1-adamantyl)-2,2,5,5-tetramethylhexane *40*[39, 74]. Results of force field calculations [32] in parentheses

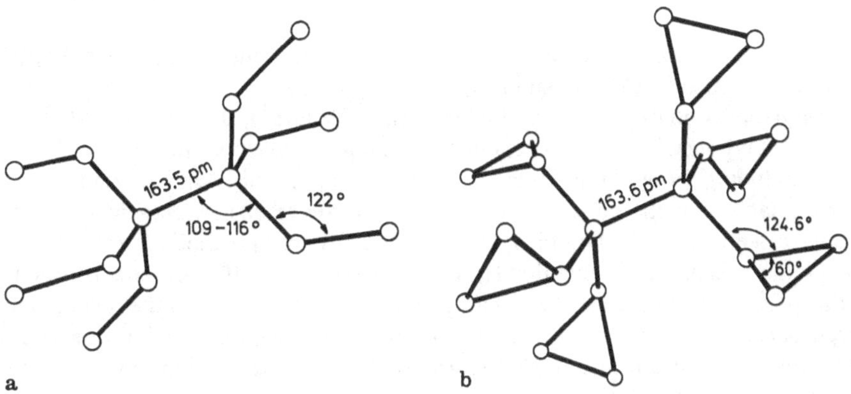

Fig. 6. Crystal structure of hexycyclopropylethane *41* [21] and force field structure [32] of hexyethylethane *43*

agreement has been found for many other highly strained compounds of the different series discussed in this review [20,25,31,55,56,78,79].

The structure of hexacyclopropylethane *41* which decomposes at ~305 °C with $t_{1/2}$ = 1 h, and its comparison with hexaisopropylethane *42* which is too unstable to be isolated (see above) is of particular interest. A comparison of the X-ray structure of hexacyclopropylethane *41* [21] with the EFF structure of hexaethylethane *43* [14b] is shown in Fig. 6. It reveals that in the latter the preferred way of escaping strain is to increase the bond angles at the α positions of the side-chains to values higher than 120°. In hexycyclopropylethane *41* the corresponding twelve bond angles are larger than 124° anyway as a consequence of the small inner angles of the cyclopropane rings. Therefore the cyclopropyl rings behave sterically more like ethyl side-chains than like isopropyl groups in which bond angle deformation is even more difficult.

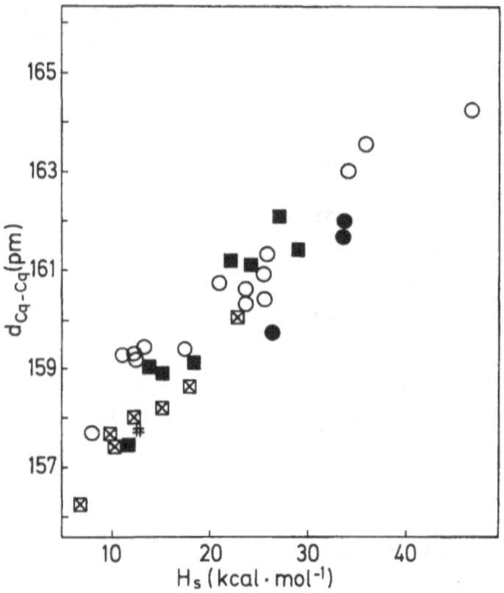

Fig. 7. Relationship between the central C_q–C_q bond lengths and the strain enthalpies in several C_q–C_q series; ○ C_q-C_q alkanes *11*, ⊠ C_q–C_q dinitriles, ■ C_q–C_q phenyl series *27*, ● C_q–C_q tetraphenylethanes ⧣ captodative dimer *39*

The final point of interest, the relationship between bond lengths and bond strengths or ground state strain H_s, is shown in Fig. 7.

A reasonably good direct correlation between the length of the central C—C bond and the ground state strain is observed for a large number of substituted C_q—C_q systems, including those of the benzhydryl type [70] and the capto-dative system *39* [74]. It is remarkable that the bond-weakening effect of the resonance stabilizing substituents has no effect on the bond length. Bond lengths are apparently a ground state phenomenon as is the strain enthalpy H_s, while the resonance effect of the substituents is a transition state effect as suggested long ago [7]. On the other hand the relationship in Fig. 7 is certainly partly fortuitous. It requires that in all series included in Fig. 7 the distribution of strain into angle deformation and bond streching must be very similar. For this reason it was not unexpected that compounds of the C_t—C_t series, which allow much better angle deformation [9b] in the central region of their structures [16, 20, 49, 54, 55, 79], do not obey the simple correlation of Fig. 7. It is restricted to the C_q—C_q compounds.

Despite this better mode of strain dissipation in the C_t—C_t series the central C—C bond length in 2,3-dimethylbutane, the parent compound, is shorter than the extrapolated bond length of a hypothetical strainless parent C_q—C_q compound [79]. This supports the proposal made before, that C_i—C_i bonds are intrinsically stronger than C_q—C_q bonds. If the difference in resonance stabilization of tertiary and secondary alkyl radicals was the dominating factor determining these bond strengths it should have no consequences for the bond lengths [80].

The separate and quantitative evaluation of accelerating steric effects on the homolytic cleavage of C—C-bonds allowed the definition of intrinsic barriers for the bond dissociation reaction, i.e. $\Delta G^{\neq}$ or $\Delta H^{\neq}$ at $H_s = 0$ in Fig. 2 or 3 or in Eq. (1–11) in general. $\Delta H^{\neq}$ is equal to or slightly than the corresponding bond dissociation energy (BDE) [6a, 81] as pointed out earlier.

Intrinsic BDE's can be determined alternatively from appropriate group increments, which define strain free heats of formation of educts and radical products. Using Schleyer's increments for alkanes [29] (CH_3 −10.05; CH_2 −5.13; CH −2.16 and C −0.30 kcal · mol^{-1}) and increments for radicals obtained by the same formalism [46] (CH_3 34.3; CH_2 36.6; CH 37.7 and C 38.6 kcal · mol^{-1}) the following two equations give numerical values for the bond energies of strain free C_t—C_t and C_q—C_q bonds:

$$\text{eq. (12)} \qquad BDE(C_tC_t) = 2\,(37.7 + 2.16) = 79.7\ \text{kcal} \cdot \text{mol}^{-1}$$

$$\text{eq. (13)} \qquad BDE(C_q{-}C_q) = 2\,(38.6 + 0.30) = 77.8\ \text{kcal} \cdot \text{mol}^{-1}$$

Obviously there are, by definition, no individual contributions of alkyl side chains in the educt molecules because they remain unchanged in the radicals. The difference in energy of these two bonds calculated in this way

$$\Delta BDE = 1.9\ \text{kcal} \cdot \text{mol}^{-1}$$

is in reasonable agreement with the difference in activation enthalpy for the C—C cleavage reaction.

$$\Delta\Delta H^{\neq}\ (H_S = 0) = 3.2 \pm 1.7\ \text{kcal} \cdot \text{mol}^{-1}\ .$$

$\Delta\Delta H^{\neq}$ is obtained from Fig. 2a and Eq. (5) when the difference in mean entropy of activation of the C_q—C_q series (16.2 e.u.) and the C_t—C_t series (13.2 e.u.) is taken into account.

The group increment analysis, in addition, suggests an explanation for the difference in BDE for C_q—C_q and C_t—C_t bonds. The exchange of hydrogen for alkyl at a C—C bond lowers the enthalpy considerably (going from a C_q—C_q bond to C_t—C_t to —CH_2—CH_2— and finally to ethane in steps, the change in ΔH is −1.8, −3.0 and −4.9 kcal · mol^{-1}, respectively). The energies of carbon radicals in contrast are decreased much less, if one goes from a tertiary to a socondary (−0.9 kcal · mol^{-1}) from there to a primary (−1.1 kcal · mol^{-1}) and finally methyl (−2.3 kcal · mol^{-1}). The large difference of these increments for secondary and tertiary hydrocarbon fragments in alkanes, mentioned above, is also the main reason for the large difference in BDE between tertiary and secondary C—H bonds.

$$BDE\ (C_{sec}{-}H - C_{tert}{-}H) = 2.1\ \text{kcal} \cdot \text{mol}^{-1}$$

This difference is therefore mainly due to a ground state phenomenon as pointed out earlier [51]. Clearly, therefore, BDE (C_{sec}—X − C_{tert}—X) depends very much on the C—X bond being broken [51].

7 Acknowledgement

Thanks are due to Deutsche Forschungsgemeinschaft and to Fonds der Chemischen Industrie for generous financial support. We also thank our colleagues and our coworkers who contributed much to this work, whose names are found in the references. Finally we thank Dr. John S. Lomas for stimulating discussions and for his help in the preparation of the manuscript.

8 References

1. Egger, K. W., Cooks, A. T.: Helv. Chim. Acta 56, 1516, 1537, (1973)
2. McBride, J. M.: Tetrahedron *30*, 2009 (1974)
3. D'iachkowski, F. S., Dubnov, M. M., Shilov, E. A.: Dokl. Akad. Nauk., SSSR *122*, 629 (1958)
4. To illustrate this factor by an example, the diameter of an atomic nucleus (10^{-12} cm) has to be compared with the distance of 1 light year (10^{18} cm)!
5. This definition of the homopolar bond dissociation enthalpy has to be distinguished from its heteropolar counterpart, the dissociation of a bond into ions [1]. An example of a particularly low heteropolar bond dissociation enthalpy for a C—C bond was recently discovered by Arnett [6].
6. Arnett, E. M., Troughton, E. B., McPhail, A. T., Molter, K. E.: J. Am. Chem. Soc. *105*, 6172 (1983), see also Chem. and Enging. News, March 5, 1984, p. 29

6a. For a recent critical discussion of this argument see A. M. DeP. Nicholas, D. R. Arnold, Can J. Chem. *62*, 1850 (1984)

7. Ziegler, K.: Angew. Chem. *61*, 168 (1949)
8. More precisely the change in strain on bond dissociation D_s determines the driving force. This has been expressed by the concept of front strain later; see e.g. Slutsky, J., Bingham, R. C., von R. Schleyer, P., Dickerson, W. C., Brown, H. C.: J. Am. Chem. Sox. *96*, 1969 (1974) and references therein

9a. For a first review see Rüchardt, C., Beckhaus, H.-D.: Angew. Chem. *92*, 417 (1980); Angew. Chem. Int. Ed. Engl. *19*, 429 (1980); see also Rüchardt, C.: Sitzungsber. Heidelberger Akad. Wiss., Math.-Nat. Klasse *1984*, 53

9b. Relationships between structure and strain in highly congested ethanes will be reviewed and discussed elsewhere, Angew. Chem. 1985 in print.

10. Beckhaus, H.-D., Schoch, J., Rüchardt, C.: Chem. Ber. *109*, 1369 (1976); Bernlöhr, W., Beckhaus, H.-D., Lindner, H. J., Rüchardt, C.: ibid. *117*, 3303 (1984)
11. De Jongh, H. A. P., De Jonge, C. R. H. I., Sinnige, H. J. M., De Klein, W. J., Huijsmans, W. G. B., Mijs, W. J., van den Hoek, W. J., Smidt, J.: J. Org. Chem. *37*, 1960 (1972)
12. Petrakis, L., Grandy, D. W.: Free Radicals in Coals and Synthetic Fuels, Elsevier, Amsterdam 1983;
    Poutsma, M. L., Douglas, E. C., Leach, J. E.: J. Am. Chem. Soc. *106*, 1136 (1984) and references therein
13. Hawkins, W. L.: Polymer Degradation and Stabilisation, Springer, Berlin 1984

14a. Beckhaus, H.-D., Rüchardt, C.: Chem. Ber. *110*, 878 (1977)

14b. Winiker, R., Beckhaus, H.-D., Rüchardt, C.: ibid. *113*, 3456 (1980)

15. Hellmann, G., Beckhaus, H.-D., Rüchardt, C.: ibid. *112*, 1808 (1979)
16. Hellmann, G., Hellmann, S., Beckhaus, H.-D., Rüchardt, C.: ibid. *115*, 3364 (1982)
17. Hellmann, S., Beckhaus, H.-D., Rüchardt, C.: ibid. *116*, 2238 (1983)
18. Kratt, G., Beckhaus, H.-D., Rüchardt, C.: ibid. *117*, 1748 (1984)
19. Bockrath, B. Bittner, E., McGrew, J.: J. Am. Chem. Soc. *106*, 135 (1984)
20. Eichin, K. H., Beckhaus, H.-D., Hellmann, S., Fritz, H., Peters, E. M., Peters, K., v. Schnering, H.-G., Rüchardt, C.: Chem. Ber. *116*, 1787 (1983)
21. Bernlöhr, W., Beckhaus, H.-D., Peters, K., v. Schnering, H.-G., Rüchardt, C.: ibid. *117*, 1013 (1984)
22. Koenig, T., Fischer, H.: Free Radicals, Vol. I (Kochi, J. K., ed.) p. 157, J. Wiley, New York 1973
23. Gibian, M. J., Corley, R. C.: Chem. Rev. *73*, 441 (1973)
24. Kaiser, J. H., Beckhaus, H.-D.: unpublished
25. Beckhaus, H.-D., Kratt, G., Lay, K., Geiselmann, J., Rüchardt, C., Kitchke, B., Lindner, H. J.: Chem. Ber. *113*, 3441 (1980)
26. Beckhaus, H.-D., Rüchardt, C., Smisek, M.: Thermochim. Acta *79*, 149 (1984)
27. Kratt, G., Beckhaus, H.-D., Bernlöhr, W., Rüchardt, C.: ibid. *62*, 279 (1983)
28. Ducros, M., Gruson, J. F., Sannier, M.: ibid. *36*, 39 (1980)
29. v. R. Schleyer, P., Williams, J. E., Blanchard, K. R.: J. Am. Chem. Soc. *92*, 2377 (1970)
30. Beckhaus, H.-D.: Chem. Ber. *116*, 86 (1983)
31. Barbe, W., Beckhaus, H.-D., Lindner, H. J., Rüchardt, C.: Chem. Ber. *116*, 1017 (1983)

32a. Allinger, N. L.: J. Am. Chem. Soc. *99*, 8127 (1977)

32b. Allinger, N. L., Yuh, Y. H.: Quantum Chemistry Program Exchange, Indiana Univ., Program No. 395
32c. Burkert, U., Allinger, N. L.: Molecular Mechanics, ACS Monograph, Series No. 177, Washington D.C. 1982
32d. Throughout this work the term "MM2 result" refers to the $\Delta H_f^\circ(g)$ values obtained from calculations using the MM2 force field for alkanes (Ref. 32a) and its extensions to alkylbenzenes (Ref. 30), succinonitriles (Ref. 31, 62), dimethoxyethanes (Ref. 62, 65, 66), succinic esters (Ref. 64) and γ-diketones (Ref. 63). The strain enthalpies derived from these values, though often called strain energy (MM2), are calculated according to Schleyer's formalism for alkanes (Ref. 29) which has been extended to the compounds under consideration (Ref. 30, 31, 62–66). Note that the formalism introduced by Allinger (Ref. 32a–c), which is included in the QCPE version of the MM2 program of Ref. 32b, normally results in somewhat higher values for the strain enthalpy than Schleyer's formalism.
33. Flamm-ter Meer, M. A., Beckhaus, H.-D., Rüchardt, C.: Thermochim. Acta *80*, 81 (1984)
34. Beckhaus, H.-D., Gleißner, R.: work in progress
35. Kratt, G., Beckhaus, H.-D., Lindner, H. J., Rüchardt, C.: Chem. Ber. *116*, 3235 (1983)
36. Beckhaus, H.-D., Rüchardt, C., Anderson, J. E.: Tetrahedron *38*, 2299 (1982); for a modified version of the MM2 force field to calculate barriers of rotation see Jaime, C., Osawa, E.: Tetrahedron **39**, 2769 (1983)
37. The activation parameters [14] and the correlation equation have been recalculated using the more advanced statistical analysis as in all other series. [38] The deviations of the results from the earlier ones [14] are only minor.
38. Program Kinetik 80, Dissertation Barbe, W., Univ. Freiburg, 1981; for the transformation of the statistical weights of the Eyring correlation, see Cvetanovic, R. G., Singleton, D. L.: Int. J. Chem. Kinetics *9*, 481 (1977)
39. Dissertation Flamm-ter Meer, M. A.: Univ. Freiburg, 1984
40a. Beckhaus, H.-D., Flamm, M. A., Rüchardt, C.: Tetrahedron Lett. *23*, 1805 (1982)
40b. Schulman, J. M., Disch, R. L.: J. Am. Chem. Soc. *106*, 1202 (1984)
40c. Clark, T., Mc O. Knox, T., Mc Kervey, M. A., Mackle, H., Rooney, J. J.: ibid. *101*, 2404 (1979)
41. Hammett, L. P.: Physikalische Organische Chemie, p. 387, Verlag Chemie, Weinheim 1973
42. Exner, O.: Progr. Phys. Org. Chem. (Streitwieser, A. S., Taft, R. W. ed.) *10*, 411 (1973)
43. Benson, S. W.: Thermochemical Kinetics, 2nd. ed., pp 21–23, J. Wiley, New York 1976
44. Glasstone, S., Laidler, K. J., Eyring, H.: The Theory of Rate Processes, 1 st ed., p. 196, McGraw-Hill, New York 1941
46a. Beckhaus, H.-D.: unpublished
46b. N. L. Allinger has independently developed a force field for radicals. In a paper submitted for publication in the Journal of Molecular Structure he has made the suggestion to use $\Delta G^{\neq}/D_s$ correlations instead of $\Delta G^{\neq}/H_s$ correlations for our kinetic results, Apparently he was not aware that this kind of analysis had already been introduced; see e.g. Ref. 17, 18. We thank Prof. Allinger for a preprint of this paper.
47. A similar approach was also suggested by Lomas, J. S., Dubois, J.-E.: Tetrahedron Lett. *24*, 1161 (1983); J. Org. Chem. *47*, 4505 (1982)
48. Mendenhall, G. D., Griller, D., Lindsay, D., Tidwell, T. T., Ingold, K. U.: J. Am. Chem. Soc. *96*, 2441 (1974) followed the rate of disappearance of di-t-butylmethyl radicals by esr but could not find the dimer [49] as a product
49. Beckhaus, H.-D., Hellmann, G., Rüchard, C.: Chem. Ber. *111*, 72 (1978)
50a. Kaiser, J. H., Beckhaus, H.-D.; work in progress and Dissertation Kaiser, J. H.: Univ. Freiburg, in preparation
50b. Schuh, H., Fischer, H.: Helv. Chim. Acta *61*, 2130, 2463 (1978)
51. Rüchardt, C.: Angew. Chem. *82*, 845 (1970); Angew. Chem. Int. Ed. Engl. *9*, 830 (1970)
52a. Benson, S. W.: Angew. Chem. *90*, 868 (1978); Angew. Chem. Int. Ed. Engl. *17*, 812 (1978)
52b. Sanderson, R. T.: J. Am. Chem. Soc. *97*, 1367 (1975)
52c. Larson, J. R., Epiotis, N. D., Shaik, S. S.: Tetrahedron *37*, 1205 (1981)
53a. Baas, J. M. A., van den Graaf, B., Beckhaus, H.-D.: paper in preparation
53b. van de Graaf, B., Baas, J. M. A., van Veen, A.: Rec. Trav. Chim. Pays-Bas *99*, 175 (1980)
54. Beckhaus, H.-D., McCullough, K. J., Fritz, H., Rüchardt, C., Kitschke, B., Lindner, H. J., Dougherty, D. A., Mislow, K.: Chem. Ber. *113*, 1867 (1980)

55. Beckhaus, H.-D., Hellmann, G., Rüchardt, C., Kitschke, B., Lindner, H.-J., Fritz, H.: ibid. *111*, 3764 (1978)
56. Baxter, S. G., Fritz, H., Hellmann, G., Kitschke, B., Lindner, H.-J., Mislow, K., Rüchardt, C., Weiner, S., J. Am. Chem. Soc. *101*, 4493 (1979)
57. McMillen, D. F., Golden, D. M.: Ann. Rev. Phys. Chem. *33*, 493 (1982)
58. Meot-Ner, M.: J. Am. Chem. Soc. *104*, 5 (1982)
59. In this paper the resonance energy H_r of a radical is defined as the increase in stabilization energy of a radical center when an α-methyl group is exchanged for a substituent, e.g. phenyl. For a definition based on methane see Ref. 60
60. Burkey, T. J., Castelhano, A. L., Griller, D., Lossing, F. P.: J. Am. Chem. Soc. *105*, 4701, (1983)
61. Barbe, W., Beckhaus, H.-D., Rüchardt, C.: Chem. Ber. *116*, 1042 (1983)
62. Zamkanei, M., Kaiser, J. H., Birkhofer, H., Beckhaus, H.-D., Rüchardt, C.: Chem. Ber. *116*, 3216 (1983); unfortunately the numerical values for ΔH and ΔS of the dissociation equilibrium between *39* and the radicals *37* were confused. They should read: $\Delta H_{diss} = 35.8 \pm 0.4$ kcal · mol^{-1} and $\Delta S_{diss} = 31.1 \pm 0.8$ e.u.
63. Diplomarbeit Gleißner, R.: Univ. Freiburg 1983 and unpublished work
64. Dissertation Rausch, R.: Univ. Freiburg 1984
65. Dissertation Birkhofer, H.: Univ. Freiburg, in preparation
66. For references see Birkhofer, H., Beckhaus, H.-D., Rüchardt, C.: Tetrahedron Lett. *24*, 185 (1983)
67. Rüchardt, C.: Mechanismen radikalischer Reaktionen, Forschungsbericht des Landes Nordrhein-Westfalen Nr. 2471, Westdeutscher Verlag, Opladen 1975
68. Fleming, I.: Grenzorbitale und Reaktionen Organischer Verbindungen, Verlag Chemie, Weinheim 1979. Fukui, K.: Topics Curr. Chem. *15*, 1 (1970). Fujimoto, H., Yamabe, S., Minato, T., Fukui, F.: J. Am. Chem. Soc. *94*, 9205 (1972)
69. Viehe, H. G., Merényi, R., Stella, L., Janusek, Z.: Angew. *Chem. 91*, 982 (1979); Angew. Chem. Int. Ed. Engl. *18*, 917 (1979)
70. Dissertation Schaetzer, J.: Univ. Freiburg, in preparation
71. Barbe, W., Beckhaus, H.-D., Rüchardt, C.: Chem. Ber. *116*, 1058 (1983)
72. Bernlöhr, W., Beckhaus, H.-D., Rüchardt, C.: ibid. *117*. 1026 (1984)
73. Andersen, P.: Acta Chem. Scand. *19*, 629 (1965); Andersen, P., Kleve, B.: ibid. *21*, 2599 (1967)
74. Peters, K., v. Schnering, H.-G.: unpublished
75. Following the equilibration of meso-*39* and *D,L-39* via nmr gave activation parameters for the dissociation of meso-*39* slightly different from those found previously by a different method [62] $\Delta G^{\neq}$ (300 °C) = 29.2 kcal · mol^{-1}, $\Delta H^{\neq} = 35.4 \pm 0.2$ kcal · mol^{-1}, $\Delta S = 10.9 \pm 0.5$ e.u. in benzene (Birkhofer, H.: unpublished, 1984). The previously used scavenger method [62] seems less reliable due to cage recombination
76. Korth, H.-G., Sustmann, R., Merényi, R., Viehe, H. G.: J. Chem. Soc., Perkin Trans. 2, *1983*, 67
77. A similar situation was described recently in an other context by Houk, K. N., Rondan, N. G., Mareda, J.: J. Am. Chem. Soc. *106*, 4291 (1984); Houk, K. N., Rondan, N. G.: ibid. 4293. We have reasons to assume that another, sandwich-like arrangement finds some stabilization by secondary valence interactions
78. Beckhaus, H.-D., Hellmann, G., Rüchardt, C., Kitschke, B., Lindner, H. J.: Chem. Ber. *111*, 3780 (1978)
79. Hellmann, S., Beckhaus, H.-D., Rüchardt, C.: ibid. *116*, 2219 (1983)
80. For the discussion of relationships between bond character and bond length see also Bartell, L. S.: Tetrahedron *17*, 1177 (1962); Szabò, Z. G., Konkoly Thege, 1.: Acta Chim. Acad. Hung. Tom. *86*, 127 (1975)
81. Hase, W. L.: Acc. Chem. Res. *16*, 258 (1983)

Selective Hydroboration and Synthetic Utility of Organoboranes Thus Obtained

Akira Suzuki[1] **and Ranjit S. Dhillon**[2]

1 Department of Applied Chemistry, Faculty of Engineering, Hokkaido University, Sapporo 060, Japan

2 Department of Chemistry, Punjab Agricultural University, Ludhiana 141004, India

Table of Contents

1 Introduction

The first practical synthesis of organoboranes involved the reaction of organometallic derivatives with boron esters or halides [1]. As this reaction required the prior formation of reactive organometallics followed by the conversion to organoboranes, there was little incentive to explore the synthetic utility of organoboranes [2]. Fortunately, the remarkable facile reactions of alkenes, alkynes, dienes, and enynes with a wide variety of hydroboration agents (Chart 1) [7] have opened a new and convenient route to a large variety of organoboranes. The resulting organoboranes have proven to be valuable intermediates in the formation of C—H, C—O, C—N, C-Halogen, C-Metal, and C—C bonds. For example, the facile hydroboration of alkynes with sterically demanding hydroborating agents gives vinylic boranes, which provide substituted olefins, dienes, and enynes of defined stereochemistry [3]. Moreover, the asymmetric hydroboration is now an important and simple tool available to organic chemists for asymmetric synthesis [4,5]. With the in situ synthesis of organoboranes from haloboranes via hydridation have widened the scope of hydroboration reaction [6]. Via hydroboration-hydridation-hydroboration, it is now possible to prepare otherwise inaccessible organoboranes. The hydroboration reactions are chemo-, regio-, and stereoselective. Thus, the convenience of these reactions makes the organoboranes one of the most versatile reagents available.

A series of books [7] and reviews [8] written on the utility of these reactions highlights the synthetic applicability of this field. However, there was no attempt to systematize the otherwise vast literature available for selective hydroboration, e.g., hydroboration of double bond in the presence of triple bond and vice versa, hydroboration of one of the double bonds of dienes, monohydroboration of alkynes, and selective hydroboration of carbon-carbon multiple bonds in the presence of many other functional groups. These topics are the aim of this review. Included here also is a brief account of the developing field of haloboration.

Diborane B_2H_6

Monoalkylboranes

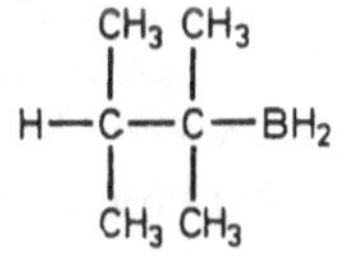

2,3-Dimethyl-2-butylborane (Thexylborane, $ThxBH_2$, ⫲BH_2)

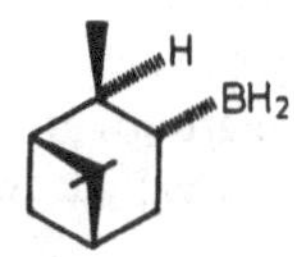

Monoisopinocampheylborane

Dialkylboranes

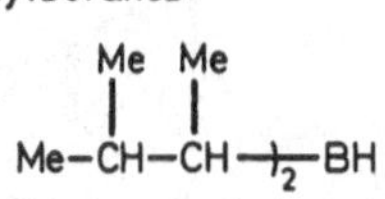

Bis(3-methyl-2-butyl)borane (Disiamylborane, Sia_2BH)

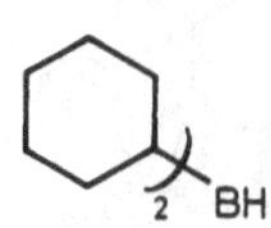

Dicyclohexylborane (Chx_2BH)

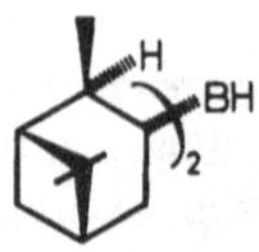

Diisopinocampheylborane
(IPC_2BH)

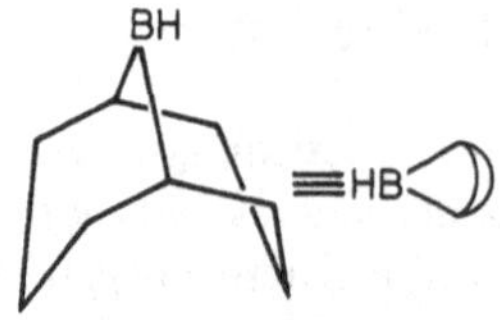

9-Borabicyclo(3,3,1)nonane
(9-BBN)

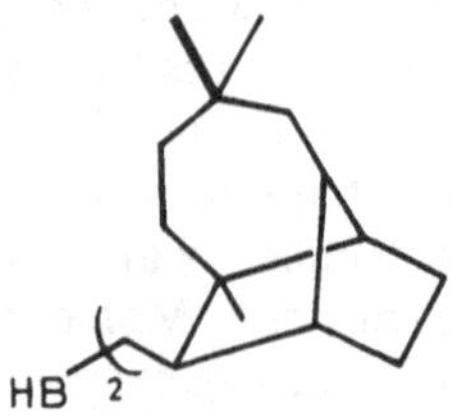

Dilongifolylborane
(Lgf_2BH)

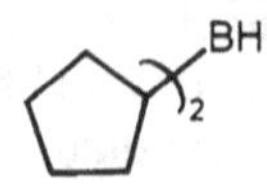

Dicyclopentylborane

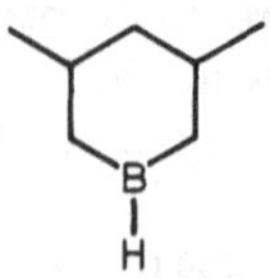

3,5-Dimethylborinane
(3,5-DMBN)

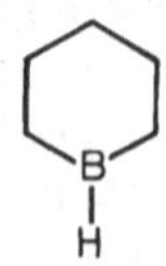

Borinane

Heterosubstituded boranes

Monohaloboranes	H_2BX	
Dihaloboranes	HBX_2	X = Br or Cl

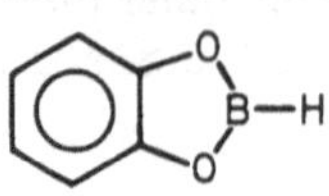

1,3,2-Benzodioxaborole
(Catecholborane)

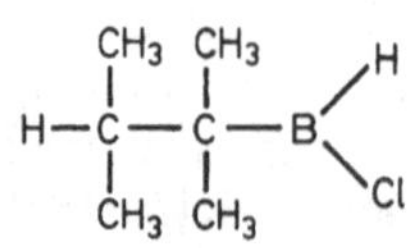

2,3-Dimethyl-2-butylchloroborane
(Thexylchloroborane, ThxBHCl)

Chart 1

2 General Remarks

The hydroboration of carbon-carbon double bonds with diborane leads conveniently to trialkylboranes. The presence of second unhindered double or triple bonds, or functional groups generally affords a complicated mixture of products when diborane is employed as hydroborating agent, arising from competing hydroboration Chart 2).

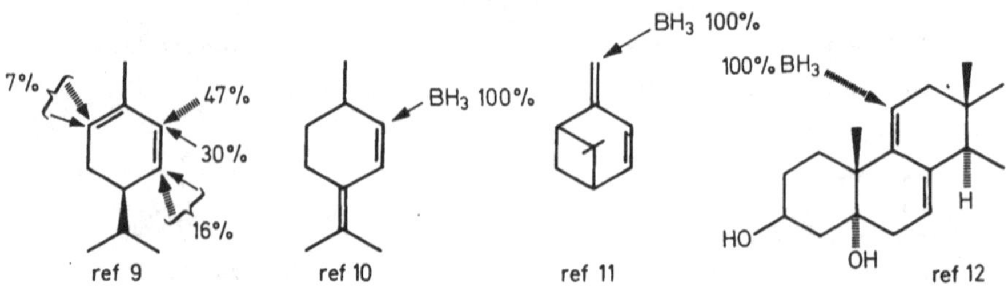

Chart 2

It has been observed that some conjugated polyenes can be selectively hydroborated with borane [13] (Chart 3) probably due to the low reactivity of conjugated system in the intermediate organoborane. In some cases, the diene moiety within the polyene system has been protected by complexation [14–16].

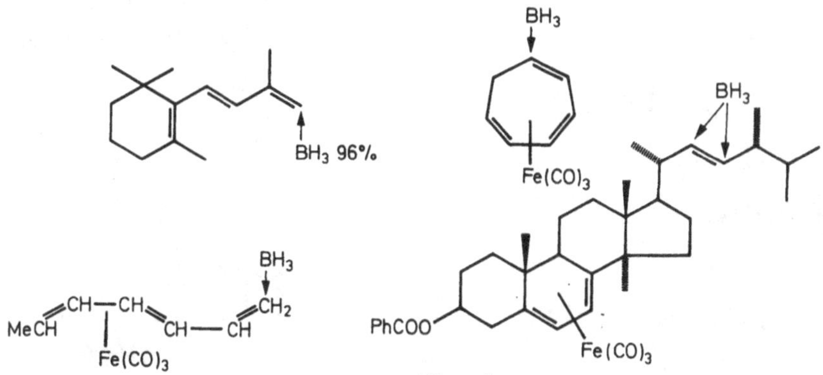

Chart 3

With the availability of a variety of sterically demanding hydroborating agents (Chart 1) it is now possible to hydroborate one of the double or triple bonds as shown in Charts 4 and 5.

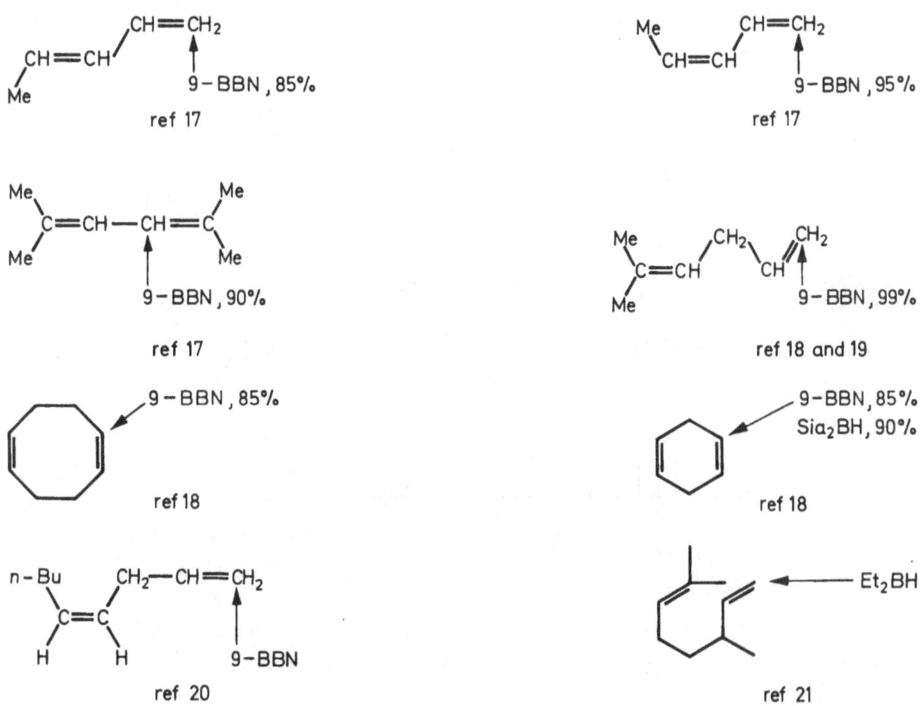

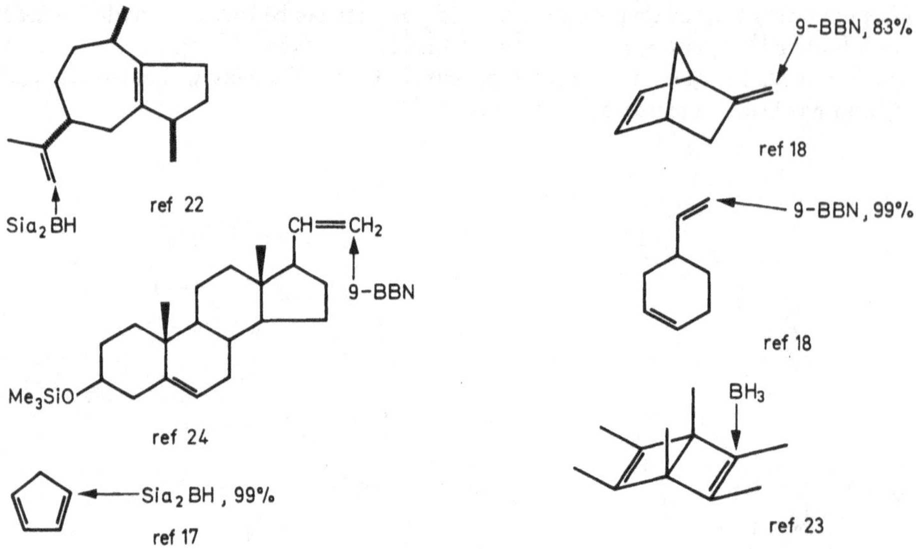
9-BBN, 83%
ref 18
ref 22
Sia_2BH
$CH{=}CH_2$
9-BBN
9-BBN, 99%
ref 18
Me_3SiO
ref 24
BH_3
Sia_2BH, 99%
ref 17
ref 23

Chart 4

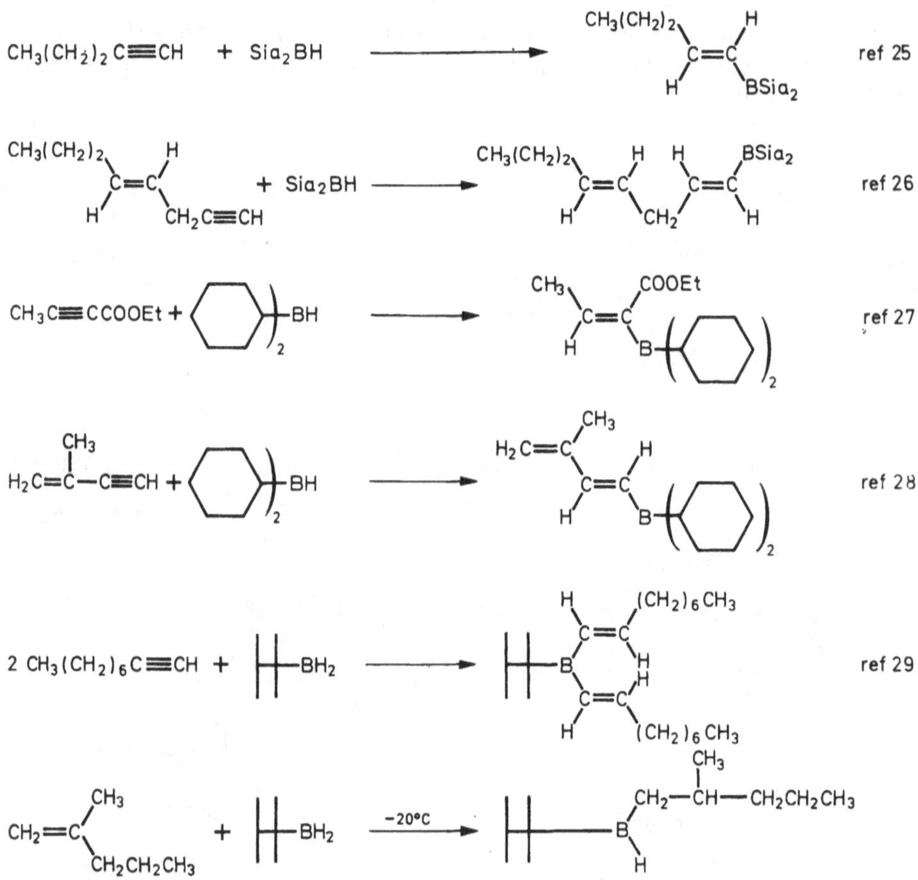
ref 25
ref 26
ref 27
ref 28
ref 29
−20°C

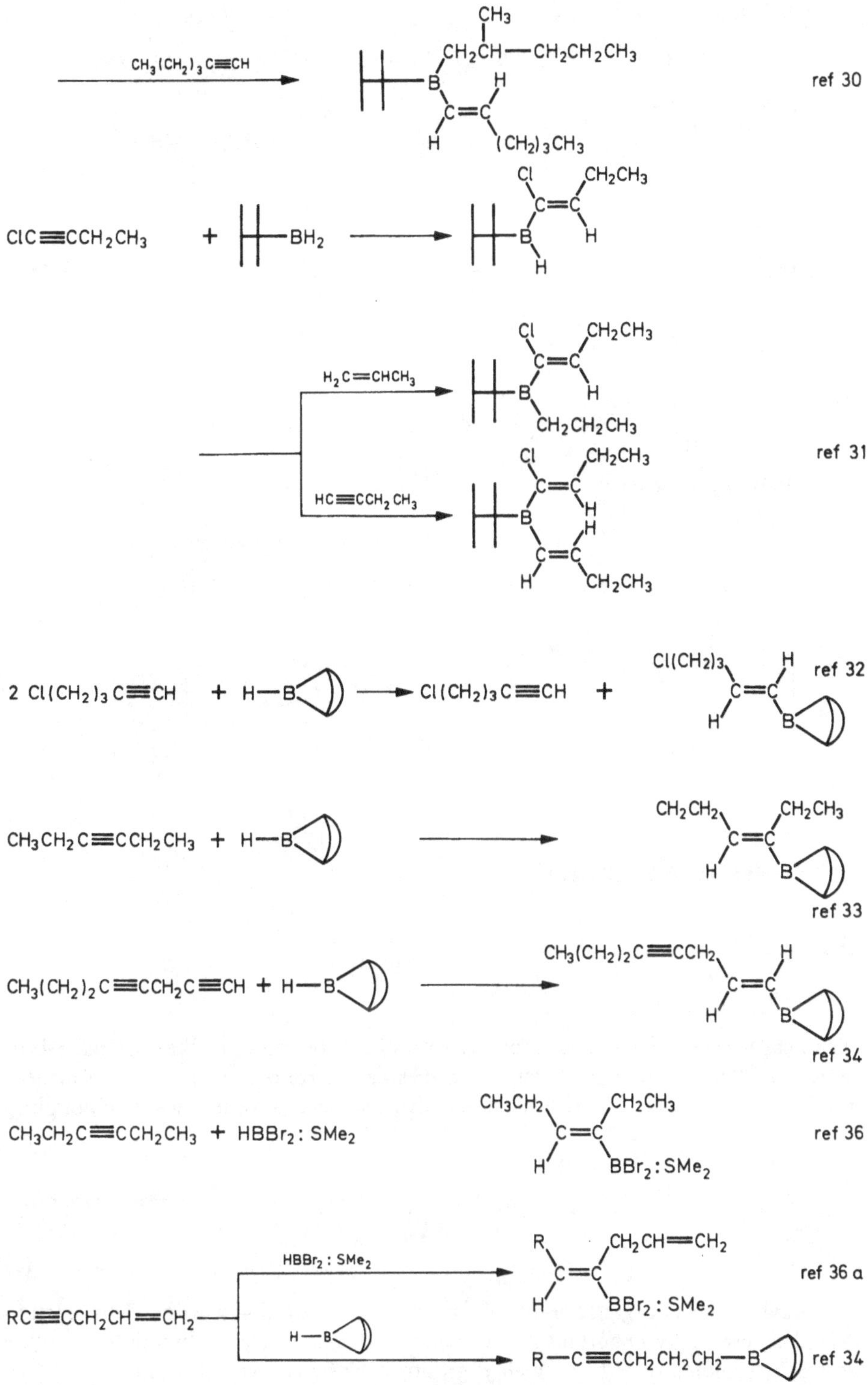
ref 30
ref 31
ref 32
ref 33
ref 34
ref 36
ref 36 a
ref 34

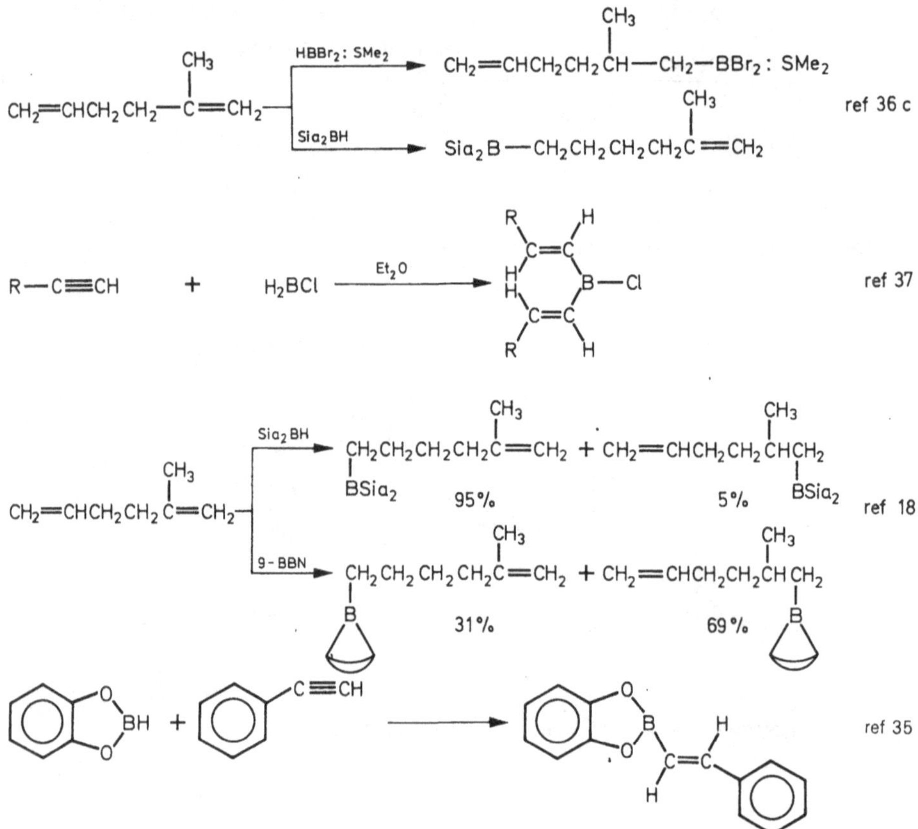

Chart 5

3 Synthetic Applications

3.1 Synthesis of Alcohols

3.1.1 Via Bromination

The light induced reactions of bromine with organoboranes proceed through α-bromination [38] followed by facile migration of an alkylgroup in the presence of nucleophiles, such as water [39,40] (Eq. 1). Thus, this sequence provides means of coupling

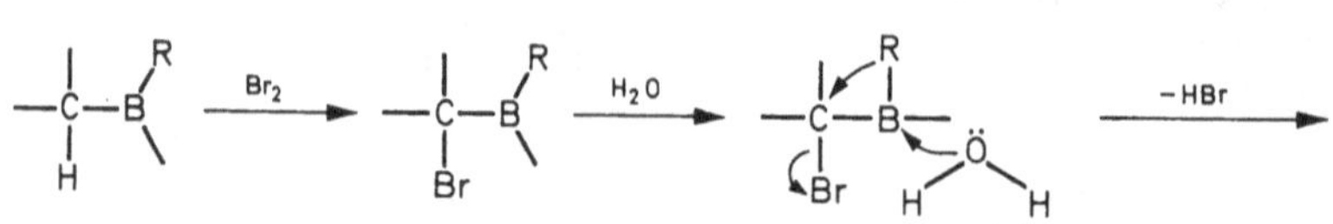

(1)

two alkyl groups of organoboranes via bromination-oxidation reaction. This reaction would be especially important if it would be possible to couple two different alkyl groups and moreover, when the alkyl groups have functional substitutes.

Thexyldialkylboranes appeared to be especially promising as these are readily available from olefins via selective stepwise hydroboration [29,40] and can easily accommodate functional derivatives [41]. Furthermore, the thexyl group does not have an α-hydrogen to be abstracted and it exhibits low migratory aptitude in this reaction. Consequently, the totally "mixed" thexylboranes are ideal intermediates for coupling of two different groups with simultaneous introduction of a hydroxy group at the point of coupling [42]. Representative thexyldialkylboranes react readily with bromine under the influence of normal laboratory light. An alkyl group capable of yielding the more stable free radical in the α-position is preferentially brominated. Alkaline hydrogen peroxide oxidation of the products gave the corresponding alcohols in good yields (Eqs. 2 and 3).

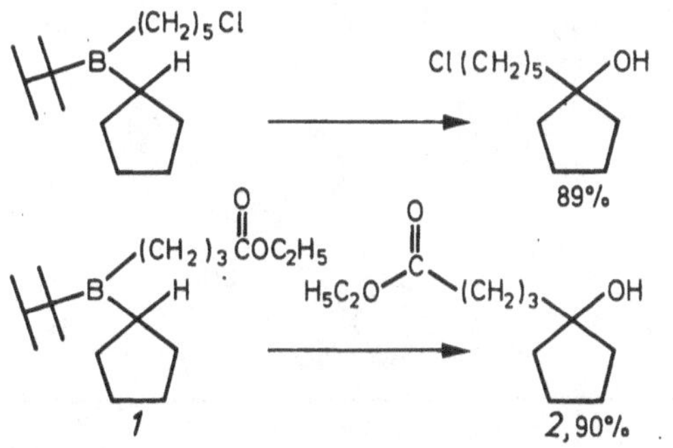

(2)

(3)

Interestingly, an excess of bromine (3 equivalents) was required to obtain the maximum yield of 1-(3-ethoxycarbonylpropyl)cyclopentanol (**2**) from the corresponding thexyldialkyborane (**1**). When one equivalent of bromine was used, 2,3-dimethyl-2,3-dibromobutane, thexylalcohol and cyclopentanol were formed together with the desired product (**2**). However, the bromination of thexyl(10-methoxycarbonyldecyl)-cyclopentylborane (**3**) with one equivalent of bromine occurred in the normal manner to give, upon oxidation, an excellent yield of 1-(10-methoxycarbonyldecyl)cyclopentanol. These results suggest that the bromine attacks both α (cyclopentyl) and β (thexyl) tertiary hydrogens in **1**, but only the α (cyclopentyl) tertiary hydrogens in **3**. This loss of regiospecific nature of the α-bromination reaction could possibly be due to an intramolecular boron-oxygen coordination as represented in structures (**2a**) and (**2b**).

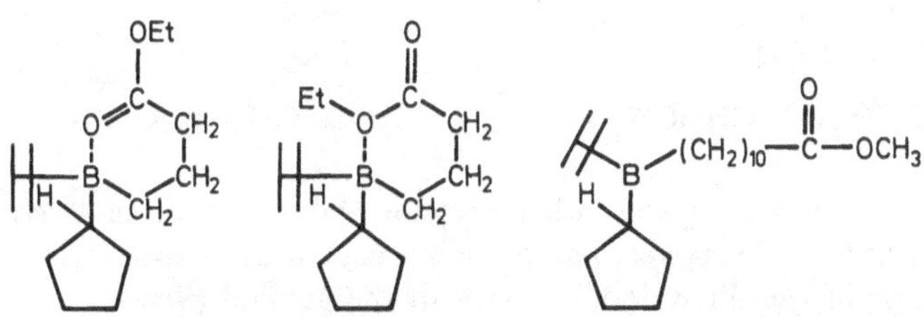

The results of coupling of two different alkyl groups via bromination-oxidation are assembled in Table 1.

Table 1. Coupling of Two Different Alkyl Groups via Bromination-Oxidation of Thexyldialkylboranes [42]

Olefin A	Olefin B	Product	Yield (%)
	$CH_2{=}CH{-}CH_2CH_3$	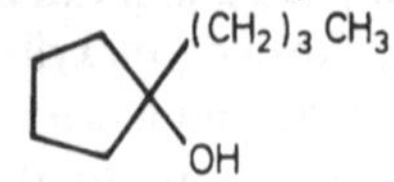	85
//	$CH_2{=}CH{-}(CH_2)_3Cl$	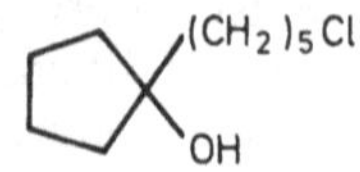	89
//	$CH_2{=}CH{-}CH_2COOC_2H_5$	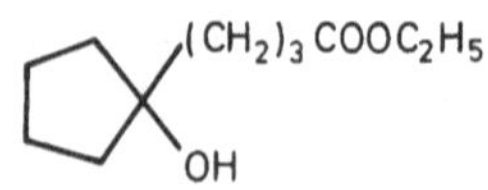	72
//	$CH_2{=}CH{-}(CH_2)_8COOCH_3$	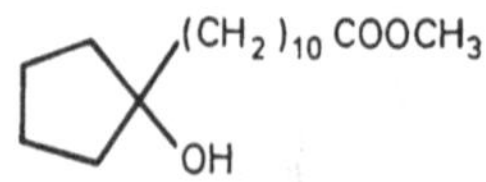	75
//	$CH_2{=}CH{-}(CH_2)_3CH_3$	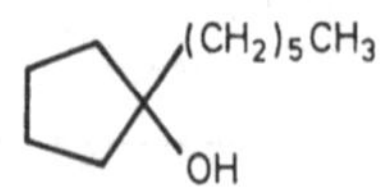	94

3.1.2 Synthesis of Unsaturated Alcohols

Disiamylborane is a highly selective hydroborating agent with excellent steric control. Thus, it reacts selectively with less substituted olefins in the presence of more substituted [43] as:

$$\underset{\longleftarrow\ \text{Reactivity}}{RCH{=}CH_2 \quad R_2C{=}CH_2 \quad R_2C{=}CHR}$$

This generalization is true for both simple as well as conjugated dienes (Eqs. 4 and 5)[44].

$$CH_2{=}\overset{CH_3}{\overset{|}{C}}{-}CH_2CH_2CH{=}CH_2 \xrightarrow[\text{ii) [O]}]{\text{i) Sia}_2\text{BH}} CH_2{=}\overset{CH_3}{\overset{|}{C}}{-}(CH_2)_4OH \tag{4}$$

$$CH_2{=}\overset{CH_3}{\overset{|}{C}}{-}CH{=}CH_2 \xrightarrow[\text{ii) [O]}]{\text{i) Sia}_2\text{BH}} CH_2{=}\overset{CH_3}{\overset{|}{C}}{-}CH_2CH_2OH \tag{5}$$

This feature has been clearly extended to the hydroboration of myrcene (**4**) which has three reactive double bonds. The use of one equivalent of disiamylborane converts selectively one of the three double bonds to the alcohol (**5**) while two equivalents afford a diol (**6**) (Eq. 6) [45,46].

Grandisol (**8**), one of the synergistic mixture of the four compounds (collectively called "grandlure") of sex pheromones of male boll weevil (*Anthonomus grandis)* has been synthesized by the following two-step procedure (Eq. 7) [47]. Racemic grandisol

(**8**) was obtained in 52% yield. This synthetic process has now been modified and scaled up to produce commercial quantities of grandisol [47].

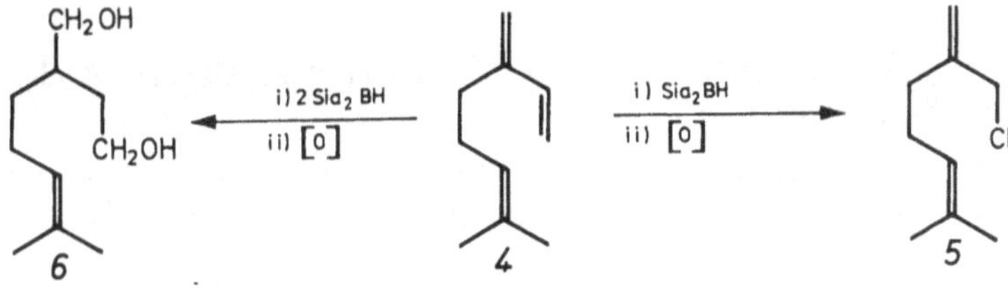

(6)

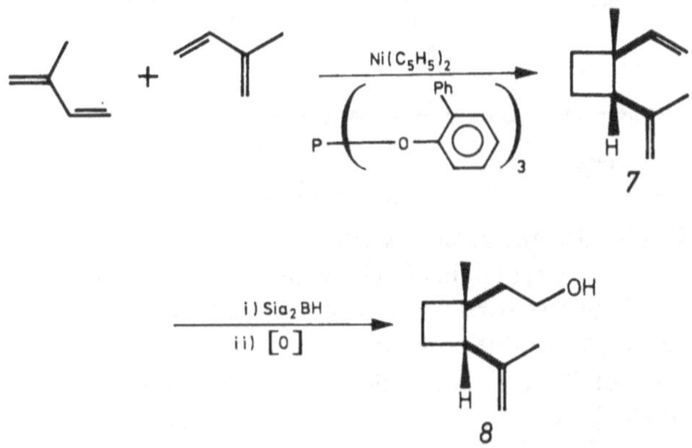

(7)

The sluggish reaction of disiamylborane with tri- and tetrasubstituted alkenes as compared to terminal and most internal disubstituted alkenes, makes it one of the most versatile hydroborating agents. The relative reactivity of disiamylborane with alkenes, dienes, and alkynes varies over a range of 10^4 (Table 2) [26], whereas that of diborane varies in the range of 20–30.

Table 2. Relative Reactivities of the Hydroboration of Various Alkenes, Dienes, and Alkynes with Disiamylborane in THF at 0 °C [26]

1-Hexyne	373	(*Z*)-3-Hexene	2.0
3-Hexyne	225	Cyclopentene	1.4
1,6-Hexadiene	176	2,4,4-Trimethyl-1-pentene	0.8
1-Octene	108	(*Z*)-4-Methyl-2-pentene	0.5
1-Pentene	105	(*E*)-2-Butene	0.4
1-Hexene	100	(*E*)-2-Pentene	0.3
3-Methyl-1-butene	57	(*E*)-3-Hexene	0.2
Cyclooctene [48]	26	(*E*)-4-Methyl-2-pentene	0.1
Cycloheptene [48]	7	(*Z*)-2,4-Dimethyl-2-pentene	0.1
3-Methyl-1-pentene	4.9	(*Z*)-4,4-Dimethyl-2-pentene	0.08
3,3-Dimethyl-1-butene	4.7	(*E*)-2,4-Dimethyl-2-pentene	0.04
β-Methylstyrene	2.3	Cyclohexene	0.01
(*Z*)-2-Butene	2.3	(*E*)-4,4-Dimethyl-2-pentene	0.01
(*Z*)-2-Pentene	2.0		

Disiamylborane has been demonstrated to be a highly selective hydroborating agent of C=C bonds in the presence of functional groups like carboxylic acids which react readily with diborane (Eq. 8). The hydroboration of 3-chloro-1-alkynes using a similar

$$CH_2{=}CH(CH_2)_8COOH \xrightarrow[\text{ii) [O]}]{\text{i) Sia}_2\text{BH}} HOCH_2(CH_2)_9COOH \qquad (8)$$

reaction sequence gives mainly the product arising from the migration of one of the siamyl groups from boron to the adjacent carbon atom (eq. 9)[49].

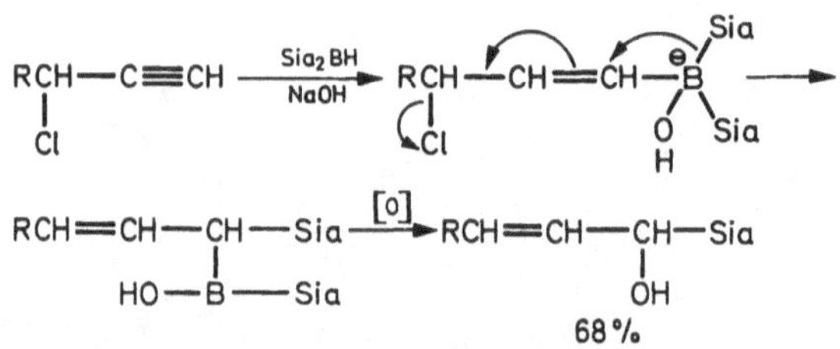

(9)

Double bonds with different geometry behave in a unique way. For example, the monohydroboration of caryophyllene (**9**) with dicyclohexylborane followed by oxidation brought about the participation of the (*E*)-trisubstituted double bond in preference to the exocyclic double bond during hydroboration to give the corresponding alcohol (**10**) (Eq. 10)[50]. However, under similar conditions, isocaryophyllene (**11**), which has a (*Z*)-endocyclic double bond, affords a mixture of unsaturated alcohols

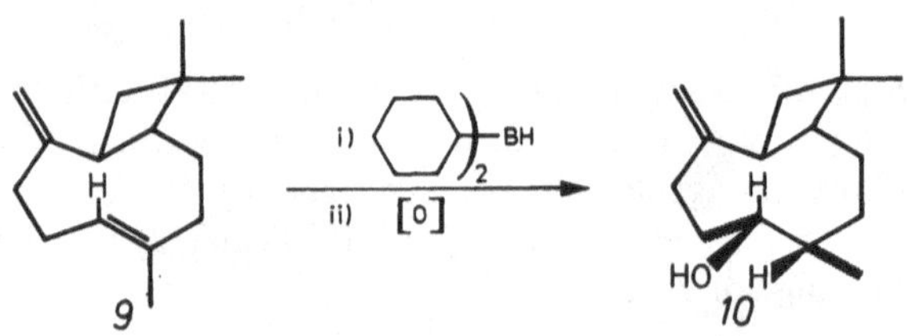

(10)

(**12**) and (**13**) in a ratio of 4:1 (Eq. 11). A complete selectivity for the exocyclic double bond has been achieved using thexylborane. This reagent gives **12** as the sole product. The selectivity as well as reactivity of caryophyllene may be attributed to the strain

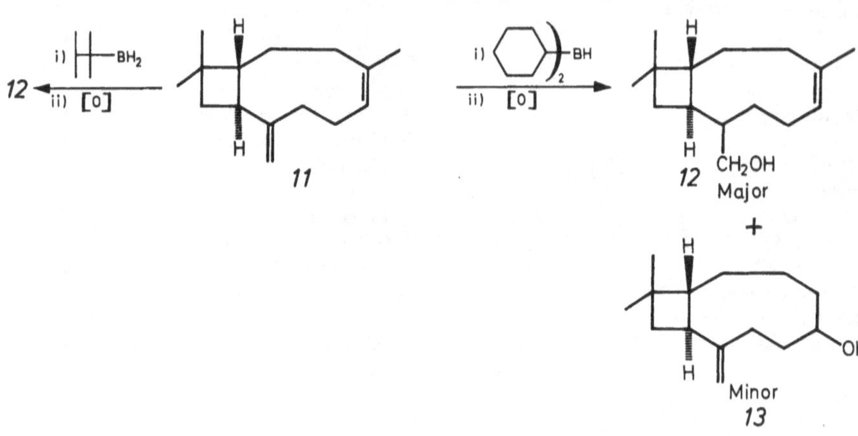

(11)

in the (*E*)-double bond in the 9-membered ring. The exocyclic bouble bond remains unchanged even though the molar ratio of caryophyllene to dicyclohexylborane is increased from 1 : 1 to 1 : 2. There was no contamination of the dihydroboration product in spite of 100% excess of dicyclohexylborane present.

3.1.3 Synthesis of 1,3-Enynols and 1,2,4-Trienols

The stereochemically defined 1,3-enynols and 1,2,4-trienols are not readily available by conventional methodologies. These compounds have now been synthesized from organoboranes [51]. The hydroboration of alkynes with thexylmonochlorobrane (**14**) affords the corresponding thexylalkenylchloroboranes (**15**). The addition of **15** to two equivalents of lithium chloropropargylide at –78 °C gave, via the intermediacy of the "ate" complexes (**16**), the organoboranes (**17**). The reaction mixture containing **17** upon treatment with an aldehyde yielded the 1,3-enynols (**18**) after oxidation. However, if the initially formed organoboranes (**17**) were brought to room temperature, they rearranged to **19**, the reaction of which with aldehydes gives 1,2,4-trienols (**20**). The reaction sequence is indicated in Scheme 1.

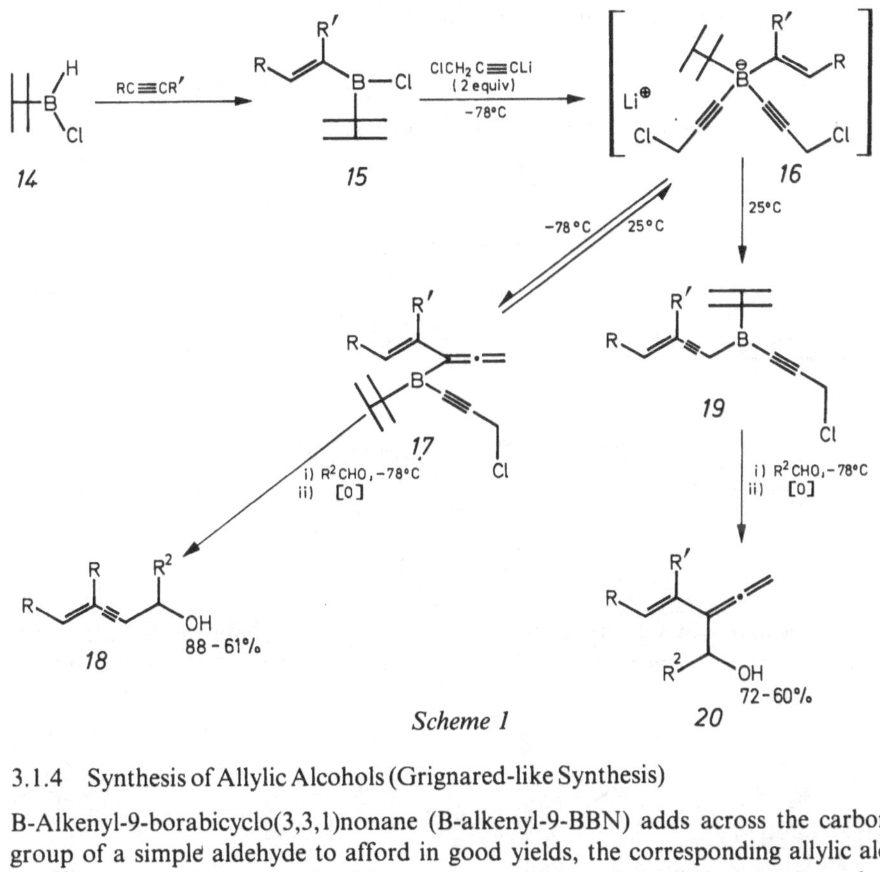

Scheme 1

3.1.4 Synthesis of Allylic Alcohols (Grignared-like Synthesis)

B-Alkenyl-9-borabicyclo(3,3,1)nonane (B-alkenyl-9-BBN) adds across the carbonyl group of a simple aldehyde to afford in good yields, the corresponding allylic alcohol [52]. The reaction proceeds with complete retention of vinylic borane stereochemistry resulting in the *trans*-disubstituted olefin linkage in the final product. The reactio tolerates a wide variety of functionalities which are not possible in Grignard synthesis of alcohols. Thus 5-chloro-1-pentyne was hydroborated with 9-BBN and then reacted with propionaldehyde to give (*E*)-8-chloro-4-octen-3-ol (Eq. 12). Evi-

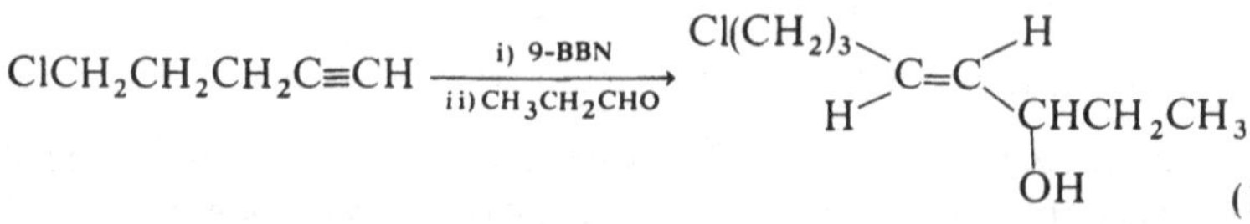

(12)

dently, this procedure proves a remarkably simple, stereospecific synthesis of allylic alcohols which was previously possible only with more reactive organometallic compounds.

3.2 Synthesis of Aldehydes

Aldehydes can be synthesized by the monohydroboration of alkynes followed by oxidation (Eqs. 13 [25], 14 [53], and 15 [54]).

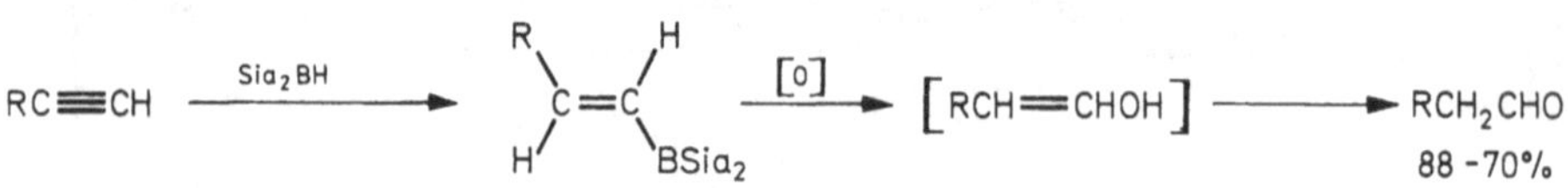

(13)

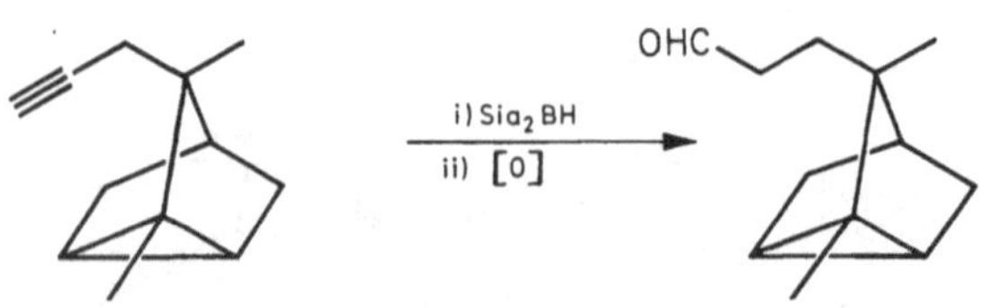

(14)

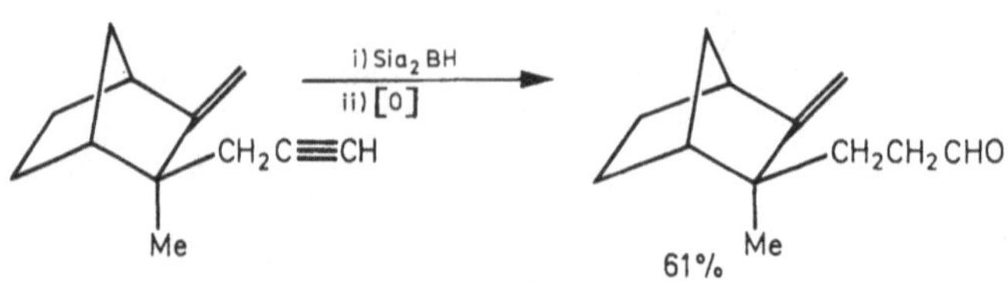

(15)

1-Alkynes add slowly to triallylborane at 20 °C to give a monoaddition product (**21**) which on treatment with methanol followed by oxidation produces 2-alkyl-4-pentenal (Eq. 16) [55].

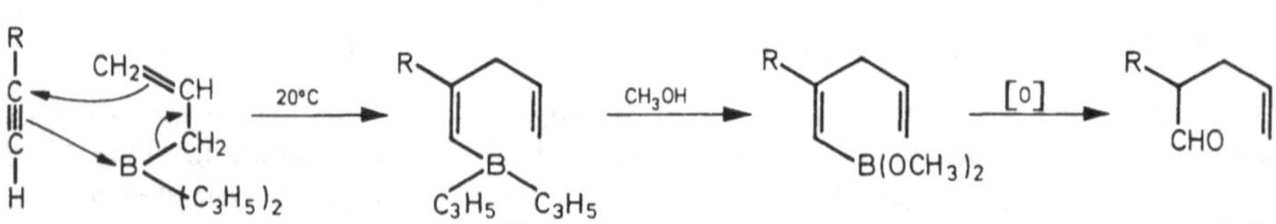

21

(16)

Most recently, the homologation of 2-alkyl-1,3,2-dioxaborinanes (**22**) readily available by hydroboration and reaction with 1,3-bis(trimethylsiloxy)propane, to α-methoxyalkyl derivatives (**23**), was achieved by reaction with LiCH(OMe)SPh (MPML) followed by treatment with $HgCl_2$. The intermediates (**23**) were smoothly oxidized with hydrogen peroxide in a pH 8 phosphate buffer to give the corresponding aldehydes in good yields (Eq. 17) [56].

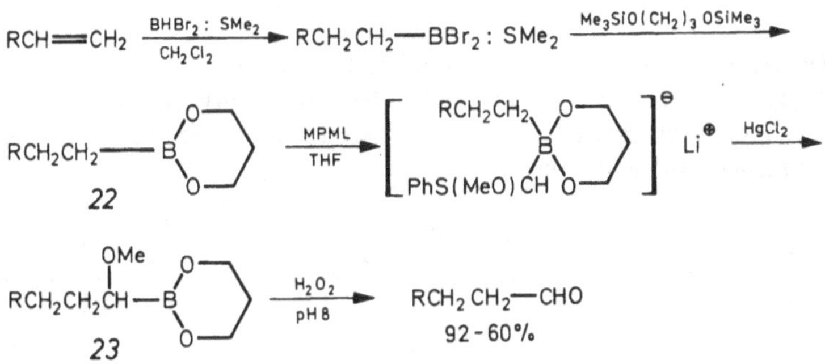

(17)

Palladium-catalyzed cross-coupling of (2-ethoxyvinyl)borane readily prepared by the hydroboration of ethoxyethyne, with aryl and benzylic halides provides a convenient method for conversion of such halides into aldehydes with two more carbon atoms (Eq. 18) [57].

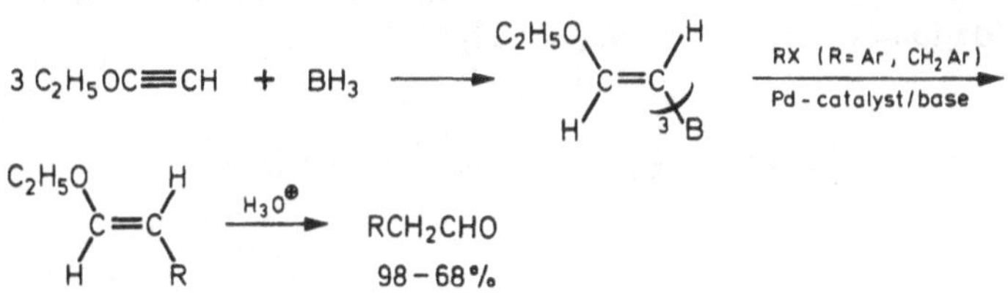

(18)

3.3 Synthesis of Ketones

Thexylborane adds to hindered olefins only once, and the product hydroborates only unhindered olefins thus leading to mixed trialkylboranes. This selective stepwise hydroboration with thexylboranes gives a simple synthesis of unsymmetrical ketones after carbonylation (Eqs. 19 and 20) [41,58,59].

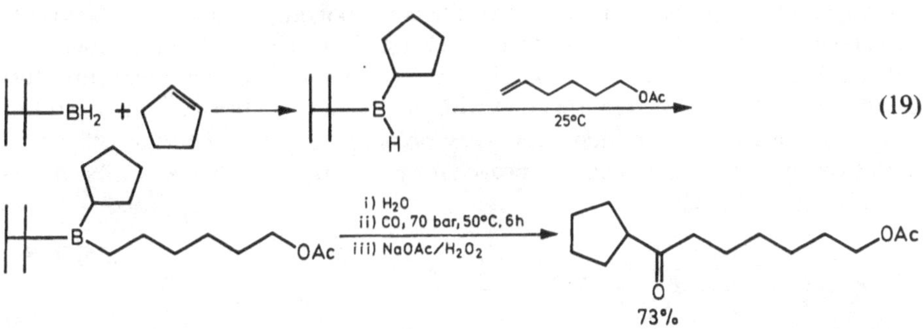

(19)

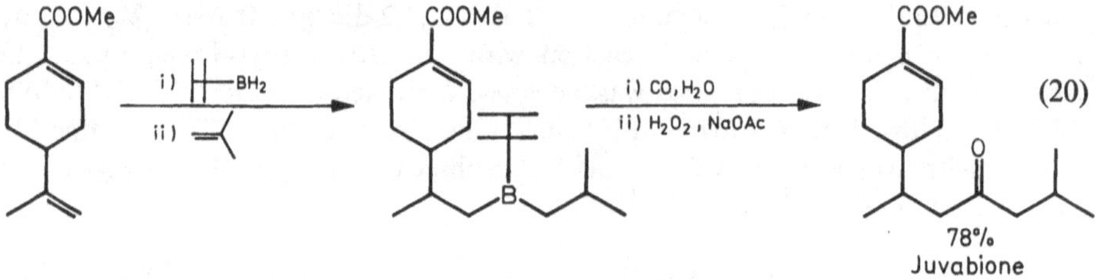

(20)

The results of these experiments have been summarized in Table 3. In no case has the migration of the thexyl group been observed to any significant extent. Although the mechanism of the reaction has not been established, the following (Eq. 21) is consistent with the experimental data.

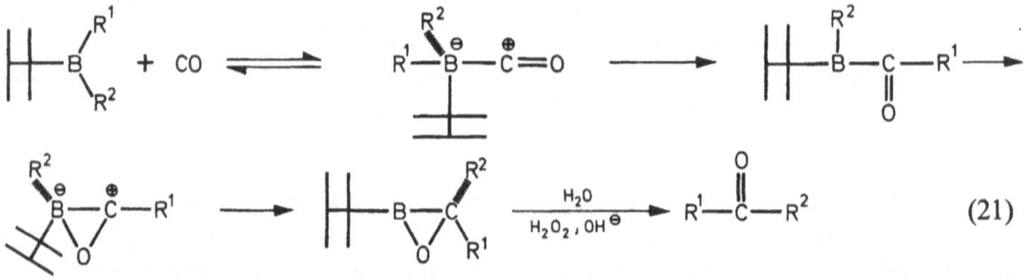

(21)

An alternative procedure to convert thexyldialkylboranes to unsymmetric ketones via cyanidation has been reported (Eq. 22) [60,61].

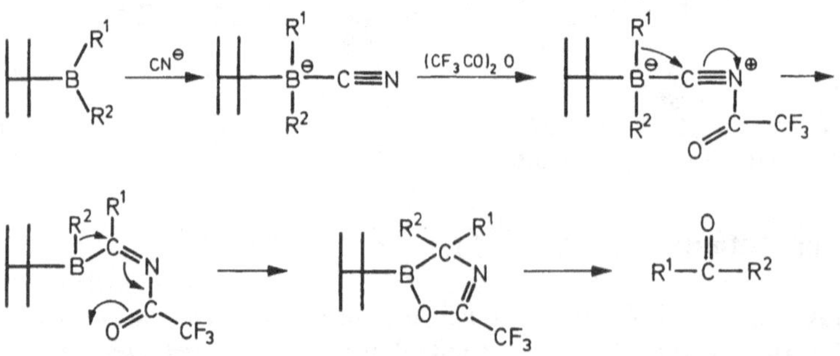

(22)

As a consequence of the low migratory aptitude of the thexyl group, thexylborane is an exceptionally valuable reagent which can be used to couple two olefins either by carbonylation or by cyanidation to form the corresponding ketones [62]. A sterically demanding alkene in the first step can be used selectively to synthesize unsymmetrical ketones. Unfortunately, the synthesis fails in the attempt to couple two monosubstituted alkenes [63]. It has been reported that with thexylchloroboranes one can avoid this synthetic limitation of thexylboranes. They provide via a hydroboration-alkylation sequence; an operationally simple approach for preparing hitherto unaccessible mixed thexylboranes (Eq. 23) [64].

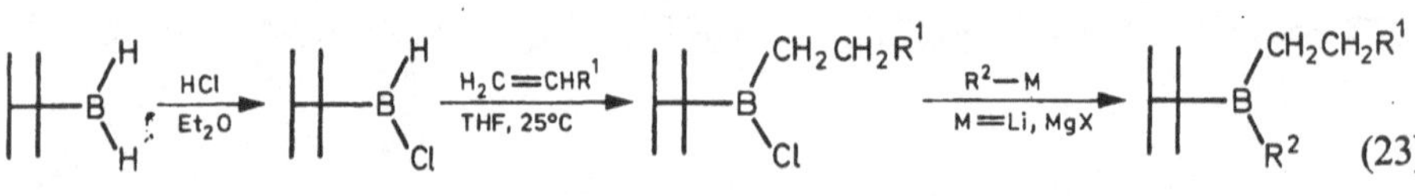

(23)

Table 3. Synthesis of Unsymmetrical Ketones via Thexyldialkylboranes

Olefin A	Olefin B	Product	Yield (%)	Ref.
$(CH_3)_2C{=}CH_2$	$CH_2{=}CHCH_2COOC_2H_5$	$(CH_3)_2CHCH_2CO(CH_2)_3COOC_2H_5$	84	[41]
	$CH_2{=}CHCH_2CN$	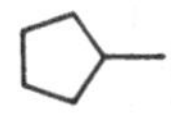 $—CO(CH_2)_3CN$	45	[41]
	$CH_2{=}CHCH_2Cl$	 $—CO(CH_2)_3Cl$	63	[41]
	$CH_2{=}CHCH_2OAc$	 $—CO(CH_2)_3OAc$	66	[58]
$CH_3(CH_2)_2C(CH_3){=}CH_2$	$CH_2{=}CH(CH_2)_2OAc$	$CH_3(CH_2)_2CH(CH_3)CH_2CO(CH_2)_4OAc$	61	[58]
$CH_3CH{=}CHCH_3$	$CH_2{=}CH(CH_2)_3OAc$	$CH_3CH_2CH(CH_3)CO(CH_2)_5OAc$	63	[58]
	$CH_2{=}CH(CH_2)_4OAc$	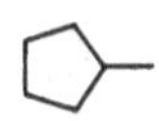 $—CO(CH_2)_6OAc$	73	[58]
$CH_3(CH_2)_2CH{=}CH_2$	$CH_2{=}CH(CH_2)_9OAc$	$CH_3(CH_2)_4CO(CH_2)_{11}OAc$	42	[58]
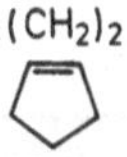	$CH_2{=}CH(CH_2)_5CH_3$	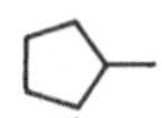 $—CO(CH_2)_7CH_3$	83	[60]
	$CH_2{=}CH(CH_2)_6Cl$	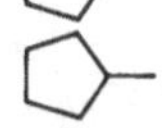 $—CO(CH_2)_8Cl$	76	[60]
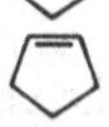	$CH_2{=}CH(CH_2)_6I$	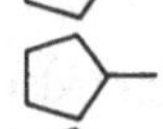 $—CO(CH_2)_8I$	76	[60]
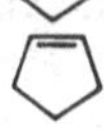	$CH_2{=}CHCH_2$— 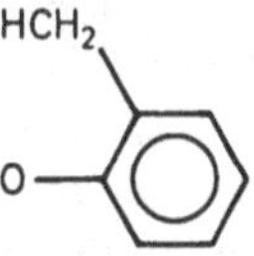 CH_3COO—	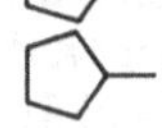 $—CO(CH_2)_3$— 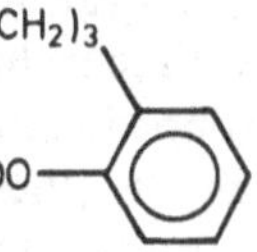 CH_3COO—	80	[60]

This novel procedure thus greatly enhances the synthetic utility of thexyl substituted organoboranes for the preparation of ketones via cyanidation reaction [65]. The utility of the procedure is exemplified by the preparation of (*Z*)-6-heneicosen-11-one (**24**) [64], a sex pheromone of Douglass fir tussock moth (Eq. 24).

The results of these experiments are summarized in Table 4.

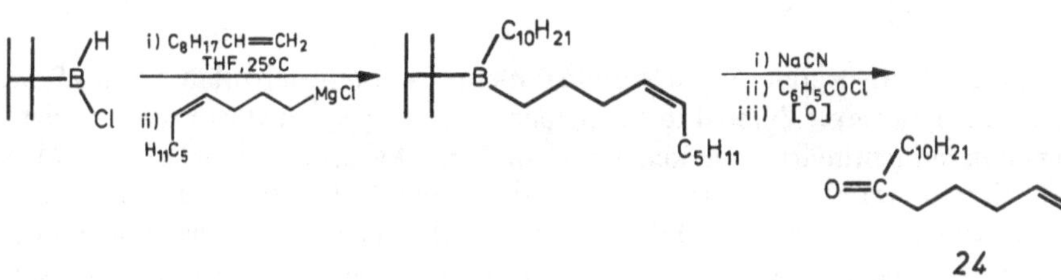

(24)

Table 4. Preparation of Mixed Thexyldiorganoboranes and their Conversion into Ketones via Cyanidation [64]

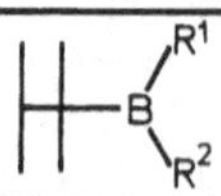

R^1 Derived from	R^2 Derived from	Structure	$R^1-\underset{\underset{O}{\Vert}}{C}-R^2$ (%)
1-Hexene	BuLi	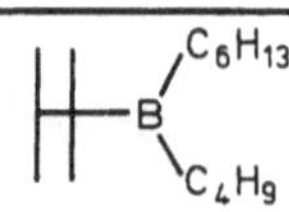	80
1-Octene	CH_3Li	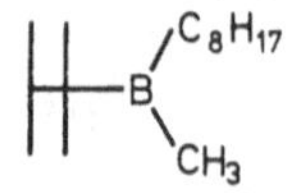	72
1-Octene	C_2H_5MgBr	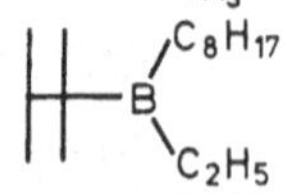	74
1-Decene	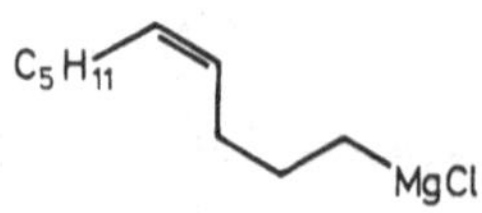	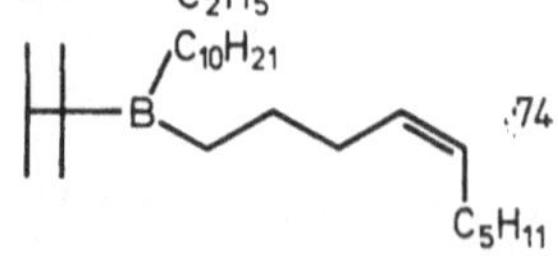	74

This reaction, however, is limited to the availability of active organometallic compounds for the alkylation reaction. Moreover, the hydroboration of olefins in THF or diethyl ether with thexylmonochloroborane at 0 °C is sluggish and at high temperature leads to disproportionation [37]. These problems, however, can be solved by changing the solvent and synthesizing thexylchloroborane from 2,3-dimethyl-2-butene and monochloroborane-methyl sulfide [66] in methylene chloride. The hydroboration proceeds smoothly forming 98% pure ThxBHCl:SMe_2 (Eq. 25). This reagent is remarkably stable at room temperature and hydroborates reactive olefins (terminal or simple disubstituted) rapidly and cleanly to give pure thexylmonoalkylchloroborane [66–68].

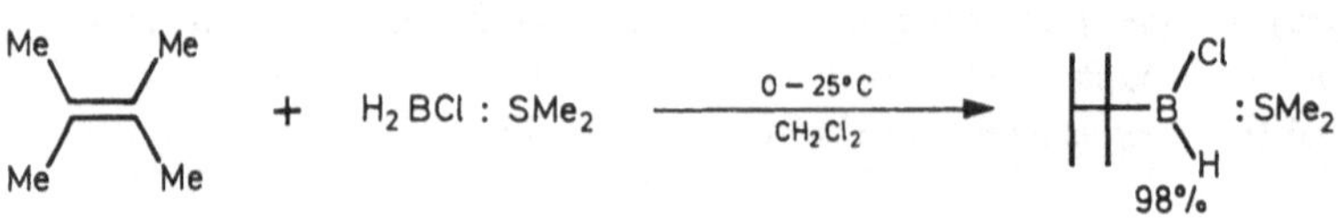

(25)

In general, the hydroboration of terminal olefins is complete within 1 h at 25 °C. The reagent is, however, limited to the preparation of thexylalkylchloroborane derivatives containing primary or unhindered secondary alkyl groups. Thus a chemist has in hand thexylborane, which cleanly leads to monohydroboration stage in case of hindered olefins, while on the other hand thexylchloroborane-methylsulfide hydroborates only unhindered olefins (Eq. 26). As the hydridation sequence provides thexylmonoalkylborane, consequently, these valuable reagents are used for selective introduction of alkenes as depicted in Eq. 26.

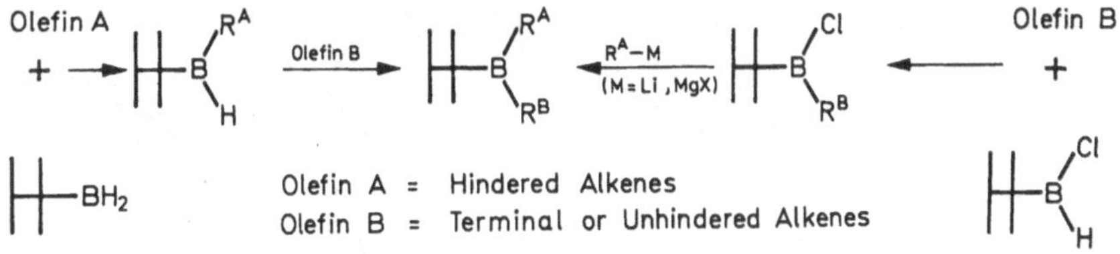

(26)

The relative reactivities of the olefins with thexylmonochloroborane-methylsulfide is presented in Table 5.

Table 5. Relative Reactivities of Representative Olefins toward ThxBHCl:SMe_2 in CH_2Cl_2 at 25 °C (1-Hexene = 100) [69]

Olefins	Relative reactivity	Olefin	Relative reactivity
1-Hexene	100	*cis*-4,4-Dimethyl-2-pentene	0.28
1-Octene	98	2-Methyl-2-pentene	0.26
Cyclooctene	89	1-Methylcyclopentene	0.15
3-Methyl-1-hexene	48	*trans*-3-Hexene	0.12
2-Methyl-1-pentene	41	*trans*-2-Pentene	0.12
Cycloheptene	12	*trans*-2-Hexene	0.12
cis-3-Hexene	11	β-Methylstyrene	0.11
cis-2-Pentene	8.5	*trans*-2-Methyl-2-pentene	0.070
cis-4-Methyl-2-pentene	7.7	Cyclohexene	0.072
4-Vinylanisole	6.3	1-Methylcyclohexene	0.0016
Cyclopentene	2.2		
Styrene	1.1		
3,3-Dimethyl-1-hexene	0.96		
Norbornene	0.63		

The selectivity of ThxBHCl:SMe_2 significantly exceeds that of 9-BBN, and parallels that observed for Sia_2BH (Chart 6) [69].

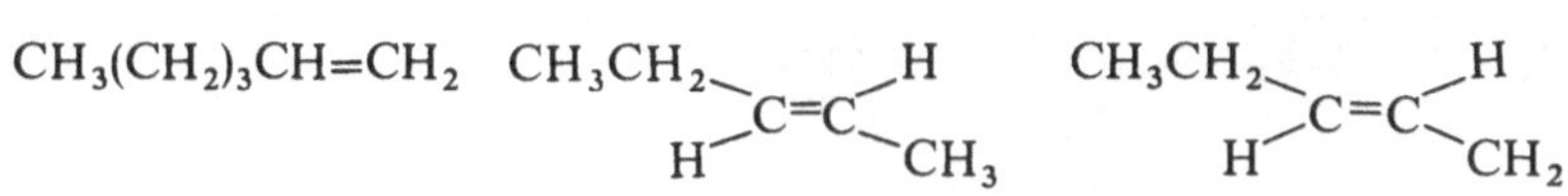

ThxBHCl:SMe_2	830	1.0	1.0
Sia_2BH	500	1.5	1.0
9-BBN	83	—	1.0

Chart 6

On the hydroboration of *cis* and *trans*-isomers, it has been observed that ThxBHCl:SMe_2 reacts at a much faster rate with *cis*-isomers than that of corresponding *trans*-isomers. This selectivity is much higher than that of Sia_2BH in the hydroboration of *cis*/*trans* pairs [69] (Chart 7).

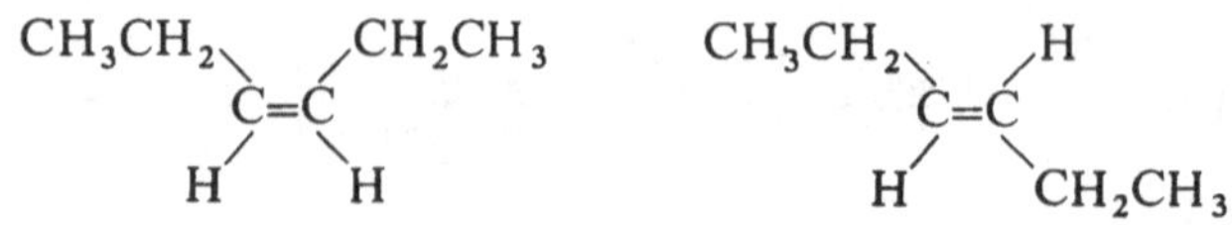

$ThxBHCl:SMe_2$	92	1.0
Sia_2BH	10	1.0
9-BBN	2	1.0

$ThxBHCl:SMe_2$	71	1.0
Sia_2BH	6.7	1.0
9-BBN	0.35	1.0

$ThxBHCl:SMe_2$	110	1.0
Sia_2BH	5.0	1.0
9-BBN	3.3	1.0

Chart 7

On the basis of symmetry, one would expect *cis*-alkenes to possess a more basic π system than their *trans*-isomers. The higher Lewis acidity of $ThxBHCl:SMe_2$ relative to Sia_2BH should cause it to interact more quickly with the *cis* π system. Thus

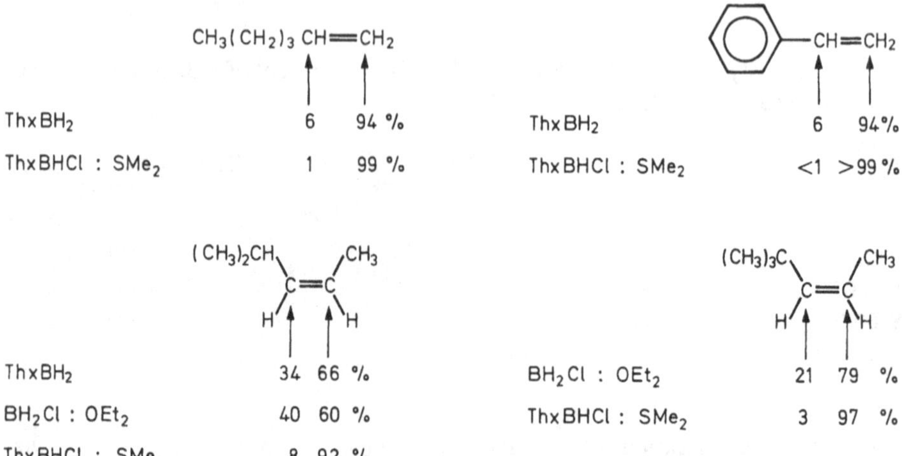

Chart 8

ThxBHCl:SMe_2 would be more effectively trapped by the *cis* isomer. The regioselectivity of this hydroborating agent is extremely high as compared to thexylborane [66,68] (Chart 8).

As pointed out earlier, thexylalkylchloroboranes are readily reduced by the reagent potassium triisopropoxyborohydride (KIPBH)[1]. The resulting thexylalkylboranes readily hydroborate a second olefin (Eq. 27). This has major advantages in some cases in directing the reaction along different paths. For example, treatment of thexylchloroborane with 4-vinylcyclohexene yields the corresponding thexylalkenylchloroborane, the reduction of which results in preferential cyclic hydroboration. Cyanidation produced a single product, bicyclo(3.3.1)nonan-2-one[73] (Eq. 28).

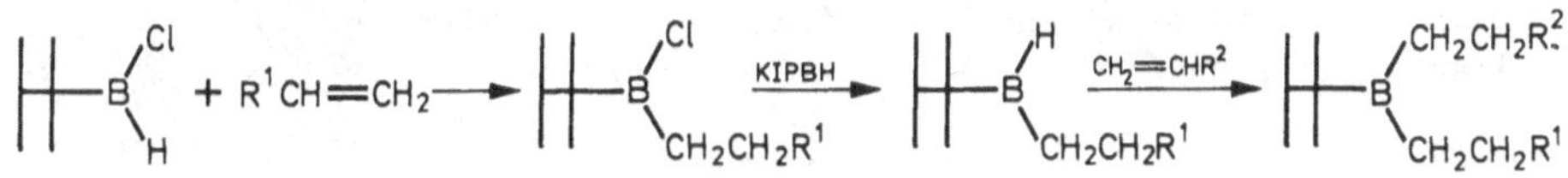

(27)

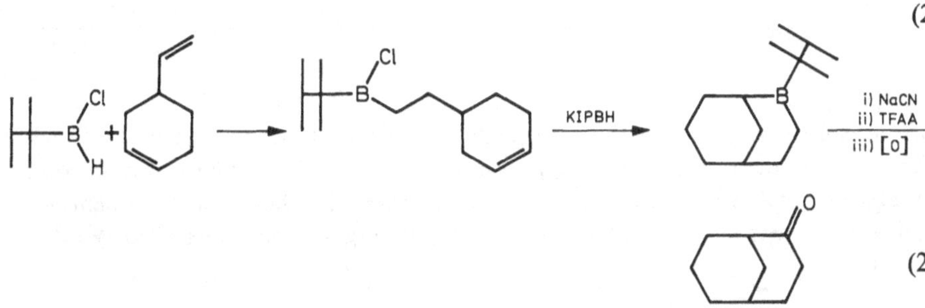

(28)

However, the following reaction gives a different type of product (Eq. 29).

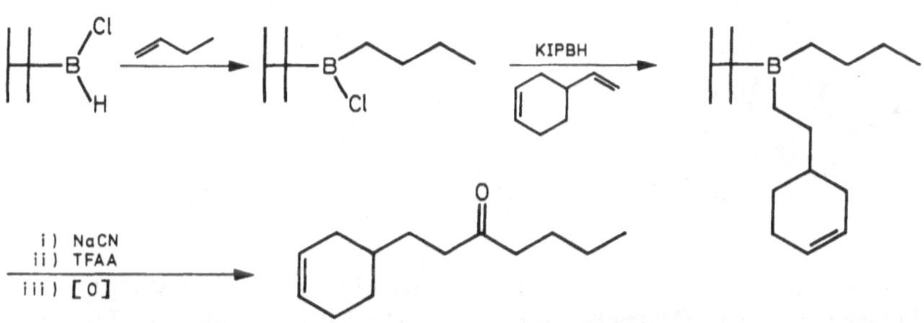

(29)

The results where two different alkenes can selectively be coupled via hydroboration-cyanidation for the preparation of a variety of unsymmetrical ketones are summarized in Table 6. Recently, this method was applied to synthesize dihydrojasmone [74].

The ketones can also be prepared from 1-halo-1-alkynes. Dialkylboranes, generated in situ via hydridation of dialkylhaloboranes, hydroborate 1-bromo-1-alkynes to

[1] Fortunately, KIPBH is a very gentle reducing agent, one which tolerates many functional groups [70]. The reagent can be prepared from potassium hydride and triisopropoxyborane in THF[71] and contains less of the undesired by-product, potassium tetraisopropoxyborate, than the commercial[72].

Table 6. Preparation of Unsymmetrical Ketones Utilizing Thexylchloroborane as Intermediate [73]

Alkene A	Alkene B	Product	Yield (%)	Purity (%)
1-Heptene	1-Pentene	6-Tridecanone	62	95
1-Octene	1-Decene	9-Nonadecanone	74	98
1-Decene	1-Octene	9-Nonadecanone	72	98
1-Dodecene	Cyclopentene	n-Dodecylcyclopentyl ketone	67	95
2-Methyl-1-pentene	1-Octene	4-Methyl-6-tetradecanone	70	97
Styrene	1-Pentene	1-Phenyl-3-octanone	59	95
Styrene	4-Methyl-1-pentene	1-Phenyl-7-methyl-3-octanone	62	96
1-Butene	4-Vinyl-cyclohexene	1-(3-cyclohexenyl)-3-heptanone	67	97
4-Pentenyl-acetate	1-Nonene	1-Acetoxy-6-pentadecanone	75	98

provide cleanly B-(*Z*-1-bromo-1-alkenyl)dialkylboranes (**25**). Treatment of **25** with sodium methoxide results in the migration of one of the alkyl groups from boron to the adjacent carbon by displacing bromine to give **26**. This upon treatment with alkaline hydrogen peroxide affords the corresponding ketones in excellent yields [75] (Eq. 30).

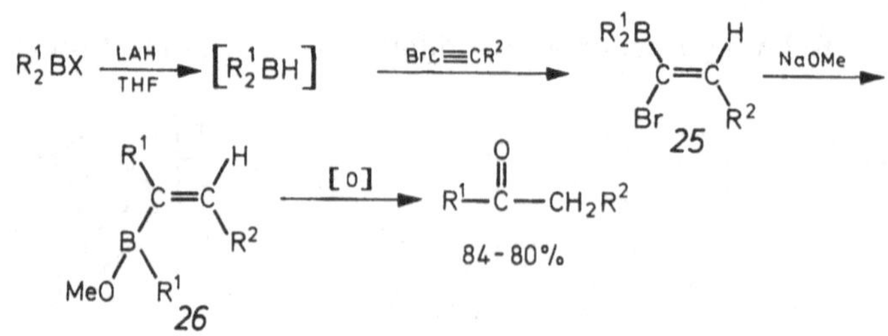

(30)

Acetylenic ketones are important synthetic intermediates in the synthesis of a variety of natural products. Optically active propargyl alcohols of either configuration are readily available in high optical purity by asymmetric reduction of propargyl ketones with B-3-pinanyl-9-borabicyclo(3.3.1)nonane [76] (Eq. 31).

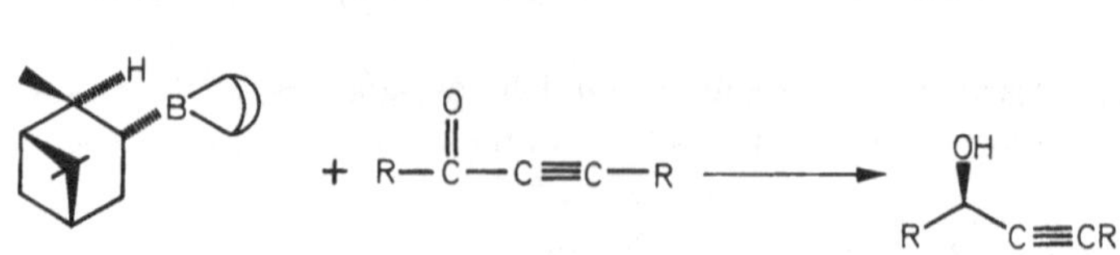

(31)

In addition, a recent discovery that potassium 3-aminopropyl amide (KAPA) [77] isomerizes internal alkynyl alcohols to the terminal alkynes without racemization, [78] provides the whole sequence a versatile process which is illustrated in the synthesis of naturally occurring optically active 7-hydroxydodecanoic acid [77] (**27**, Eq. 32).

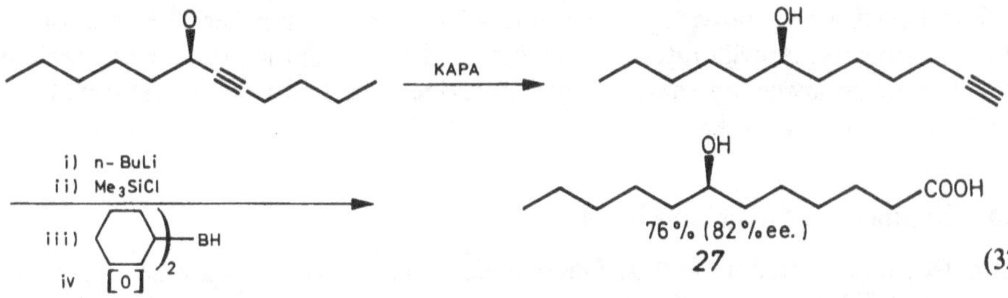

27

(32)

The α,β-acetylenic ketones can be synthesized in good yields by the selective monohydroboration-oxidation process of conjugated diynes. The monohydroboration of conjugated diynes with disiamylborane places boron preferentially at the internal triple position of the diyne system. The resultant organoboranes on treatment with sodium hydroxide and 30% H_2O_2 afforded the α,β-acetylenic ketones (Eq. 33) [79].

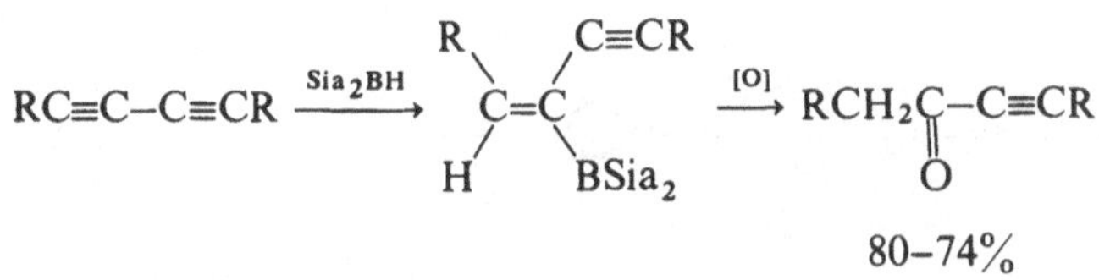

(33)

A more general method for the synthesis of α,β-acetylenic ketones has been recently reported, and is indicated in Eq. 34 [80].

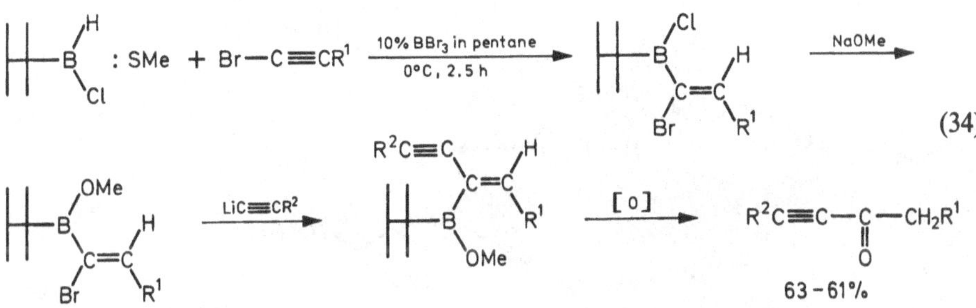

(34)

The conjugate addition of B-(1-alkynyl)-9-borabicyclo(3.3.1)nonanes (B-1-alkynyl-9-BBN) to α,β-unsaturated ketones provides a convenient procedure for the preparation of γ,δ-acetylenic ketones [81] (Eq. 35). Cuprous methyltrialkylborates react

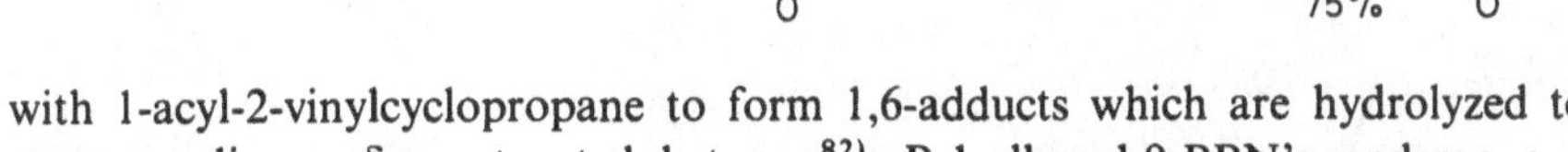

(35)

with 1-acyl-2-vinylcyclopropane to form 1,6-adducts which are hydrolyzed to the corresponding, γ,δ-unsaturated ketones [82]. B-1-alkenyl-9-BBN's undergo a facile reaction with 4-methoxy-3-buten-2-one at room temperature to give conjugated dienones in almost quantitative yields [83]. Trialkyl(1-alkynyl) boranes readily available from trialkylboranes and lithium acetylides reacts with orthoesters in the presence of titanium tetrachloride, followed by the oxidation with hydrogen peroxide and sodium hydroxide to provide α,β-unsaturated ketones [84]. Further, the reaction of lithium 1-alkynyltrialkylborates with methyl vinyl ketone in the presence of titanium tetrachloride followed by the hydrogen peroxide in a buffer solution provides a synthetic method of δ-diketones [85].

3.4 Synthesis of Carboxylic Acids

Saturated, unsaturated as well as functionalized carboxylic acids are an important class of natural products. Moreover, optically active hydroxy carboxylic acids are significant biological molecules [86], and intermediates for organic synthesis. [78] Recently, the synthesis of a wide variety of functionalized and unsaturated carboxylic acids has been accomplished through organoboranes. The reaction of organoboranes with the dianion of phenoxyacetic acid, followed by alkaline hydrogen peroxide oxidation afforded in good yields the corresponding carboxylic acids with two carbon homologation [87,88] (Eqs. 36 and 37).

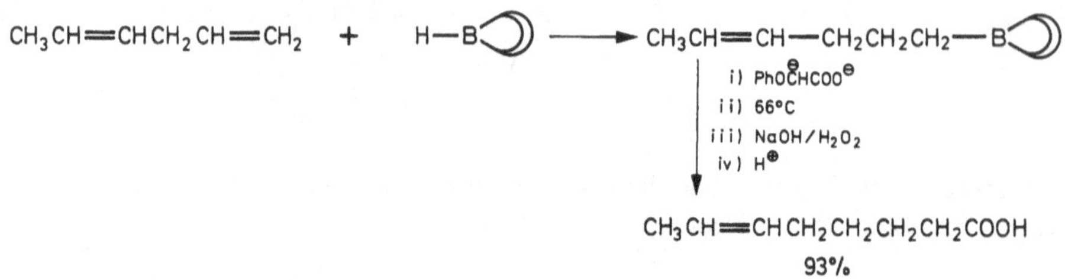

(36)

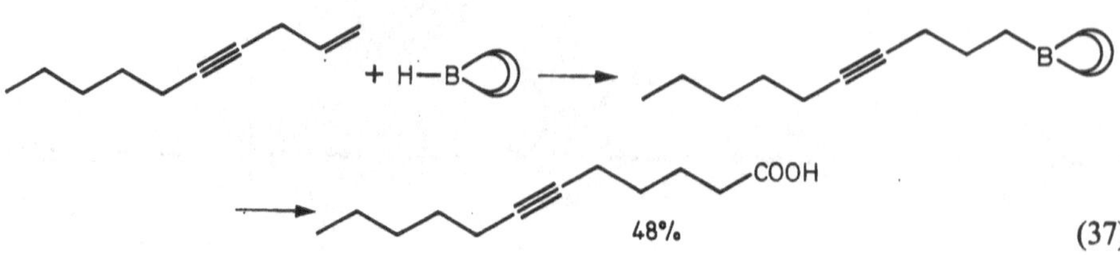

(37)

The reaction seems to involve the following path (Eq. 38):

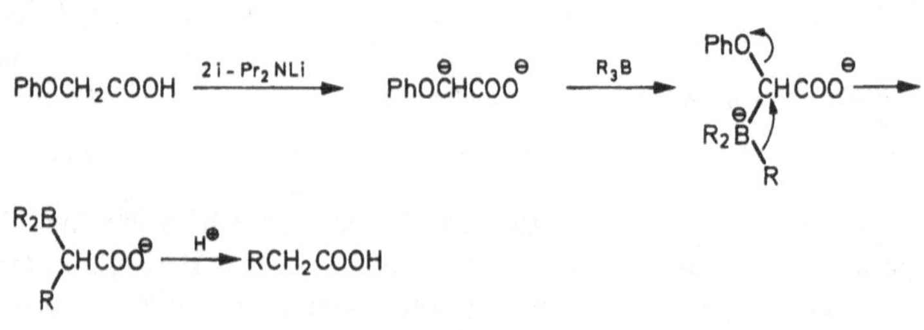

(38)

The results are summarised in Table 7.

Table 7. Synthesis of Carboxylic Acids via the Reaction of Organoboranes and Dianion of Phenoxyacetic Acid [88)]

Entry	Alkene	Product[a]	Reaction time, h	Yield %
1	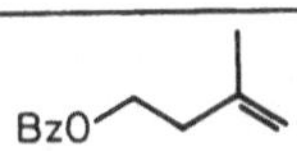	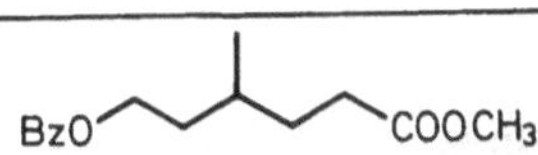	10	75[b]
2			12	86[b]
3	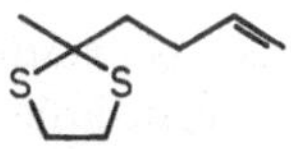	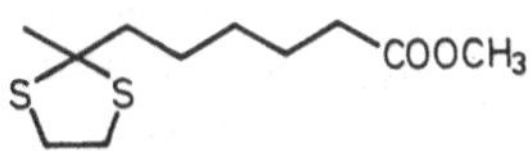	12	53[b]
4	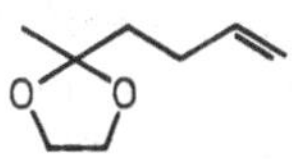	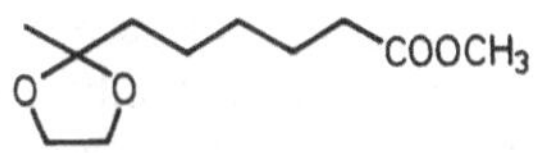	12	83[b]
5	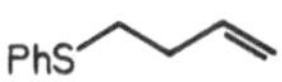		12	69[b]
6	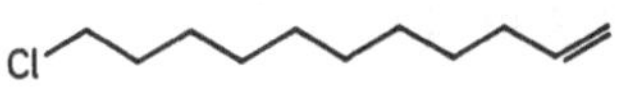	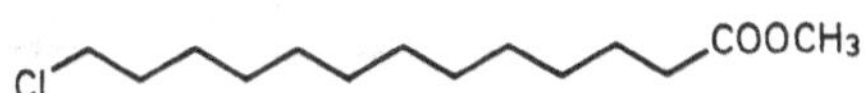	12	73[b]
7	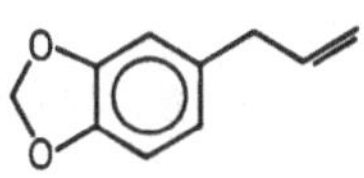	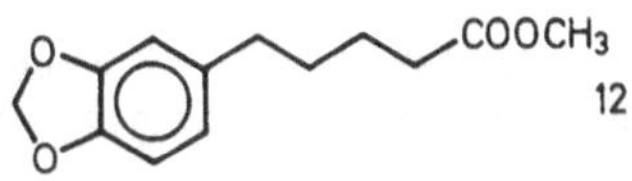	12	90[b]
8	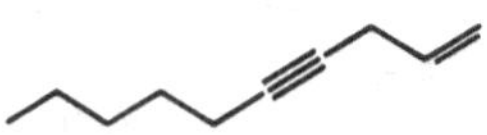	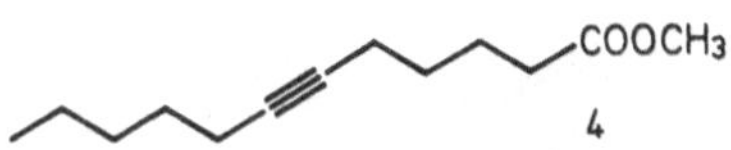	4	48
9	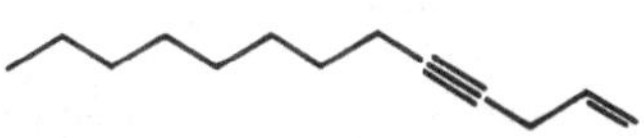		4	42
10		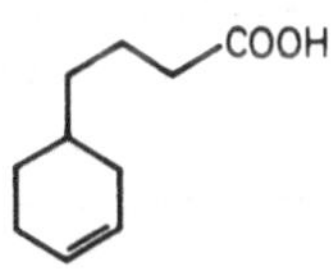	6	60

[a] Carboxylic acids entry 1–9 are characterized as their methyl esters, after treating with diazomethane;
[b] Yields based on trialkylboranes used.

An alternative preparation of one-carbon-extended carboxylic acids from alkenes via hydroboration has been most recently reported. Various 2-alkyl-1,3,2-dithiaborolanes (**28**) readily prepared from alkenes by the hydroboration with $BHBr_2:SMe_2$ followed by treatment with a suspension of lithium alkylthiolate in pentane, are converted to the corresponding carboxylic acids by using $LiCCl_3$ in THF, followed by alkaline hydrogen peroxide oxidation [89] (Eq. 39).

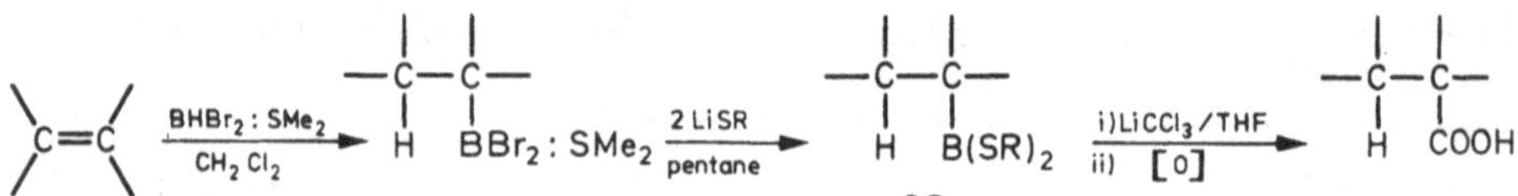

28

(39)

3.5 Allylic Boranes

3.5.1 Introduction and Preparation

The chemistry of unsaturated organoboranes often differs markedly from that of their saturated analogues. Both vinylic and allylic boranes react readily with many substrates toward which trialkylboranes are inert [90]. Allylic boranes can be synthesized selectively via direct hydroboration of an appropriate allene or conjugated diene and are of immense synthetic importance [91–93]. Mikhailov in his book and review [7,94] has documented the synthetic applicability of allylic boranes with caution of high thermal reactivity with respect to allylic rearrangement. For example, (1-methyl-2-propenyl)dialkylboranes rearrange spontaneously to the 2-butenyl isomer even at –78 °C (Eq. 40).

$$R_2B\overset{\overset{CH_3}{|}}{C}H{-}CH{=}CH_2 \xrightarrow{-78\,°C} R_2B{-}CH_2CH{=}CH{-}CH_3 \qquad (40)$$

Hydroboration of conjugated diene, e.g., monohydroboration of 1,3-cyclohexadiene with disiamylborane [95] yields the corresponding allylic boranes. A similar reaction with 9-BBN affords B-2-cyclohexenyl-9-BBN in 95% yield (Eq. 41).

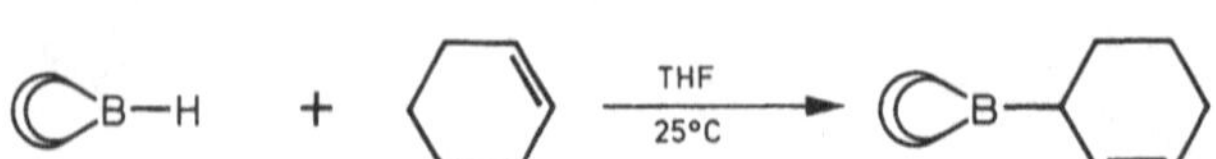

95%

(41)

The boron atom of 9-BBN is exclusively placed at one of the internal position when 2,5-dimethyl-2,4-hexadiene is hydroborated with 9-BBN (Eq. 42) [96].

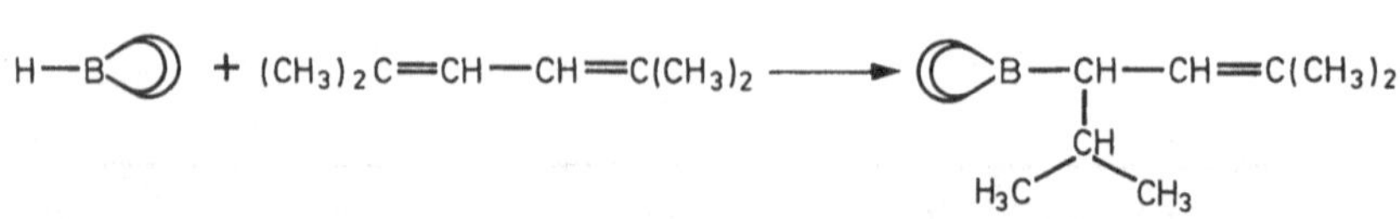

(42)

Monohydroboration of allenes with 9-BBN places the boron atom exclusively at the terminal position. Thus, 3-methyl-1,2-butadiene gives B-3-methyl-2-butenyl-9-BBN as the only product (Eq. 43) [96]. However, such a procedure for the preparation of allylic boranes by the hydroboration of conjugated dienes and allenes is not a general

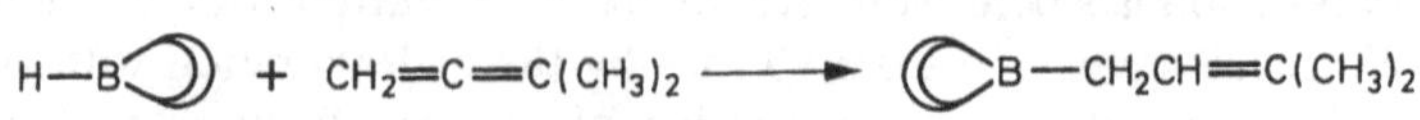

(43)

and convenient method. The reaction of allylic organometallics with boron halides or boronic esters is used as a preparative method of allylic boron derivatives [7].

One of alternative syntheses of allylic boranes is their synthesis via hydroboration of alkynes followed by one carbon chain homologation using thiomethoxymethyllithium. This provides exclusive placement of boron atom at the terminal position as shown in Eq. 44 [97].

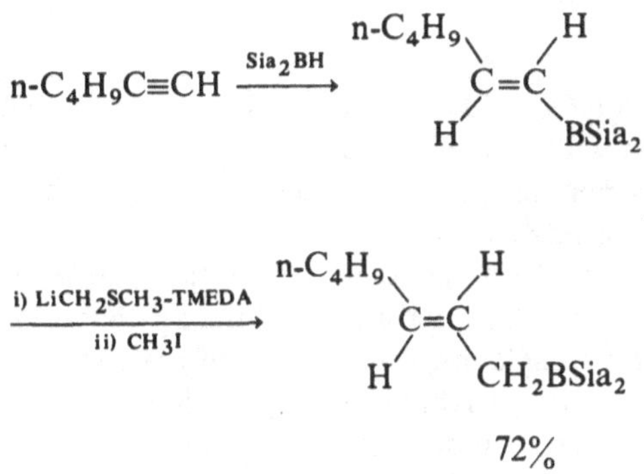

(44)

3.5.2 Synthetic Applications

Triallylborane reacts with aldehydes to form esters of diallylboronic, allylboronic or boric acids depending upon the stoichiometry of the reactants (Chart 9) [92]. Similarly, ketones afford esters of either diallylboronic or allylboronic acids. The third allyl moiety fails to react with ketones as a result of severe steric crowding around the boron atom as shown in Chart 9[92].

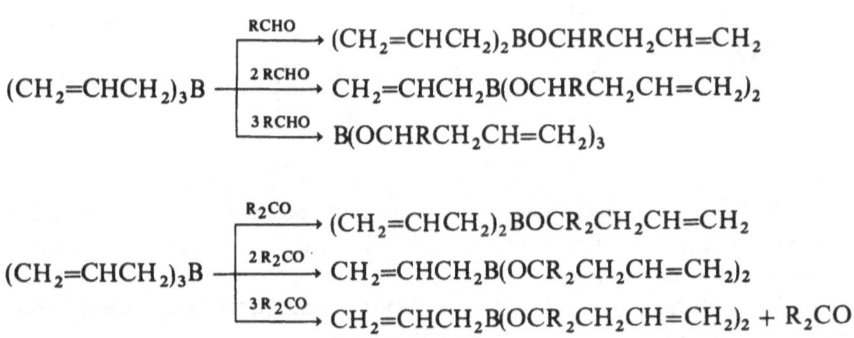

Chart 9

The boronic esters (Chart 9) are easily hydrolyzed to the corresponding homoallylic alcohols using triethanolamine [98]. Consequently, the allylboration sequence provides a synthesically useful alternative to the familiar Grignard synthesis of homoallylic alcohols. However, the protonolysis by triethanolamine causes a problem in the isolation of homoallylic alcohol from the thick, sticky, air sensitive boron-containing mixture. Fortunately, treatment of a pentane solution of borinate esters of 9-BBN with 1-equivalent of ethanolamine results in the rapid formation of a fluffy white precipitate, which can be easily removed. This air stable solid was characterized as the ethanolamine of 9-BBN (**29**, Eq. 45).

(45)

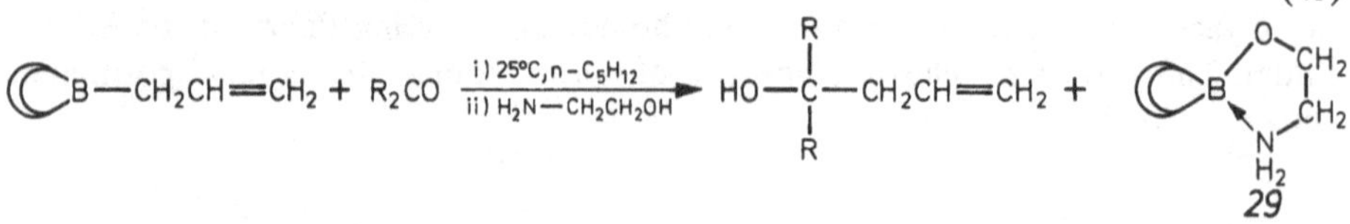

The stoichiometry of the allylboration of aldehydes and ketones with B-allyl-9-BBN is 1:1, and the reaction with aldehydes and unhindered ketones is very rapid, being complete within a few minutes at room temperature. However, as the steric crowding around the carbonyl group increases, the reaction rate decreases as is evident from di-tert-butyl ketone, which affords less than 25% of the expected product after 5 days in refluxing n-octane. The steric sensitivity of the allylboration is in sharp contrast with the corresponding allylation using Grignard reagent [99]. Even di-tert-butyl ketone is completely allylated by allylmagnesium bromide within 6 h in refluxing ether. Thus, this remarkable selectivity of allylboration offers an advantage, to selectively allylate the unhindered carbonyl moiety present in the same molecule.

Mikhailov [92] has pointed out that allylic boranes add selectively in 1,2-fashion to α,β-unsaturated carbonyl compounds unlike their counterparts trialkylboranes which add in 1,4-manner [100].

When unsymmetrical allylboranes are used, a complete rearrangement of the allylic moiety takes place [98] in a six-membered transition state mechanism (Eq. 46) [92].

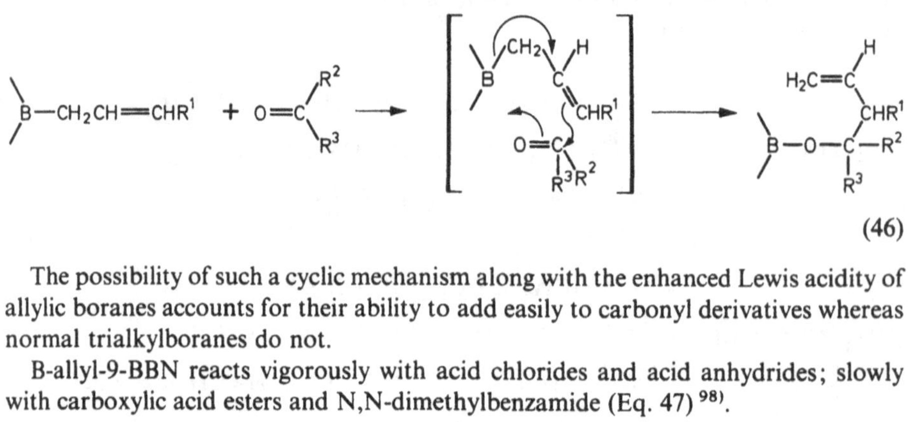

(46)

The possibility of such a cyclic mechanism along with the enhanced Lewis acidity of allylic boranes accounts for their ability to add easily to carbonyl derivatives whereas normal trialkylboranes do not.

B-allyl-9-BBN reacts vigorously with acid chlorides and acid anhydrides; slowly with carboxylic acid esters and N,N-dimethylbenzamide (Eq. 47) [98].

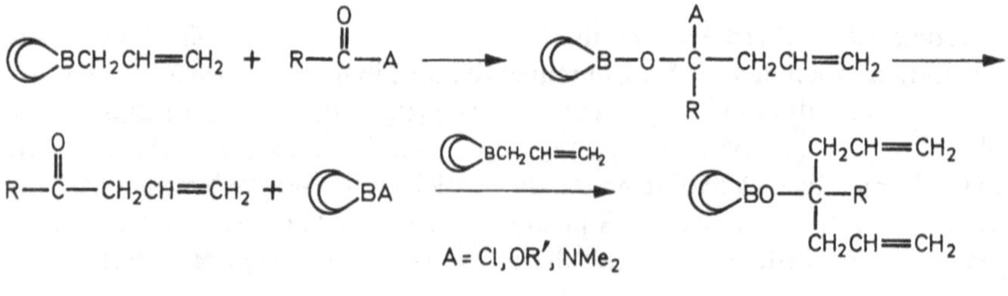

(47)

Thus allylboration of carbonyl compounds with B-allylic derivatives of 9-BBN, followed by transesterification with ethanolamine, provides a simple, convenient method for the synthesis of homoallylic alcohols [98].

It has been reported that enantiomerically enriched allylic boranes provide homoallylic alcohols with high degree of enantio- and diastereoselectivity. The scheme is illustrated in Eqs. 48 [101] and 49 [102].

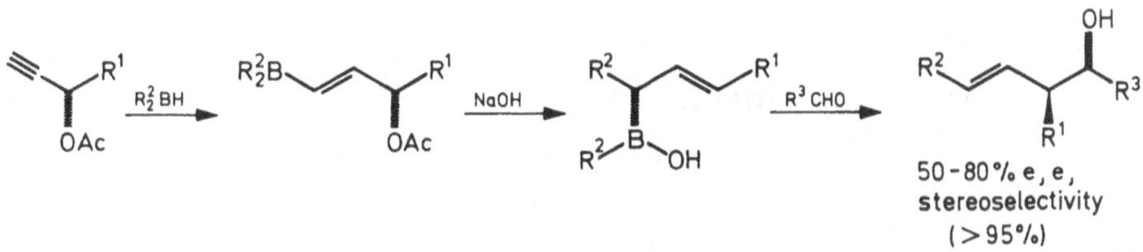

(48)

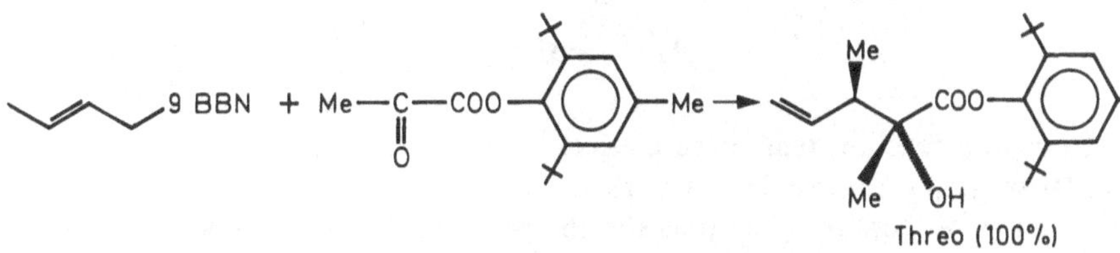

(49)

As the double bond in cyclopropene adds readily to the Grignard reagent [103], tri(2-buten-1-yl)borane adds to the double bond of cyclopropene in *cis*-fashion with rearrangement as shown in Eq. 50 [104].

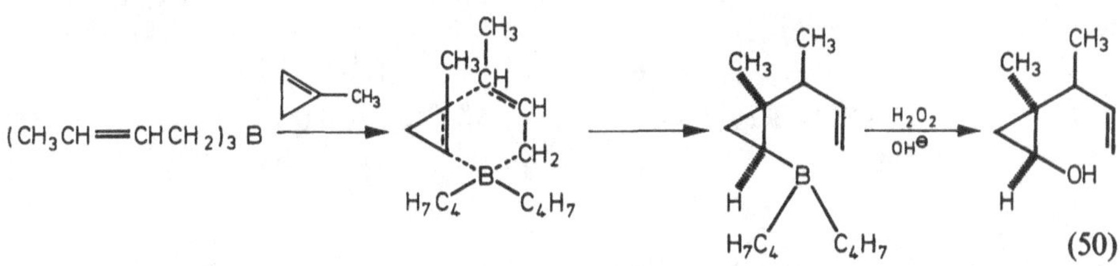

(50)

3.6 Vinylic Boranes

The reaction of internal alkynes with diborane provides the trialkenylboranes in moderate yields (Eq. 51). Terminal alkynes yield little or none of the desired alkenyl-

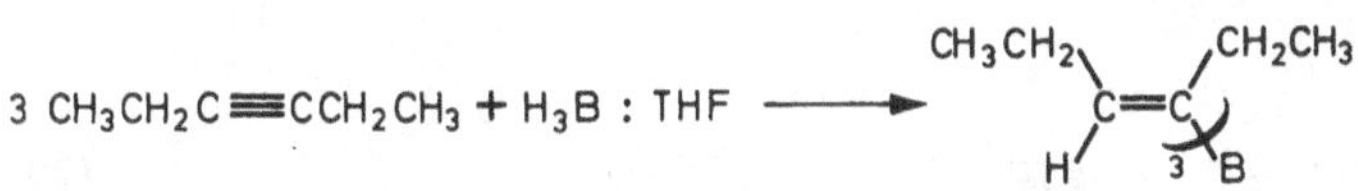

(51)

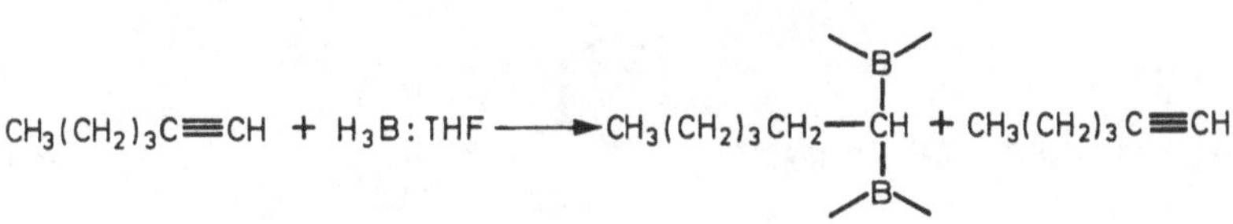

(52)

boranes [25], because apparently dihydroboration competes with monohydroboration in the case of terminal acetylenes (Eq. 52).

However, many mono-, dialkylboranes and heterosubstituted boranes (Chart 1) exhibit highly selective behavior in hydroboration of alkynes [7]. Thus, they appeared attractive for the controlled monohydroboration of alkynes for the synthesis of the desired vinylic boranes from terminal and internal alkynes (Eqs. 53 and 54).

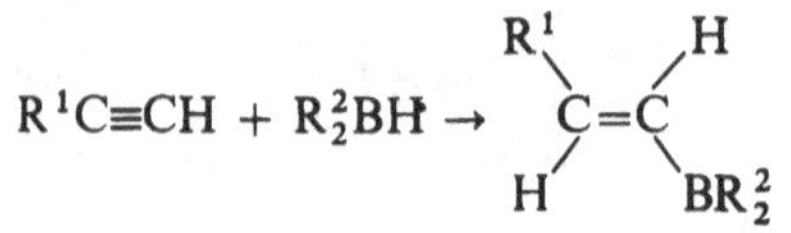
(53)

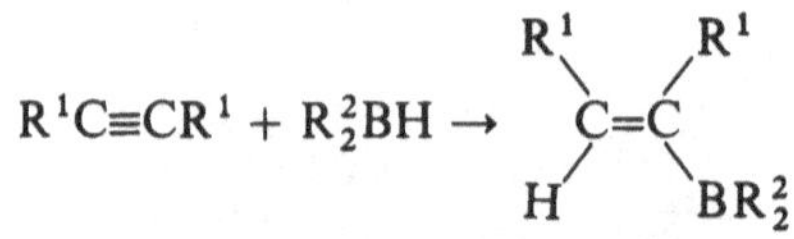
(54)

The steric and reaction temperature effects in the hydroboration of alkynes with representative dialkylboranes in THF are given [105].

The mildness of these reagents tolerates the presence of various functional groups such as ester, ether, halogen, and nitrile. The stereospecific *cis* nature of hydroboration gives exclusively the *trans* alkenylboranes, often also in high regioisomeric purity (Eq. 53). On the other hand, highly pure (*Z*)-1-alkenyl-dialkylboranes are prepared without any difficulty via the monohydroboration of 1-halo-1-alkynes with disiamylborane or dicyclohexylborane, followed by treatment with t-butyllithium (Eq. 55) [106].

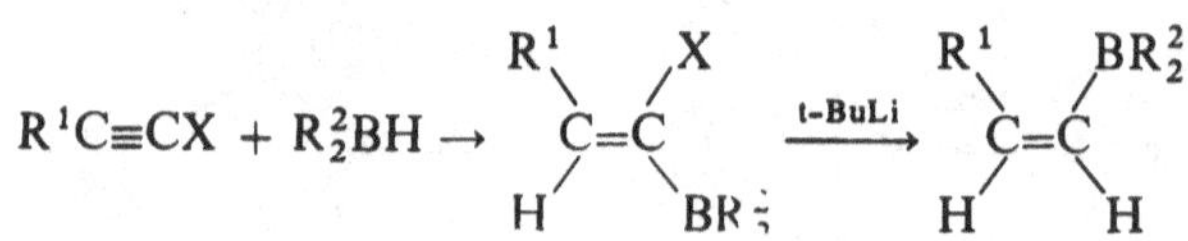
(55)

Subsequent reactions of the alkenylboranes usually proceed by stereodefined pathway, thus allowing highly stereo- and regiospecific syntheses.

The monohydroboration of alkynes can also be achieved with $BH_2Cl:OEt_2$ at 0 °C using an internal alkyne or a terminal alkyne (Eqs. 56 and 57, Table 8) [37], while dihydroboration is a significant reaction when THF is used as a solvent. This is due to the reason as the nucleophilic character of THF is higher as compared to diethyl ether.

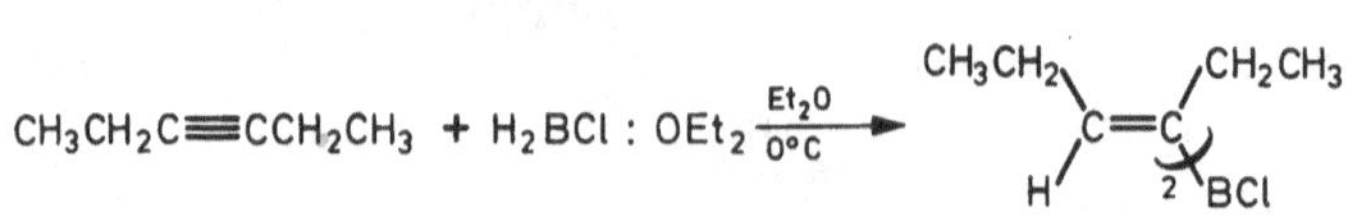
(56)

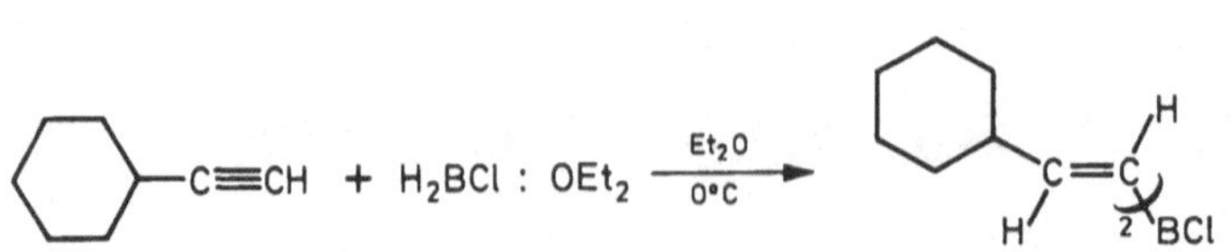
(57)

Table 8. Reactions of Alkynes with BH_2Cl (5 mmol) in Diethyl Ether at 0 °C (0.87 M BH_2Cl)

Alkyne	Amount of alkyne used (mmol)	Time h	Yield of dialkenyl-chloroborane (%)
1-Hexyne	14	0.25	100
Cyclohexylethyne	14	0.25	100
Phenylethyne	14	1.00	96
3-Hexyne	10	0.25	95

The dialkenylchloroboranes undergo the usual reactions of vinylic boranes, e.g., protonolysis with acetic acid gives olefins, oxidation with alkaline hydrogen peroxide provides the corresponding carbonyl compounds. However, the most useful reactions of these compounds are their ready conversion to the stereochemically pure (*E*, *Z*)-1,3-dienes by the Zweifel reaction with I_2-NaOH [37,107,108] and into the symmetrical (*E*,*E*)-1,3-dienes [109,110], mono-olefins [111] and 1,4-dienes (Chart 10).

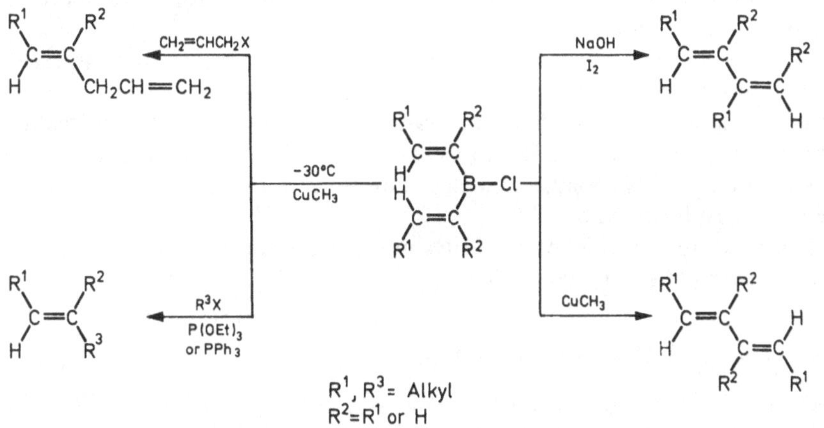

Chart 10

Earlier, alkenyldichloroboranes have been synthesized by the transmetallation of zinc [112] or tin [113] compounds. The simple preparation of alkenyldichloroborane via hydroboration of alkynes with dichloroborane has been reported [114]. Unlike monochloroborane, the reactions of $BHCl_2:THF$ and $BHCl_2:Et_2O$ are slow and incomplete, and gives a mixture of products [115]. This is due to the reason that $BHCl_2$ is a stronger Lewis acid, more strongly complexed than BH_2Cl. When the reaction was carried out with neat $BHCl_2:Et_2O$ in the absence of excess diethyl ether, the major product formed was dialkylchloroborane. Again in the presence of benzene, the sole product isolated after 6 h was R_2BCl[114] and not monoalkyldichloroborane. These results suggest that all of the $BHCl_2:Et_2O$ disproportionates to BH_2Cl and BCl_3 prior to the reaction with olefins [115] (Eq. 58). Fortunately, $BCl_3:Et_2O$ is insoluble in pentane, and complexation of $BHCl_2$ with diethyl ether can be broken by adding the stronger Lewis acid BCl_3 (Eq. 59) [116]. Thus, the free $BHCl_2$ formed in situ (Eq. 59) reacts cleanly with terminal and internal alkynes to form the corresponding alkenyldichloroboranes (Eqs. 60 and 61) [116] in pentane.

$$2\,BHCl_2{:}OEt_2 \rightleftarrows BH_2Cl{:}OEt_2 + BCl_3{:}OEt_2 \quad (58)$$

$$BHCl_2{:}OEt_2 + BCl_3 \xrightarrow{\text{Pentane}} BHCl_2 + BCl_3{:}OEt_2\downarrow \quad (59)$$

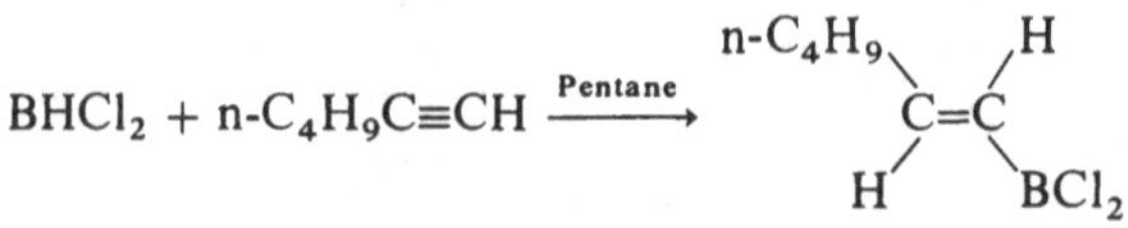

(60)

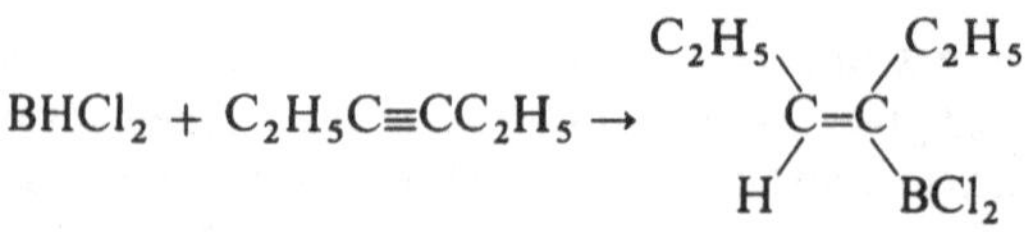

(61)

To avoid the formation of small amounts of the dihydroborated derivatives, it is desirable to use 10% excess of alkyne. Hydrolysis or alcoholysis of the resulting alkenyldichloroborane provides a simple preparation of the desired alkenylboronic acid or esters via hydroboration.

Among the haloboranes, dibromoborane-methylsulfide has proven to be an excellent hydroborating reagent. Its reaction with terminal and internal alkynes cleanly affords the corresponding vinylic dibromoboranes without the concomitant formation of 1,1-dibora derivatives. This reagent reacts only with internal triple bond in the presence of terminal double bond, and with disubstituted terminal double bond in the presence of monosubstituted one [36c)].

Vinylic boranes are versatile intermediates in organic synthesis, such examples of which will be exemplified in later sections.

3.7 Synthesis of *Z*-Disubstituted Olefins

The synthesis of *Z*-disubstituted double bond was accomplished by Zweifel. The Zweifel's (*Z*)-alkene synthesis involves the iodination of dialkylvinylboranes in the presence of sodium hydroxide which results in the transfer of an alkyl group from boron and the formation of the (*Z*)-alkene (Eq. 62) [117,118)], although the introduction of alkyl groups is limited.

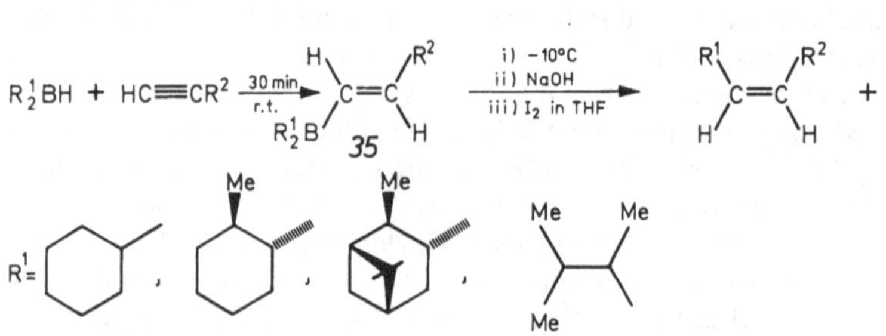

(62)

The reaction is believed to proceed through:

(i) coordination of the base with the boron atom,

(ii) formation of iodonium ion,

(iii) *trans*-migration of one of the R groups, and
(iv) trans-elimination of the boron-iodine moiety (Eq. 63).

The versatility of this reaction is further demonstrated by the fact that the alkyl groups on vinylic dialkylboranes migrate with retention of configuration (Eq. 64)[118]; thus, greatly extending the scope of this reaction.

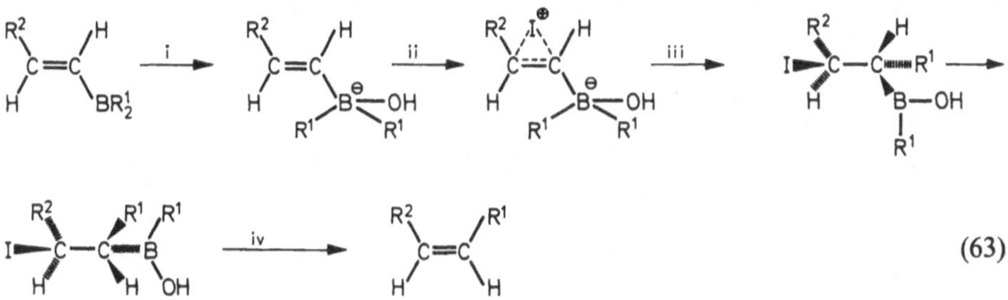

(63)

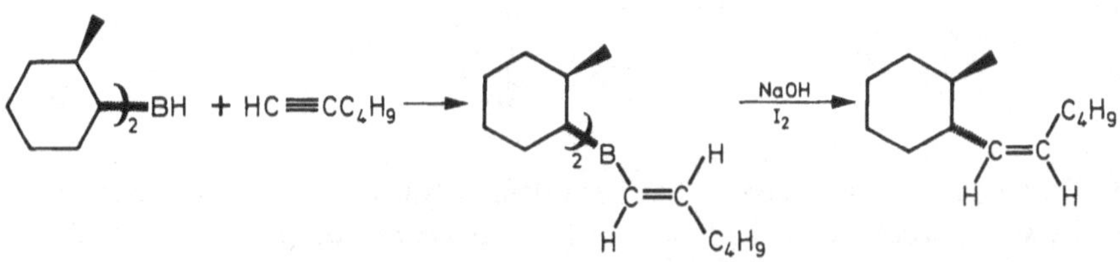

(64)

However, the utility of this Zweifel *cis*-olefin synthesis was severely handicapped in the past on account of the following difficulties:

(1) Monohydroboration of terminal alkynes with relatively unhindered dialkylborane is accompanied by competing dihydroboration, while for thexylmonoalkylboranes both thexyl and R group migrates competitively.
(2) The iodination of dialkylvinylboranes in the presence of an appreciable amount of 1,1-dibora derivatives produces considerable amounts of alkyl iodide along with the desired *cis*-alkenes[119,120].

Recently, a wide variety of dialkylboranes have been readily prepared in situ, via hydridation of dialkylhaloboranes[6,105]. And the first diffidulty is overcome by carrying out the hydroboration of 1-alkynes with dialkylboranes at lower temperatures[105,107]. Fortunately, the second difficulty, the concurrent formation of alkyl iodides, can be circumvented by carrying out the reaction of dialkylvinylboranes with sodium methoxide and iodine at lower temperature (–25 to –78 °C, Eqs. 65 and

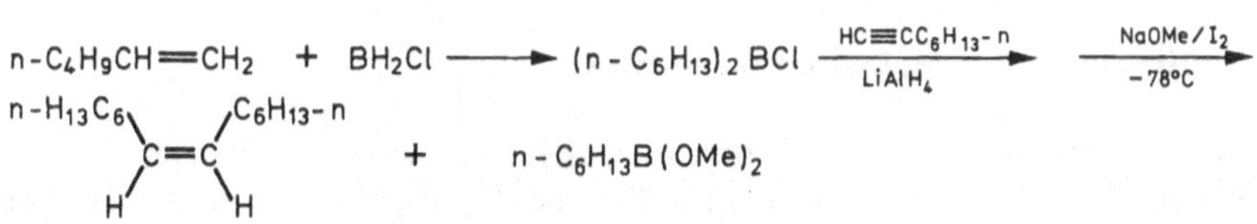

(65)

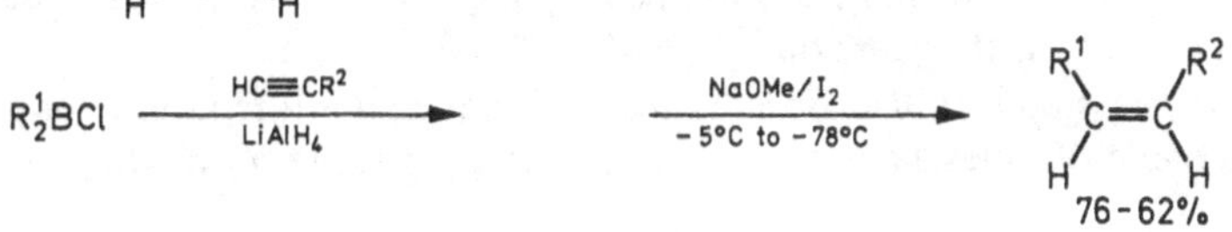

(66)

66)[121]. Thus, this procedure extends the range of applicability of the synthesis of *cis*-olefins, as now the alkenes of low steric requirements like 1-hexene are readily utilized (Eq. 65). This particular class of alkenes had previously been inaccessible via the iodination reaction.

However, this method has an obvious disadvantage: one of the R group is wasted. The final solution to this problem has now been provided with dibromoborane-methyl sulfide (Eq. 67) [122].

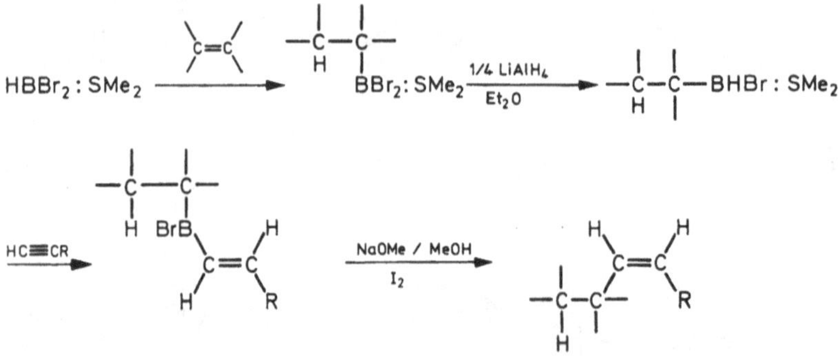

(67)

The utility of this new procedure has been demonstrated by the synthesis of muscalure, the sex pheromone of the house fly (*Musca domestica*) (Eq. 68) [122] and disparlure, the sex pheromone of the Gypsy moth (*Porthetria dispar L.*) [123], and the sex attractant of soybean loopers (*Pseudoplusia includens*) [124].

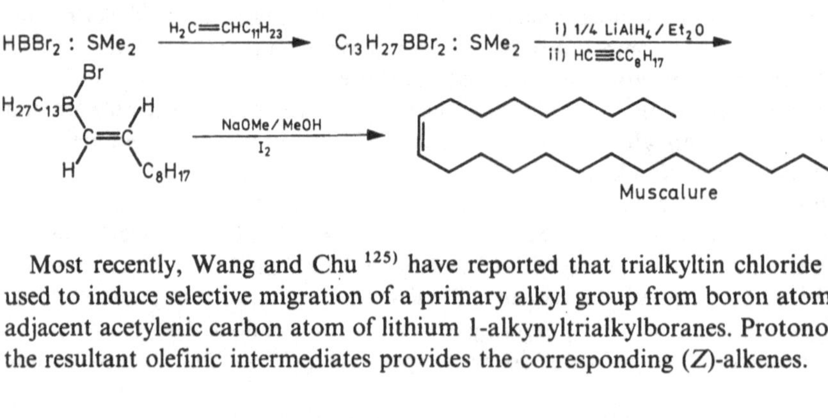

(68)

Most recently, Wang and Chu [125] have reported that trialkyltin chloride can be used to induce selective migration of a primary alkyl group from boron atom to the adjacent acetylenic carbon atom of lithium 1-alkynyltrialkylboranes. Protonolysis of the resultant olefinic intermediates provides the corresponding (*Z*)-alkenes.

3.8 Synthesis of *E*-Disubstituted Alkenes

The Zweifel *trans*-alkene synthesis from dialkyl-1-alkenylboranes involves the hydroboration of 1-halo-1-alkynes with dialkylborane (Eq. 69) [126]. It is of interest that the base produces a transfer with inversion [126, 127].

Protonolysis in the last step can also be achieved with aqueous silver ammonium nitrate for acid sensitive molecules as in a simple stereoselective route to prostaglandins (Eq. 70) [128].

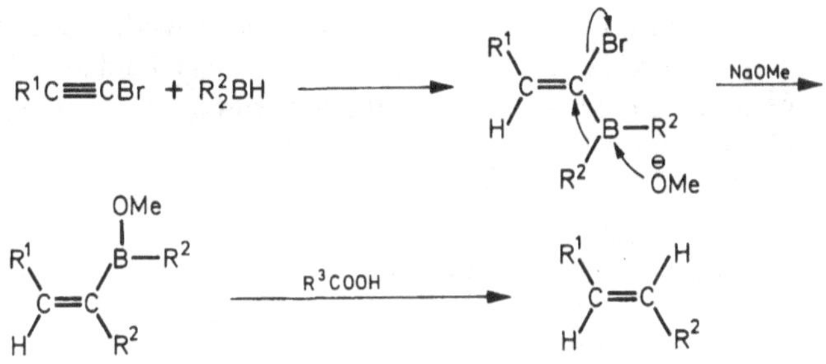

(69)

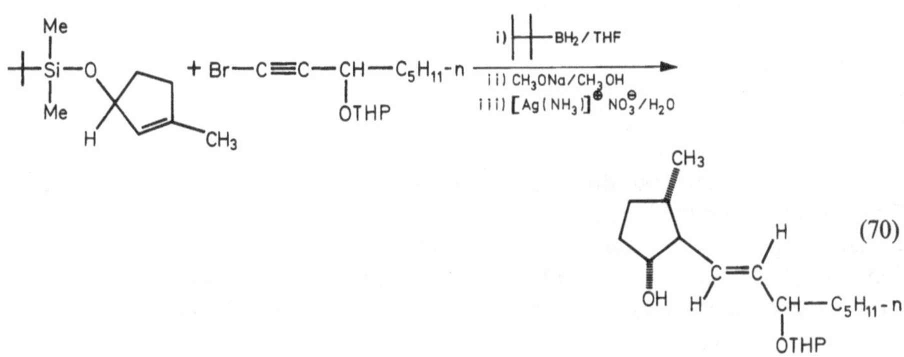

(70)

Although *trans*-disubstituted olefins are readily prepared by the partial reduction of acetylenes with sodium in liquid ammonia [129], this novel olefin synthesis via hydroboration has a unique advantage over the conventional method in which four consecutive stereocenters can be created in a predictable manner as shown by the following example (Eq. 71) [118].

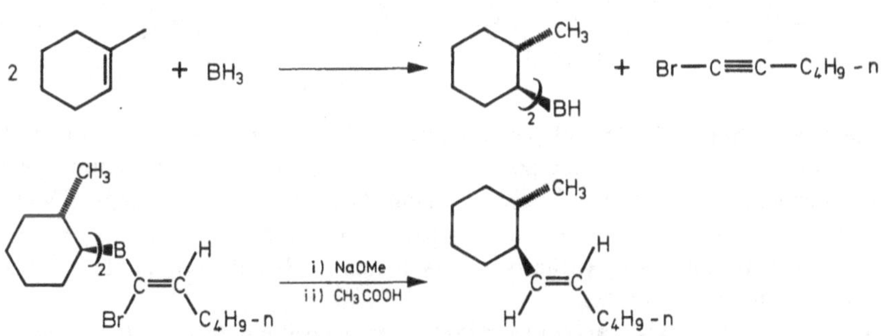

(71)

The loss of one of the valuable alkyl groups can be avoided by the hydroboration of 1-halo-1-alkynes with monoalkylthexylborane, generated in situ via the hydridation of monoalkylthexylchloroborane which produces the corresponding thexylalkyl-1-halo-1-alkenylboranes (**30**) in excellent yields. However, one difficulty is encountered, where the hydridation with potassium triisopropoxyborohydride (KIPBH) in the presence of 1-halo-1-alkyne affords the undesired product. This difficulty, fortunately, can be overcome by generating the thexylmonoalkylborane first at low temperature followed by rapid addition of the haloalkynes. The treatment of resulting thexylalkyl-

(1-halo-1-alkenyl)borane (**30**) with sodium methoxide leads with selective migration of an alkyl group from boron to give 1,2-disubstituted vinylic borane (**31**) [130]. Protonolysis with 2-methylpropanoic acid affords the desired (*E*)-alkene in high chemical yield in essentially pure form (Eq. 72) [131].

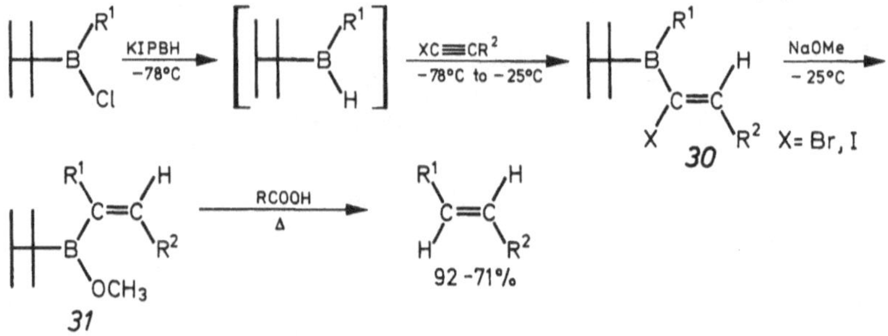

(72)

A more convenient procedure for the synthesis of (*E*)-alkenes has been reported [75]. This involves the hydridation step with lithium aluminium hydride at 0 °C. The loss of thexyl group is also undesirable and it can be avoided by use of dibromoborane-dimethylsulfide (Eq. 73) [25,122]. This procedure appears to be general for essentially all R groups with no disadvantages known. There is no loss of R groups.

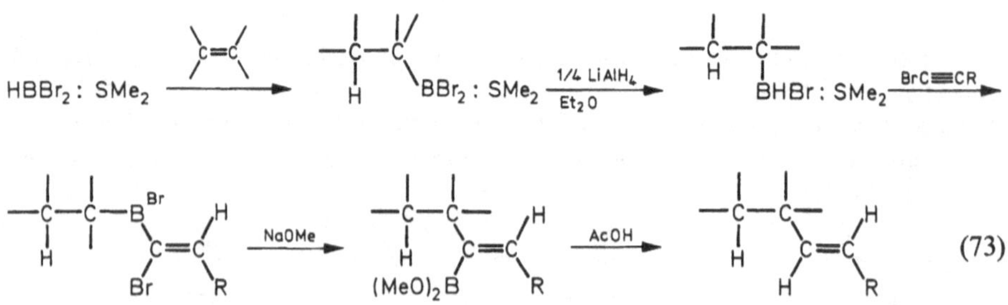

(73)

Both *Z*- and *E*-configurations of carbon-carbon double bond are an important structural moiety of insect pheromones. The synthesis of unsaturated alcohols has attracted considerable attention of organic chemists in recent years because such alcohols and their acetates are known to be insect sex attractants. Recently, a very convenient, general, one pot synthesis of *Z*-alkenols [132] and *E*-alkenols [133] via organoboranes has been reported.

trans-Double bond can also be converted in a stereospecific manner, where vinylic boranes add to α,β-unsaturated carbonyl compounds in a 1,4-fashion (Eq. 74) [134].

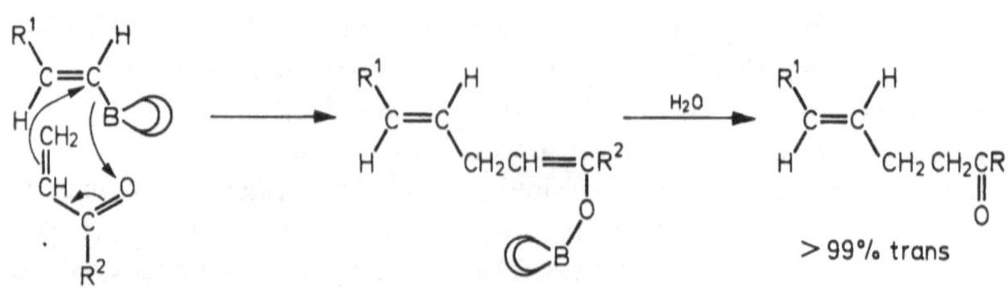

(74)

3.9 Synthesis of Trisubstituted Alkenes

Zweifel and his co-workers have also introduced a stereospecific synthesis of trisubstituted alkenes (Eq. 75) [117,135], although it has limitations of the use.

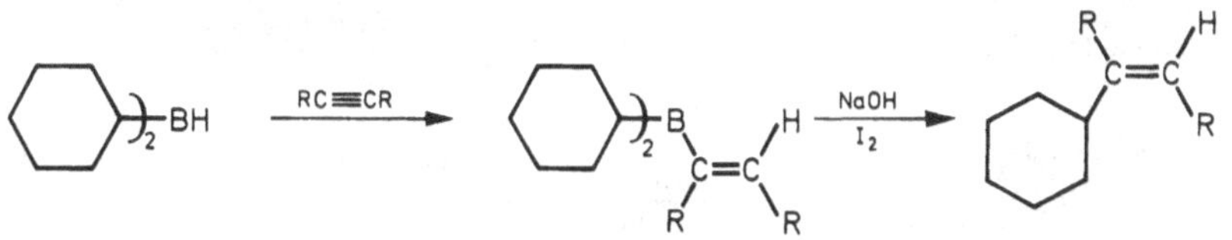

(75)

The other methods which involve organoboranes for the synthesis of trisubstituted alkenes employ either trialkyl(1-alkynyl)borates [136] or trialkyl(1-alkenyl) borates [137].

Unfortunately, the Zweifel's hydroboration procedure suffers from the same disadvantage as his synthesis of *cis* and *trans* alkenes. It requires R_2BH as a reagent. Application of the new synthesis of R_2BH via hydridation [105] of R_2BX made the procedure general for essentially all R groups available through hydroboration (Eq. 76) [138].

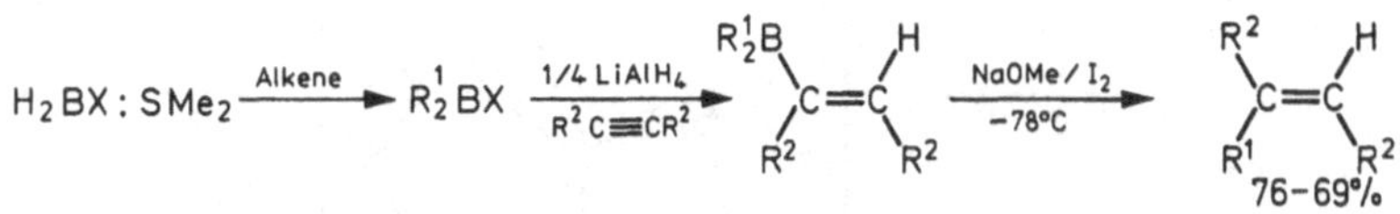

(76)

These results indicate a mechanism analogous to that involved in the formation of *cis*-disubstituted alkenes is operative (Eq. 77) [138].

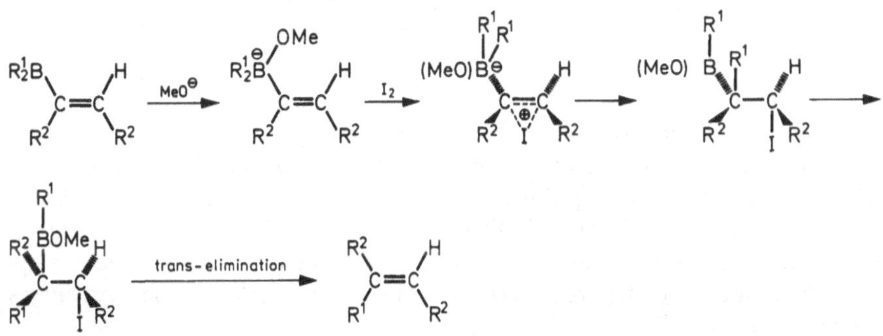

(77)

The procedure still suffered from the disadvantage that only one of the R group was utilized. However, the recent development of hydridation procedure makes it possible

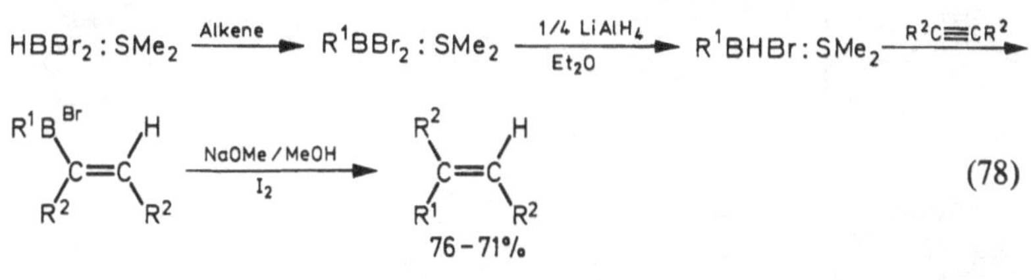

(78)

to replace only one of the bromine with hydrogen. This procedure worked ideally (Eq. 78), thus providing a general one pot and stereospecific synthesis of trisubstituted alkenes, where all R groups are utilized [139].

The following procedure was also reported (Eq. 79) [75].

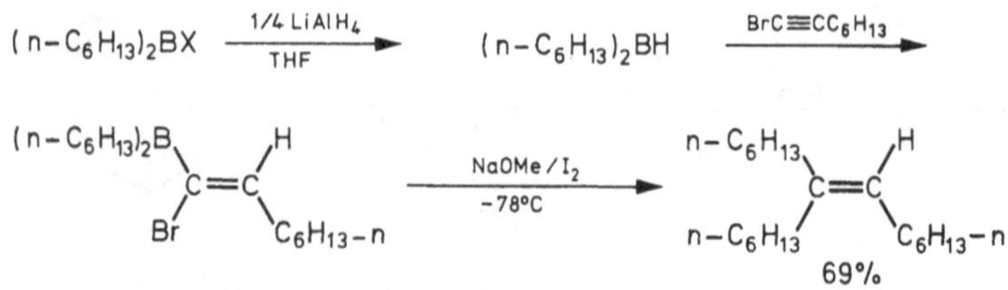

(79)

3.10 Synthesis of Conjugated (*Z*, *E*)-Dienes

Synthesis of symmetrical (*Z*, *E*)-dienes may be performed by the following procedure (Eq. 80) [108]. It has been suggested that the migration of an alkyl group proceeds with inversion at the migration terminus and deiodoboration occurs in a *trans* manner.

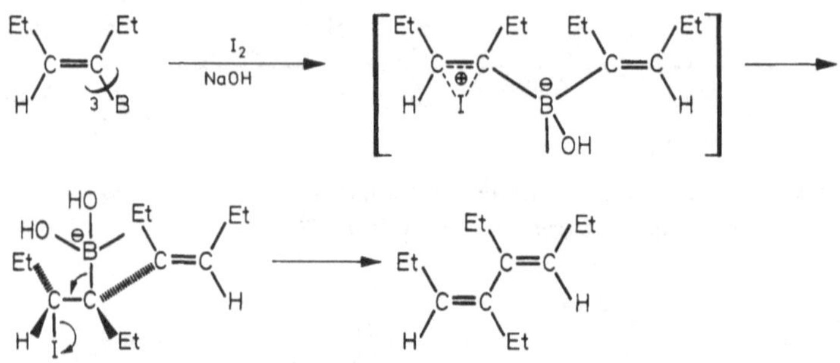

(80)

The use of thexylborane to avoid the loss of an alkenyl groups, however, leads to a mixture of products with competitive migration of thexyl moiety. This problem has been circumvented by employing trimethylamine oxide; divinylthexylboranes derived from terminal alkynes result in the selective oxidation of the thexyl moiety to yield the corresponding divinylborinate derivatives. In the presence of iodine and sodium hydroxide, these intermediates rearrange to give (*Z*, *E*)-1,3-butadiene derivatives (98% pure) (Eq. 81) [140].

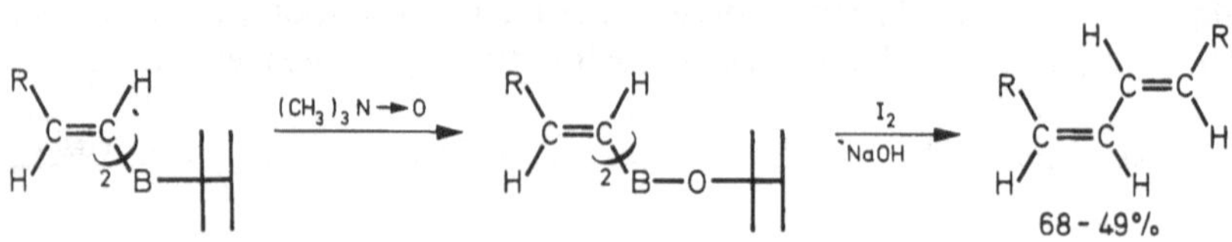

(81)

Another method to symmetrical (*Z*, *E*)-dienes is through dialkenylchloroboranes (Eq. 82) [107].

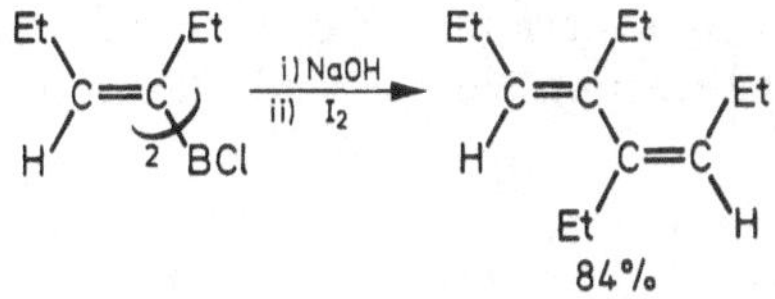

(82)

The synthesis of unsymmetrical derivatives has been reported, which involves stepwise treatment of an alkenyldialkylborane with lithium alkynide followed by boron trifluoride. Protonolysis of the intermediate affords in a good yield the isomerically pure unsymmetrical (*Z*, *E*)-diene (Eq. 83) [141].

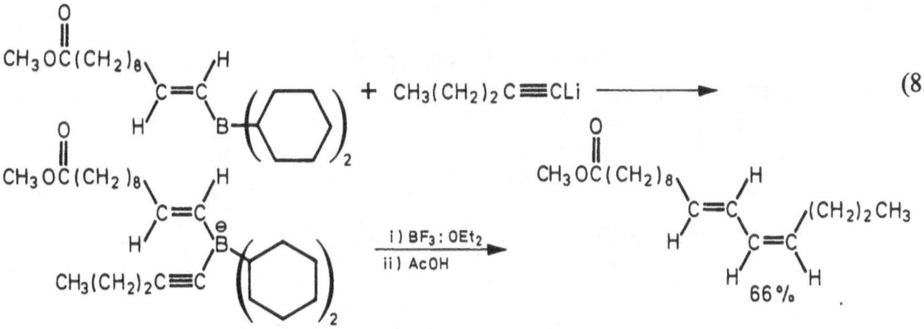

(83)

Most recently, a new and general synthesis of conjugated alkadienes has been reported by the reaction of 1-alkenyldiorganoboranes with 1-alkenyl halides in the presence of palladium catalysts and bases [142]. According to this procedure, (*Z*, *E*)-alkadienes are readily obtained, as revealed in Eq. 84.

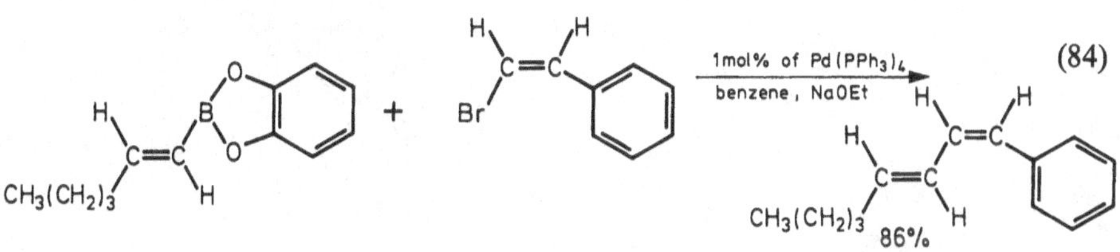

(84)

3.11 Synthesis of Conjugated (*E*, *E*)-Dienes

Recently, dialkenylchloroboranes were reported to undergo methylcopper-induced coupling to produce symmetrical (*E*, *E*)-1,3-dienes in excellent yields and high stereochemical purity (Eq. 85) [37, 143]. In this reaction, it is essential to utilize 3 molar

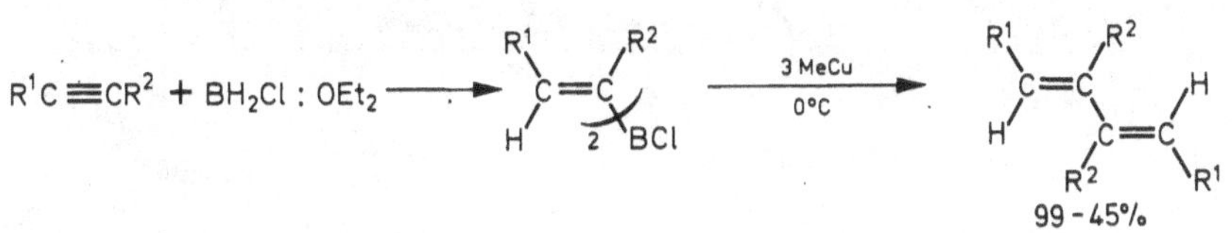

(85)

equivalents of methylcopper in order to achieve the effective conversion of organoboranes into the dienes. The proposed reaction mechanism is described in Chart 11, where R means 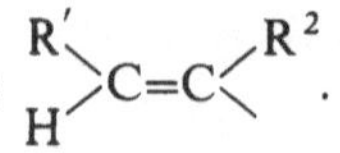 .

$$R_2BCl + MeCu \longrightarrow R_2BMe + CuCl$$

$$R_2BMe + MeCu \rightleftharpoons [R_2\overset{\ominus}{B}Me_2]\,Cu^{\oplus} \rightleftharpoons RCu + RBMe_2$$

$$RBMe_2 + MeCu \rightleftharpoons [R\overset{\ominus}{B}Me_3]\,Cu^{\oplus} \rightleftharpoons RCu + Me_3B$$

$$RCu \longrightarrow 1/2\ R{-}R + Cu$$

Chart 11

An alternative procedure where alkenyldibromoboranes, without isolation, are directly converted to dienes is reported (Eq. 86) [36a].

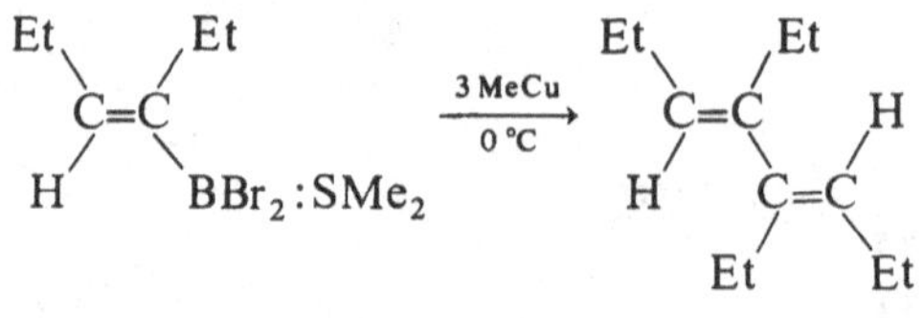

(86)

90%

A more milder procedure is, where sodium methoxyalkenyldialkylborates obtained by the simple treatment of alkenyldialkylboranes with sodium methoxide, react readily with cuprous bromide-methylsulfide at 0 °C to afford symmetrical conjugated dienes (Eq. 87) [144]. Although the intermediacy of a copper(I) borate complex formed

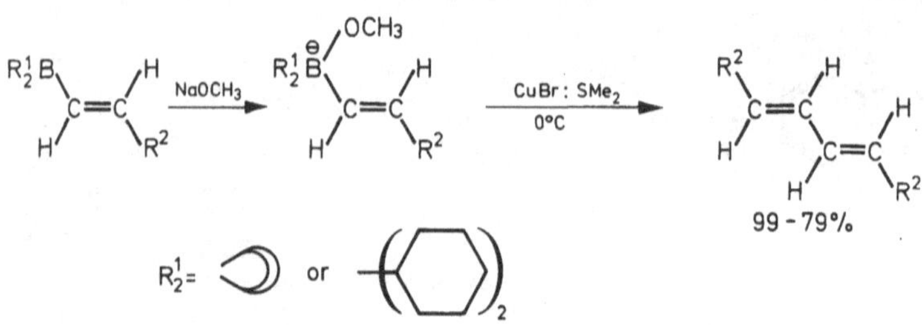

(87)

by cation exchange with sodium appears likely, it remains to be established whether this intermediate decomposes directly to give a diene or dissociate to yield alkenylcopper compound.

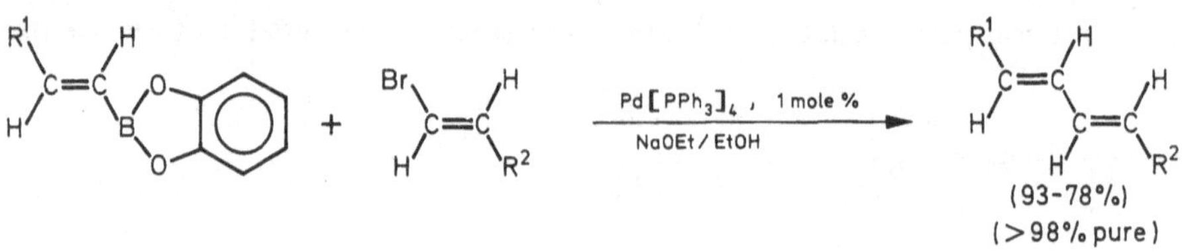

(88)

Conjugated (*E*, *E*)-dienes can also be synthesized from (*E*)-1-alkenyl-1,3,2-benzodioxaboroles by palladium catalyzed reactions (Eq. 88) [142,145] and (Eq. 89) [146] in high regio- and stereospecificity.

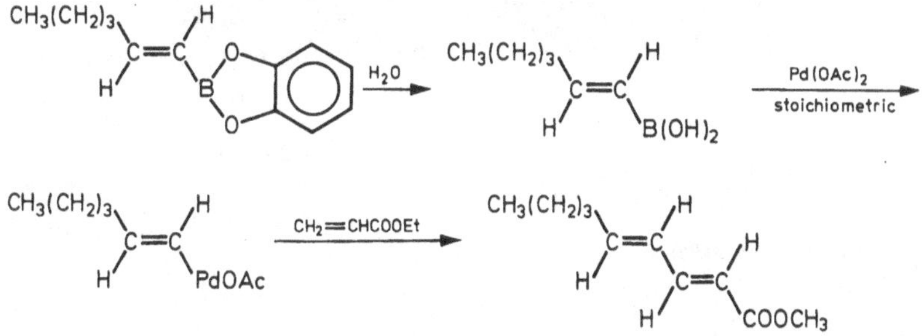

(89)

Unsymmetrical (*E*, *E*)-dienes are also available by the following procedure using 1-halo-1-alkynes (Eq. 90) [31].

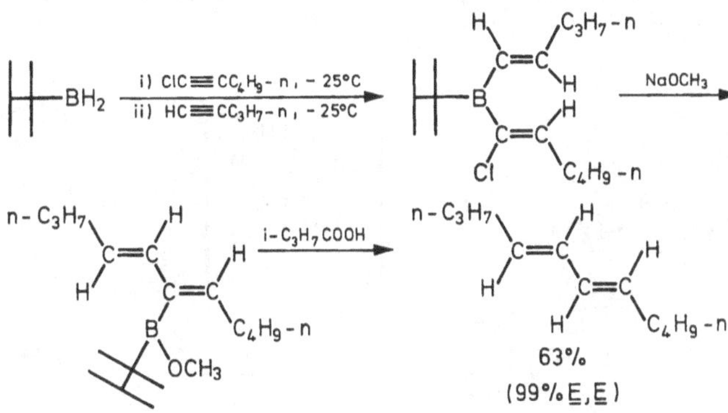

(90)

3.12 Synthesis of Conjugated (*Z*, *Z*)-Dienes

By an analogous cuprous bromide-methyl sulfide catalyzed procedure for the synthesis of (*E*, *E*)-dienes; conjugated (*Z*, *Z*)-dienes can also be prepared by substituting the initial alkyne with 1-halo-1-alkyne. For example, (5*Z*, 7*Z*)-5,7-dodecadiene is prepared in 85% yield in high stereochemical purity (Eq. 91) [144].

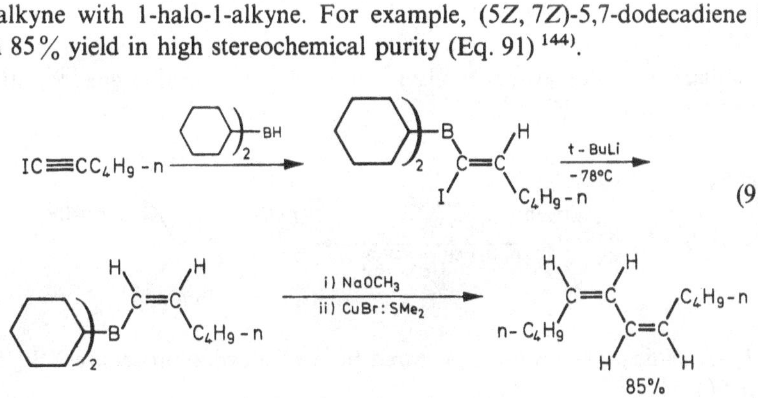

(91)

An alternative method is also reported (Eq. 92) [142,147], although the yields are comparatively low.

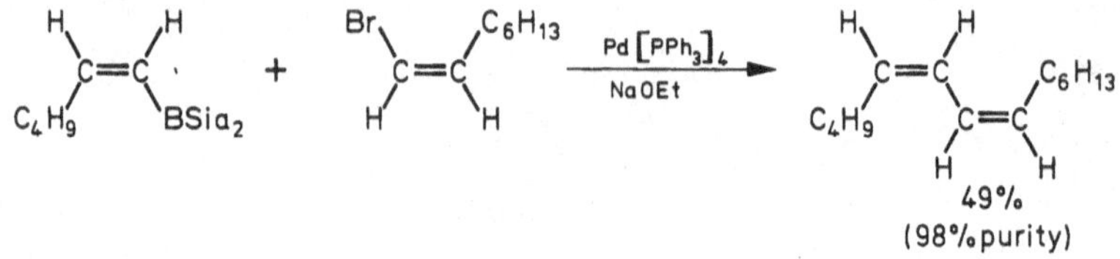

(92)

3.13 Synthesis of 1,4-Dienes

As discussed in the last sections, addition of sodium methoxide-alkenylborane complexes to copper bromide-dimethyl sulfide at 0 °C immediately gives conjugated dienes. However, if the temperature is lowered to −15 °C, the dark blue-black complex formed is stable and can be trapped by allylic halides to afford a stereochemically defined synthesis of (4*E*)-1,4-dienes (Eq. 93) [148].

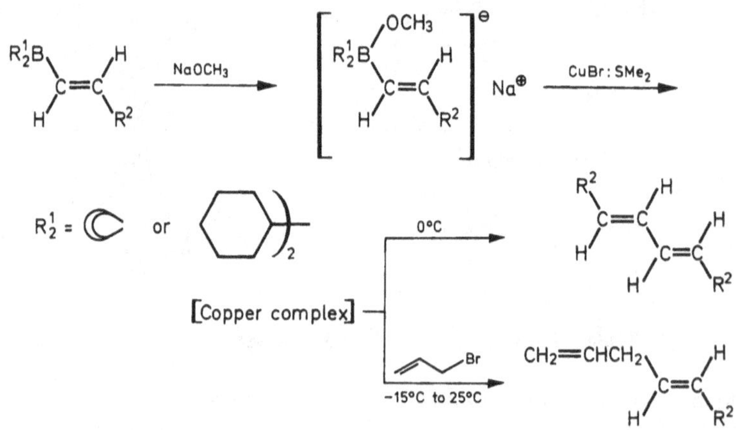

(93)

Similarly, readily available (1*Z*)-1-hexenyldicyclohexylborane [106] affords (4*Z*)-1,4-nonadiene (Eq. 94) [148].

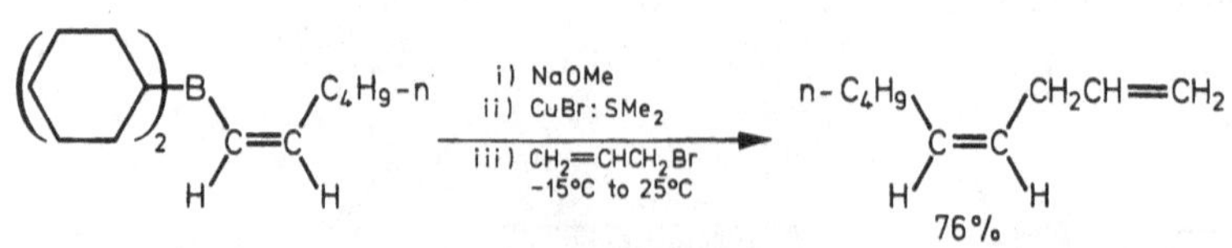

(94)

1,4-Alkadienes can also be prepared by the following procedures (Eq. 95) [149] and (Eq. 96) [150].

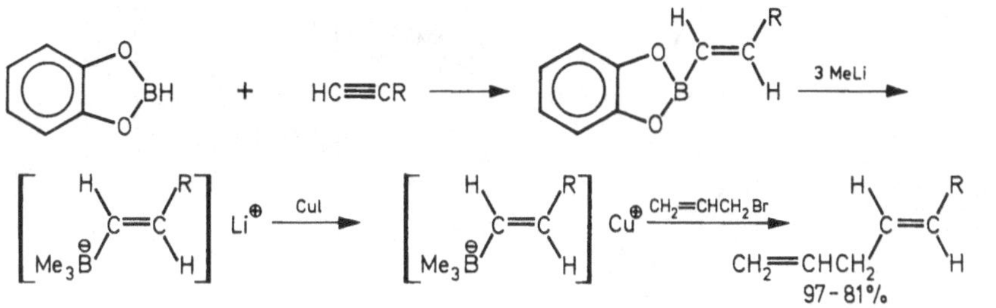

(95)

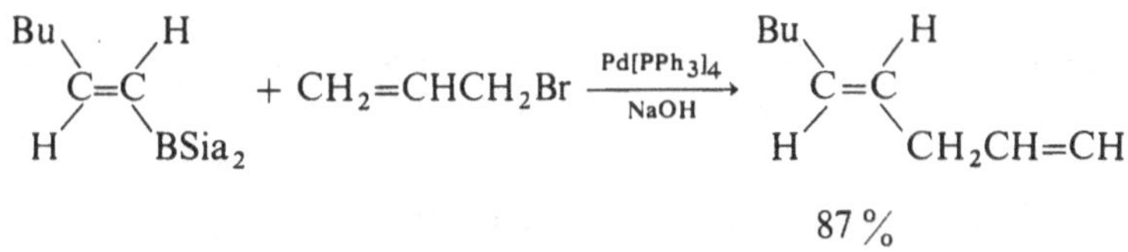

(96)

3.14 Synthesis of Conjugated Enynes

It has been previously discussed that alkenylcopper compounds prepared from alkenylboranes can be converted in a stereodefined manner to 1,3- and 1,4-dienes. Recently, it has been reported that alkenylcopper reagents could be coupled to 1-halo-1-alkynes in the presence of 1–2 equivalents of TMEDA to provide excellent yields of conjugated enynes [151]. Alkenylcopper reagents generated via organoboranes also react with 1-halo-1-alkynes in a stereodefined manner to afford conjugated *trans*-enynes (Eq. 97) [152]. As the alkenylcopper reagents prepared from organoboranes have different stereochemistry and substitution pattern from those prepared via carbometalation reaction, thus this method is stereochemically complementary to the one developed by Normant [151]. In this reaction, cuprous iodide is the reagent of choice as compared to cuprous bromide-dimethyl sulfide.

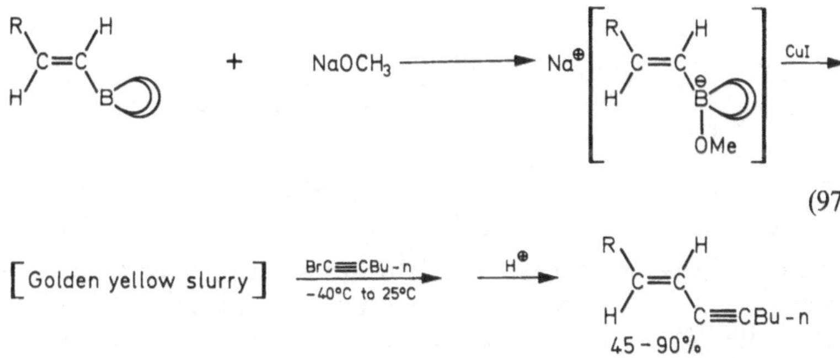

(97)

Another important synthesis of 1,3-*trans*-enynes has also been reported (Eq. 98)[153].

Many insect pheromones have conjugated *cis, trans*-diene system [152, 153]. The versatility of these reactions is exemplified in the synthesis of many natural products [153, 154, 155]. In Eq. 99, the four step synthesis of a natural sex pheromone of the European grape vine moth (*Lobesia botrana*), (*7E*, *9Z*)-7,9-dodecadien-1-yl acetate, is illustrated [155].

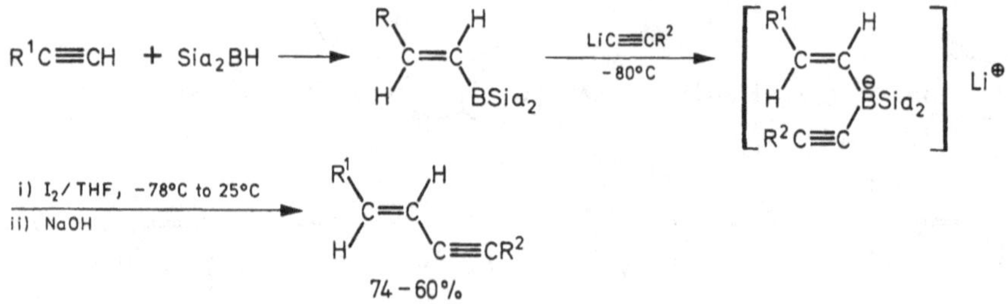

(98)

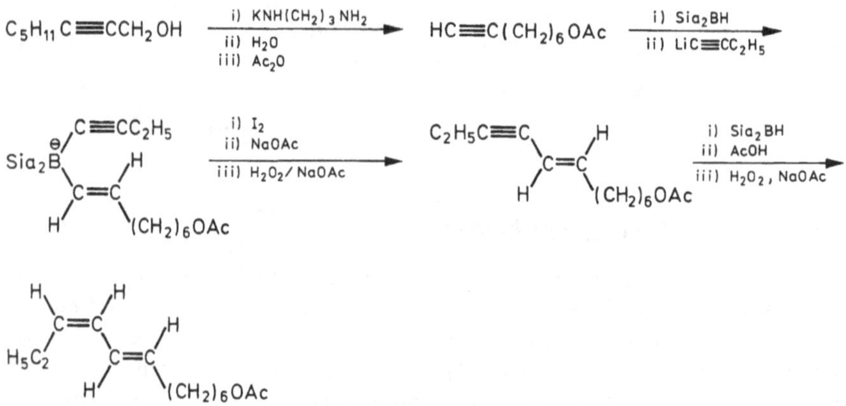

(99)

(*E*)-Enynes can also be synthesized in excellent yields with high regio- and stereoselectivity by transition metal catalyzed cross-coupling reaction. Thus, (*E*)-1-alkenyldisiamylboranes react with 1-halo-1-alkynes in the presence of catalytic amount of tetrakis(triphenylphosphine)palladium to afford conjugated *trans*-enynes (Eq. 100)[145].

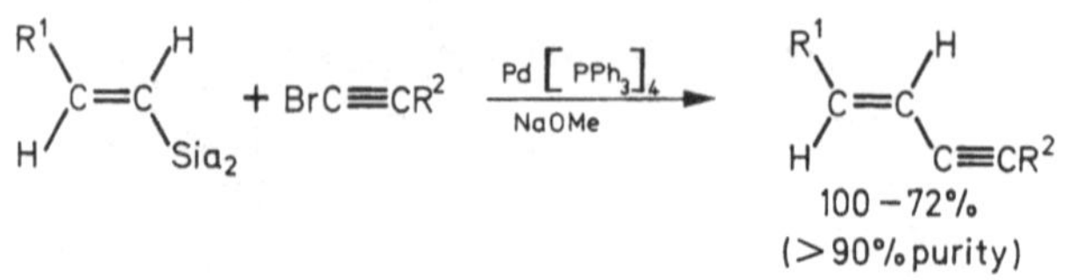

(100)

Recently, Rossi et al. have applied the cross-coupling of 1-alkenylboranes with 1-alkynyl halides for the synthesis of natural products. Thus, (7*E*, 9*Z*)-7,9-dodecadien-1-yl acetate (**32**), the sex pheromone of *Lobesia botrana*, has been synthesized by the following sequence involving:

(a) the cross-coupling of (*E*)-8-(2-tetrahydropyranyloxy)-1-octenyldisiamylborane with 1-bromo-1-butyne in the presence of tetrakis(triphenylphosphine)palladium and sodium methoxide;
(b) the acetylation of the crude product of the reaction;
(c) *cis*-stereoselective reduction of the conjugated (*E*)-enynyl acetate (Eq. 101)[156].

The versatility of such cross-coupling reactions of 1-alkenylboranes has been amply demonstrated by the stereospecific synthesis of *trans*-(C_{10})allofarnesene [158], bombykol and its three geometrical isomers [159], and humulene [160].

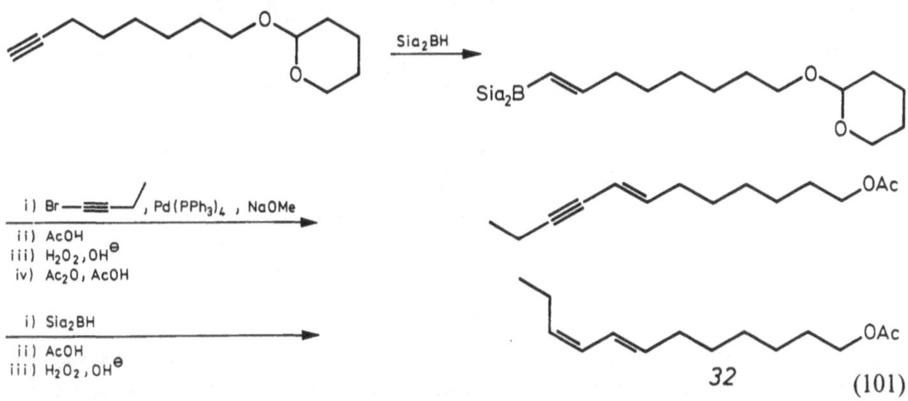

(101)

3.15 Synthesis of (*E*)-1,2,3-Butatrienes

The stereoselective synthesis of the title compounds has been achieved. Thexylborane on reaction with 2 molar equivalents of 1-iodo-1-alkyne at 0 °C proceeds to near completion (88% for 1-iodo-1-hexyne) to form fully substituted organoborane (**33**), which upon treatment with 2 molar equivalents of sodium methoxide at 0 °C readily produces *trans*-1,2,3-butatrienes (Eq. 102) [157]. The same reaction, however, with either 1-chloro or 1-bromo-1-alkynes is sluggish to form the thexyl-1-halo-1-alkenylborane [31].

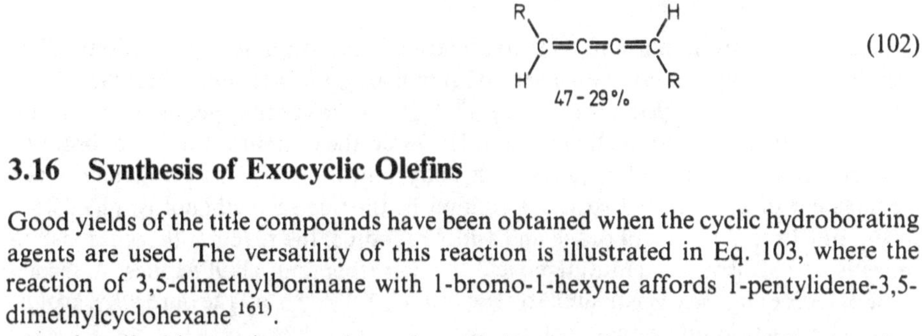

(102)

3.16 Synthesis of Exocyclic Olefins

Good yields of the title compounds have been obtained when the cyclic hydroborating agents are used. The versatility of this reaction is illustrated in Eq. 103, where the reaction of 3,5-dimethylborinane with 1-bromo-1-hexyne affords 1-pentylidene-3,5-dimethylcyclohexane [161].

As described in previous sections, the palladium catalyzed cross-coupling reaction of stereodefined 1-alkenylboranes with stereodefined 1-alkenyl halides in the presence of relatively strong bases such as sodium alkoxides and hydroxide ("head-to-head

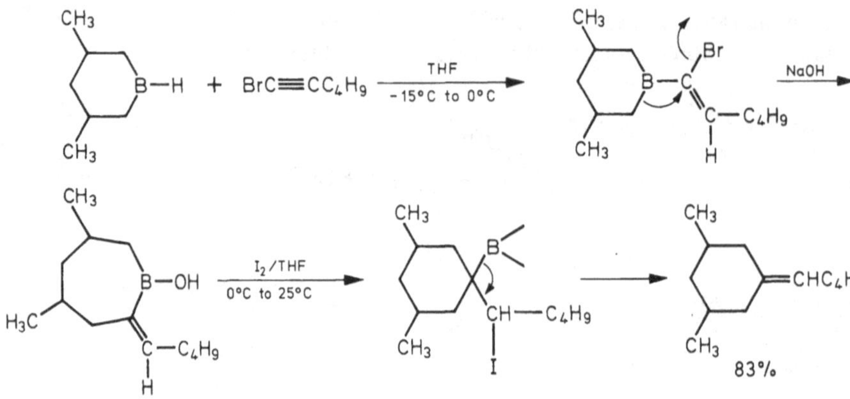

(103)

coupling") proceeds smoothly to provide various synthetic methodologies. On the other hand, it has been reported that "head-to-tail crosscoupling" products are obtained in the presence of Pd-catalyst and amines as bases, instead of sodium alkoxides and hydroxide (Eq. 104) [162]. Recently, this reaction has been applied for the synthesis of methylenecycloalkanes (eq. 105) [163].

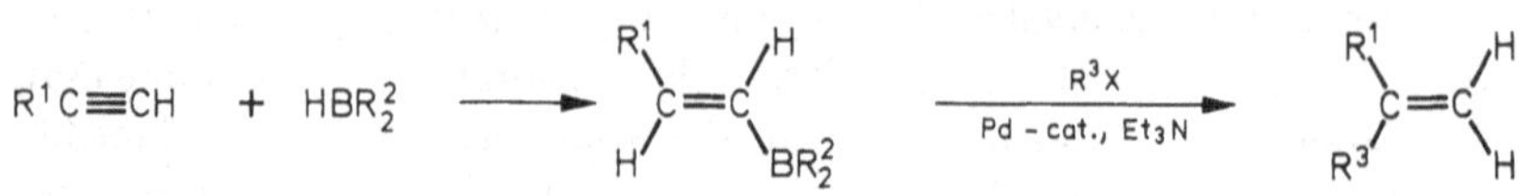

(104)

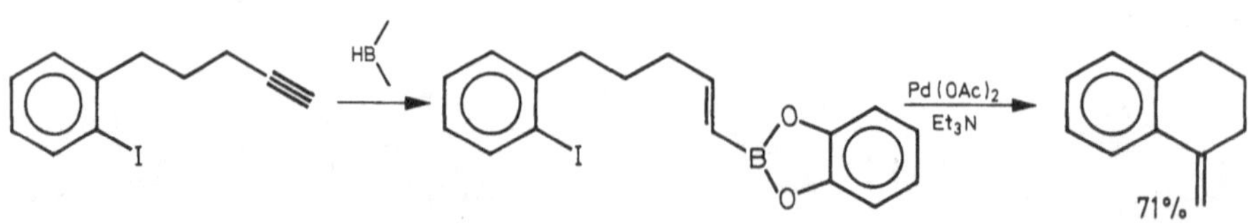

(105)

3.17 Synthesis of Organic Halides

Organic halides are important synthetic intermediates in numerous reactions. They are also important due to the variety of radiohalogen-containing pharmaceuticals which have been developed in recent years [164–166]. It has been observed that organoboranes can be converted to alkyl halides. However, the ordinary reaction of bromine and iodine is sluggish and requires high temperature (150 °C). Fortunately, when 1 equivalent of sodium hydroxide or sodium methoxide in methanol is added to a mixture of 1 mol of organoborane and 1 mol of iodine, the reaction becomes instantaneous and affords in quantitative yield the corresponding alkyl halides. A second mole of iodine and base reacts similarly (Eq. 106) [119,167,168], but the third alkyl group resists the reaction under these conditions.

$$R_3B + 2\,I_2 + 2\,NaOH \rightarrow 2\,RI + 2\,NaI + RB(OH)_2 \qquad (106)$$

4-Vinylcyclohexene is converted to give the corresponding unsaturated alkyl iodide selectively (Eq. 107)[119].

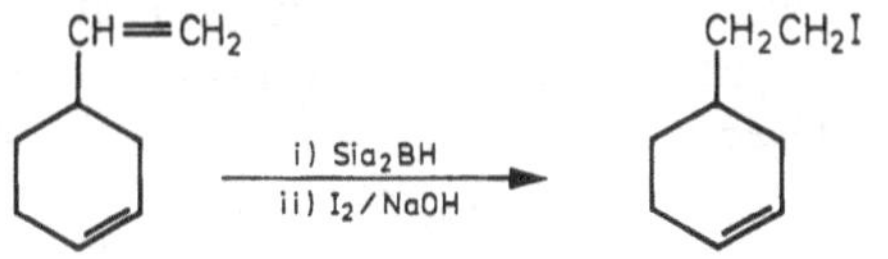

(107)

This reaction, however, has some obvious disadvantages,
(i) the strong base used in the reaction can react with sensitive functional groups or (ii) can cause dehydrohalogenation. Recently, this problem has been circumvented by the use of iodine monochloride and methanolic sodium acetate.

The reaction is instantaneous with the migration of two alkyl groups to afford the corresponding alkyl iodides (Eq. 108)[169].

$$(RCH_2)_3B \xrightarrow[\text{NaOAc/MeOH}]{\text{excess ICl}} 2\,RCH_2I \tag{108}$$

Alkyl bromides can also be obtained from organoboranes by treatment with bromine or bromine chloride[170].

The stereodefined alkenyl halides are of prime importance due to the recent developments of di- or trisubstituted alkene synthesis by cross-coupling reactions between organometallics and alkenyl halides catalyzed by transition metal compounds[171]. These alkenyl halides can be conveniently obtained from alkenylboranes or alkeneboronic acids. B-Alkenylcatecholboranes undergo rapid hydrolysis when stirred with excess water at 25 °C (Eq. 109)[102]. The alkeneboronic acids are usually crystalline solids of low solubility in water and can be easily isolated and handled in air without significant deterioration.

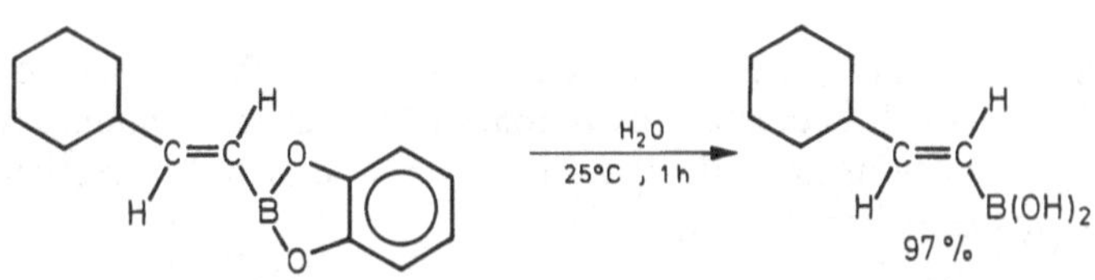

(109)

The reaction of alkeneboronic acids with iodine in the presence of base affords alkenyl iodides with retention of configuration (Scheme 2)[173] while treatment of

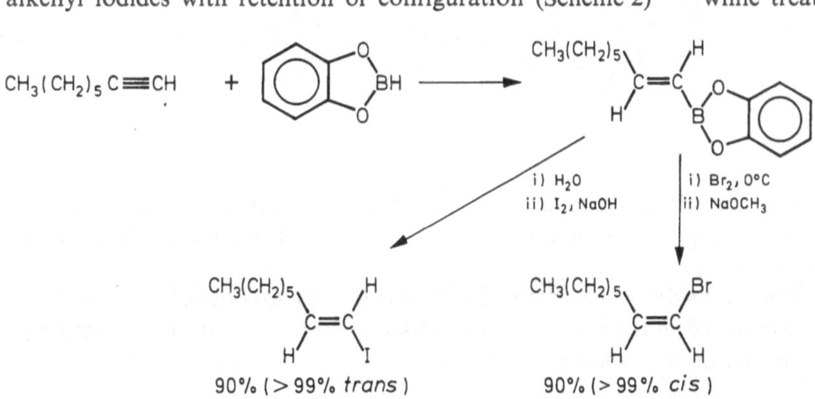

Scheme 2

B-alkenylcatecholboranes with bromine in the presence of base leads to inversion of configuration to give alkenyl bromides (Scheme 2) [174]. Thus, this method provides a convenient procedure to synthesize either *trans*-alkenyl iodides or *cis*-alkenyl bromides.

The possible mechanism for the synthesis of alkenyl bromides involves the usual *trans*-addition followed by case induced *trans*-elimination [174] as illustrated in Eq. 110.

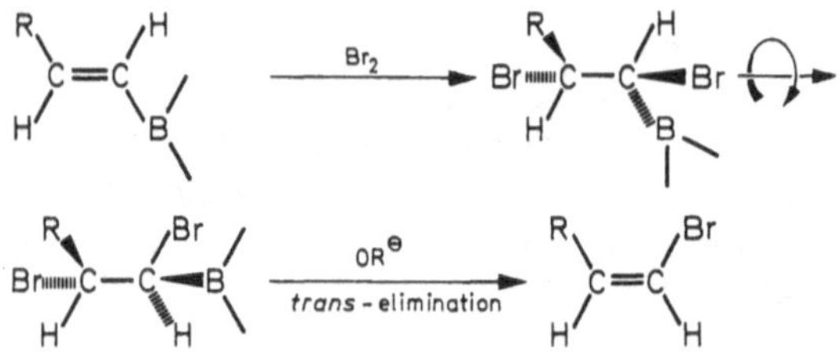

(110)

The synthesis of alkenyl iodides probably involves the *trans*-addition of hypoiodous acid via iodonium ion intermediate followed by *cis*-elimination (Eq. 111) [173].

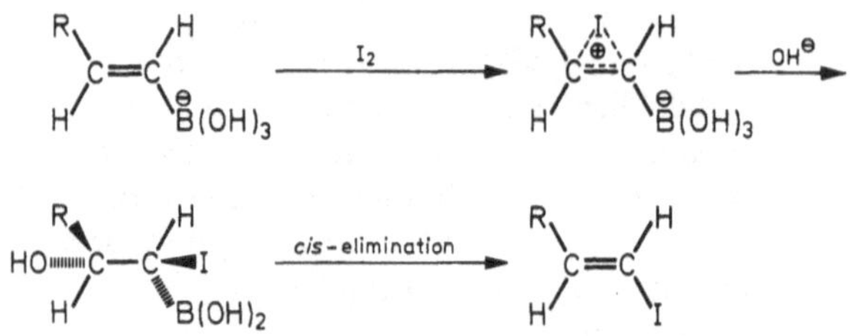

(111)

Alkenylboronic acids synthesized from internal alkynes fail to undergo the reaction with iodine and base. However, alkenylboronic acids and esters obtained from internal alkynes react with bromine in the presence of base to afford in excellent yields the corresponding alkenyl bromides (Eq. 112) [174].

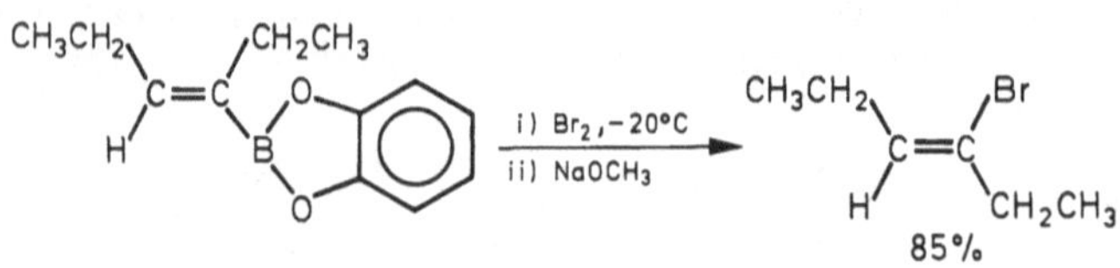

(112)

The importance of these reactions is exemplified in the synthesis of (5*Z*, 7*E*)-5,7-dodecadien-1-ol (Eq. 113), a sex pheromone component of the forest tent catapillar (*Malacosoma disstria*) [175].

As discussed already, some of the functional groups cannot tolerate the basic conditions employed in the reaction. This difficulty, however, can be obviated also by the use of iodine monochloride and methanolic sodium acetate, as illustrated in Eq. 114 [176].

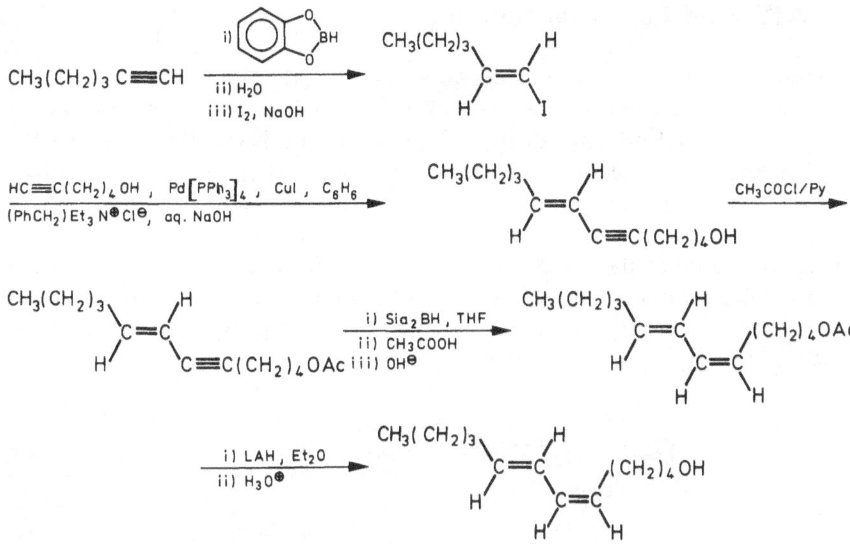

(113)

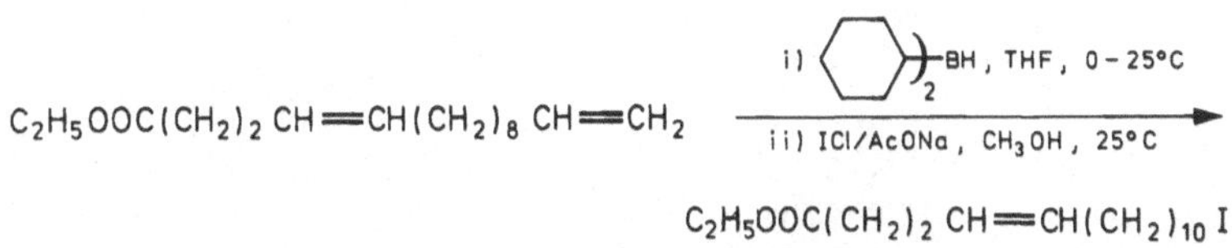

(114)

The iodine monochloride is an ideal reagent except for the necessity of its handling or preparation. This difficulty has been overcome by the reaction of organoboranes with iodide ion in the presence of mild oxidizing agent chloramine-T [177]. The reaction proceeds under extremely mild conditions to afford the corresponding alkenyl iodides (Eq. 115) [178] where it can tolerate a variety of functional groups.

(115)

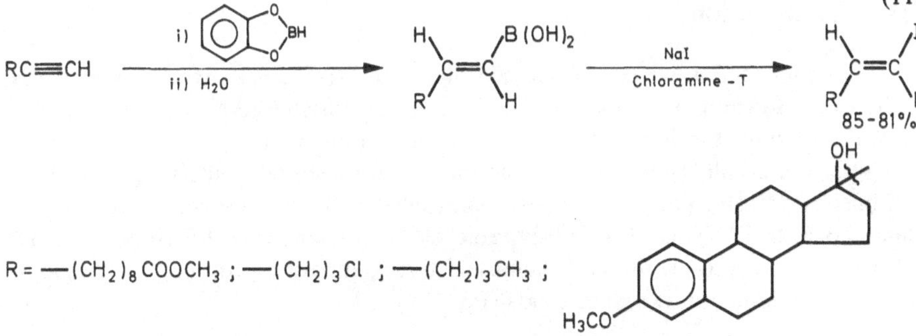

This method has been employed conveniently for labelling molecules with $Na^{125}I$ on a tracer level [179].

The haloboration reaction of terminal alkynes has been recently demonstrated to provide useful synthetic procedures for various halogen derivatives, which will be discussed later.

3.18 Synthesis of 1,1-Dihalo-1-alkenes

The currently available methods for the synthesis of the title compounds are confined to the preparation of homo-1,1-dihalo-1-alkenes [180], while only a few reports are available for mixed 1,1-dihalo-1-alkenes of defined stereochemistry [181]. As the hydroboration reaction proceeds in a stereospecific manner, the hydroboration-oxidation-bromination-debromoboration sequence of 1-chloro-1-alkynes produces selectively (*Z*)-1-bromo-1-chloro-1-alkenes (Eq. 116) [182]. The oxidation with anhydrous trimethylamine oxide of the alkenylborane prior to the addition of bromine is necessary to avoid the competing transfer of one of 1,2-dimethylpropyl group from boron to the adjacent carbon atom. Similar reaction sequence provides 1,1-dibromo-1-alkenes (Eq. 117) [182].

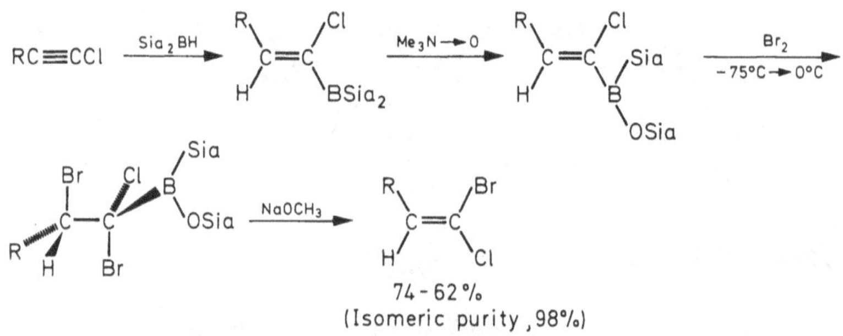

(116)

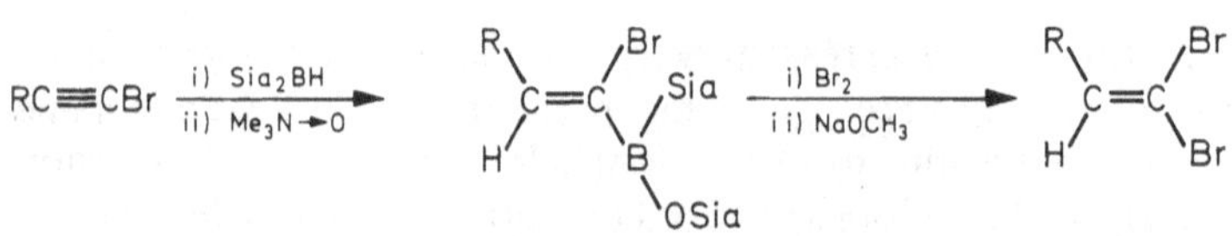

(117)

3.19 Haloboration

In the previous section we have discussed that 1-halo-1-alkenes can be conveniently synthesized. However, with these methods it is not possible to synthesize 2-halo-1-alkenes. Although the halometallation reaction would be a powerful tool for the preparation of 2-halo-1-alkenes, the reaction has not been adequately developed for such purpose [183]. Recently, it has been reported that B-bromo-9-borabicyclo[3.3.1]-nonane (B-Br-9-BBN) and B-iodo-9-borabicyclo[3.3.1] nonane (B-I-9-BBN) [184] react with 1-alkynes, stereo-, regio- and chemoselectively, and after protonolysis, 2-halo-1-alkenes are obtained in excellent yields (Eq. 118) [185].

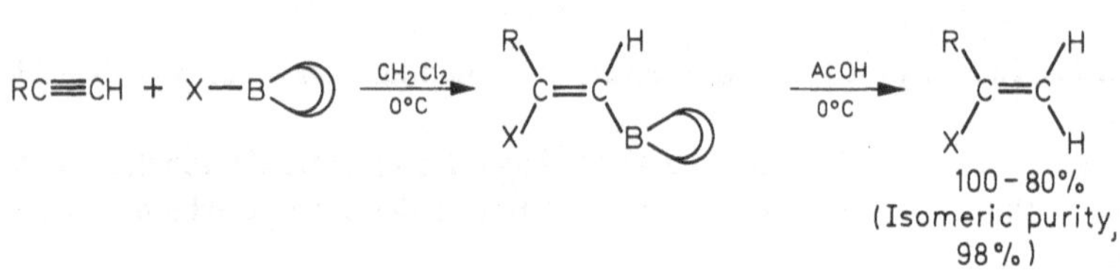

(118)

The experiments have shown that B-Br-9-BBN attacks only the terminal triple bond whereas internal triple bond, internal and terminal double bonds can withstand the bromoboration reaction conditions (Eqs. 119 and 120).

$$CH_3(CH_2)_5C{\equiv}CH + CH_3(CH_2)_4C{\equiv}CCH_3 \xrightarrow[\text{ii) } CH_3COOH]{\text{i) B-Br-9-BBN}} \underset{99\%}{CH_3(CH_2)_5CBr{=}CH_2} + \underset{97\%}{CH_3(CH_2)_4C{\equiv}CCH_3} \tag{119}$$

$$CH_3(CH_2)_5C{\equiv}CH + CH_3(CH_2)_5CH{=}CH_2 \xrightarrow[\text{ii) } CH_3COOH]{\text{i) B-Br-9-BBN}} \underset{99\%}{CH_3(CH_2)_5CBr{=}CH_2} + \underset{96\%}{CH_3(CH_2)_5CH{=}CH_2} \tag{120}$$

As the haloboration occurs stereospecifically, the reaction is extended for a stereo-defined synthesis of (Z)-1-alkynyl-2-halo-1-alkenes (Eq. 121) [186].

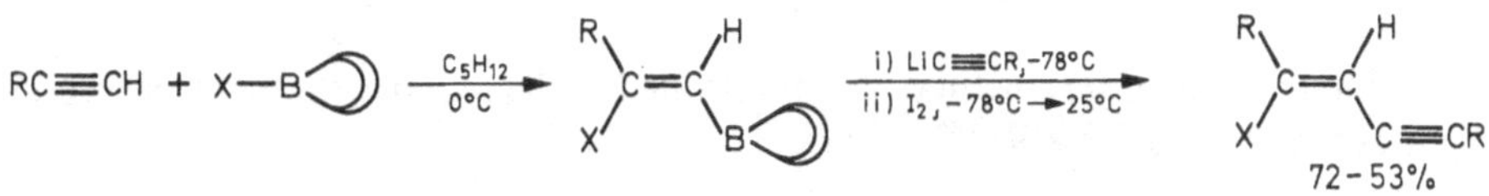

(121)

In an attempt to synthesize (Z)-1,2-dihalo-1-alkenes from bromoboration adducts of 1-alkynes with B-X-9-BBN, all efforts have been unsuccessful. Fortunately, it has found that the haloboration of 1-alkynes with tribromoborane, followed by the reaction with iodine or bromine chloride in the presence of sodium acetate gives the expected (Z)-1,2-dihalo-1-alkenes stereospecifically (>98%) in good yields (Eq. 122) [187].

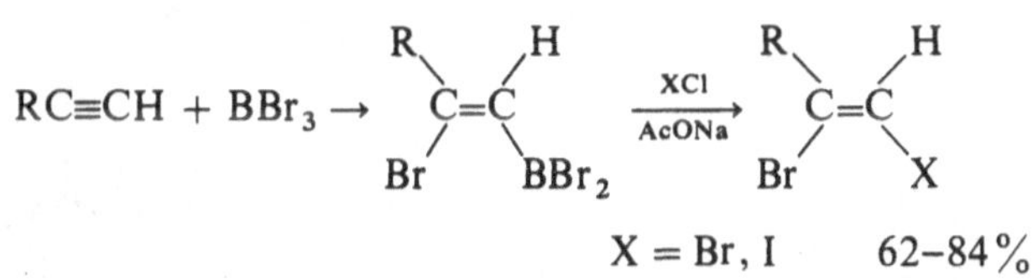

(122)

Most recently, B-halo-1-alkenyl-9-BBN's have revealed to undergo a 1,4-addition reaction to α,β-unsaturated ketones such as methyl vinyl ketone. Hydrolysis of the initially formed intermediates produces the corresponding (Z)-δ-halo-γ,δ-unsaturated ketones in good yields stereospecifically (>98%) [188]. In the reaction with methyl vinyl ketone, the major products isolated after hydrolysis are not the expected halo-enones but the aldol adducts, which are readily converted into the halo-enones by treatment with sodium hydroxide, as depicted in Scheme 3. No aldol adducts are formed in the case of enones other than methyl vinyl ketone.

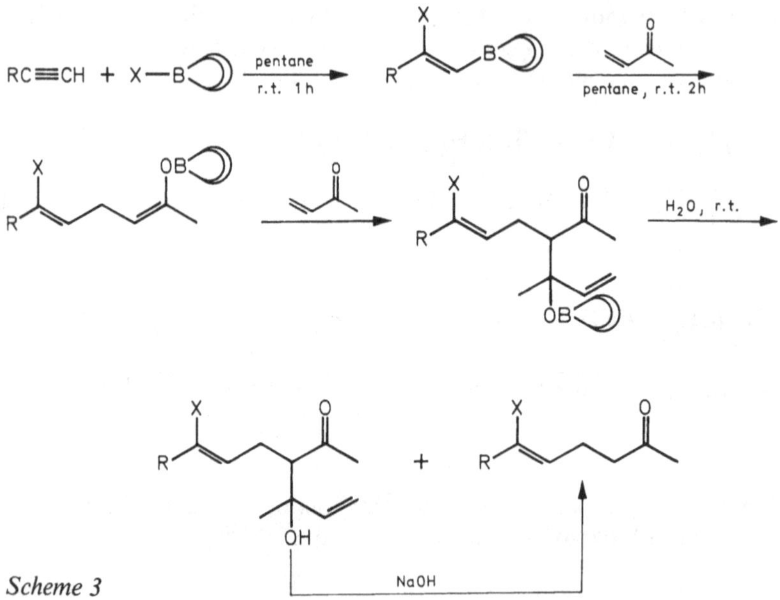

Scheme 3

Synthetic utility of this method is demonstrated in the selective synthesis of several natural products (Scheme 4) [188].

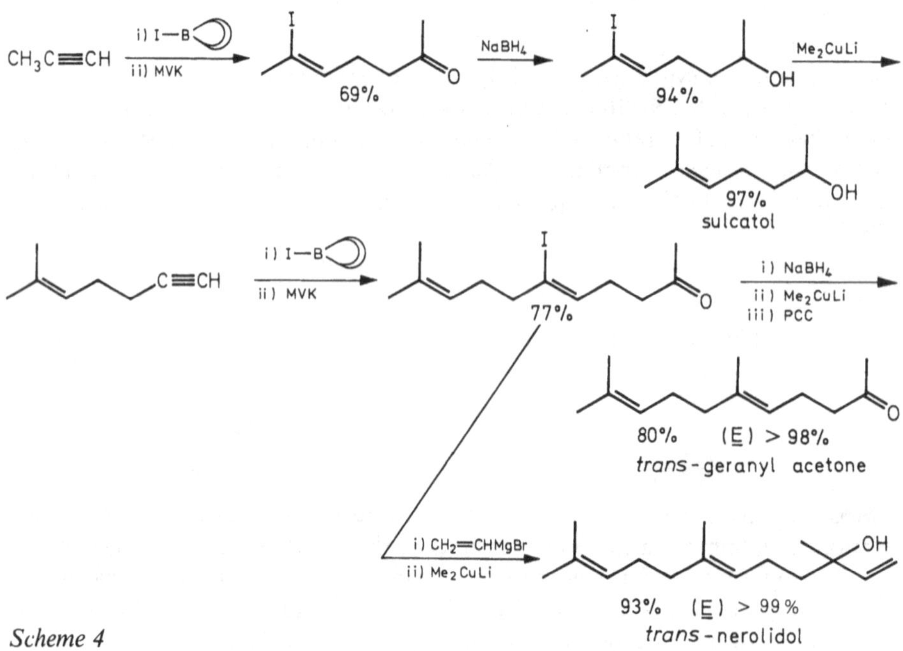

Scheme 4

The haloboration reaction are also applicable for the preparation of N-phenyl-β-bromo-α,β-unsaturated amides [189] and 2-bromoalkanals [190].

3.20 Trans-Metallation

The synthesis of stereodefined substituted alkenes is one of the major challenges in organic synthesis as many naturally occurring compounds like pheromones and hormones contain mono-, di- or trisubstituted double bonds. In many cases, a content of small amounts of the wrong isomer acts as an inhibitor of their biological activity. Thus, a highly stereospecific synthesis of such compounds is required. In addition to such methods employing hydroboration and haloboration, vinylic metal species obtained from alkynes are now widely used for the stereospecific synthesis of these compounds. The easy accessibility of alkenylboranes of known geometry makes hydroboration-metallation sequence a method of prime importance. Of particular significance is the fact that hydroboration is selective and many functional groups are readily accommodated in this reaction. Thus, through hydroboration-mercuration, organic chemists have a convenient route to a variety of organomercurials not previously available [191-193]. The synthetic applications of organomercurials have been reviewed [194].

However, for the synthesis of alkenylmercuric salts, one difficulty was encountered as mercuric acetate also reacted with the residual double bond. Fortunately, cyclohexene with mercuric acetate in tetrahydrofuran gave only 8% reaction after 42 h[195]. Thus, 2-(4-cyclohexenyl)-ethylmercuric chloride was obtained in 94% isolated yield via hydroboration-mercuration sequence of 4-vinyl cyclohexene (Eq. 123) [192].

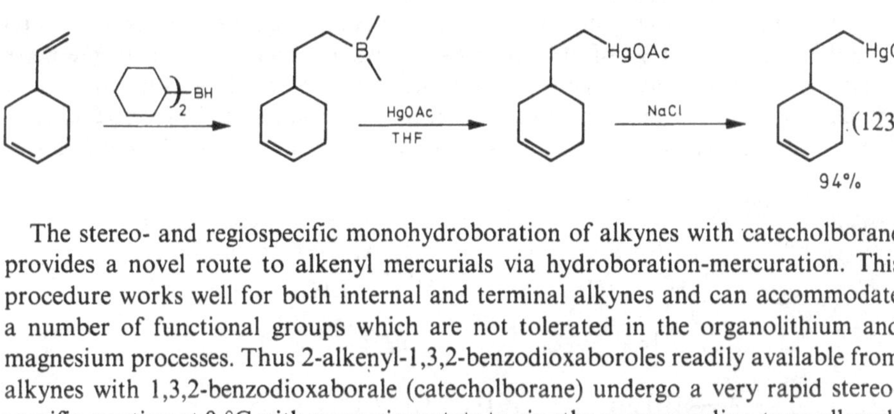

(123)

The stereo- and regiospecific monohydroboration of alkynes with catecholborane provides a novel route to alkenyl mercurials via hydroboration-mercuration. This procedure works well for both internal and terminal alkynes and can accommodate a number of functional groups which are not tolerated in the organolithium and magnesium processes. Thus 2-alkenyl-1,3,2-benzodioxaboroles readily available from alkynes with 1,3,2-benzodioxaborale (catecholborane) undergo a very rapid stereospecific reaction at 0 °C with mercuric acetate to give the corresponding *trans*-alkenylmercuric acetates. Treatment with aqueous sodium chloride provides 97–99% isolated yields of the alkenylmercuric chlorides (Eq. 124) [196].

(124)

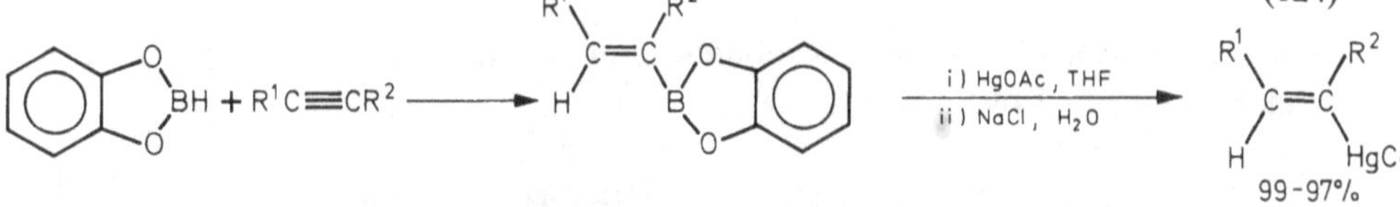

It is to be pointed out that vinylboranes synthesized via hydroboration with disiamylborane and dicyclohexylborane are less satisfactory and afforded [193] byproducts in addition to the desired *trans*-alkenylmercuric acetate.

These stable and easily handled alkenylmercurial compounds are important synthetic intermediates and can be used directly for the preparation of other alkenyl metallics. One such application is illustrated in Scheme 5 for the synthesis of a prostaglandin analogue [197].

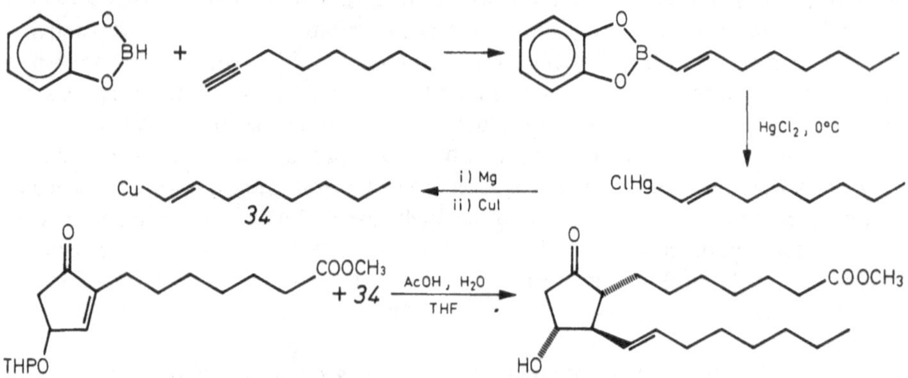

Scheme 5

3.21 Separation of Isomeric Mixtures

We have seen the relative reactivities of structurally different alkenes towards various dialkylboranes vary over a broad range of differences, e.g., disiamylborane [43] and 9-BBN[48] vary in a range of 10^5. This makes selective hydroboration of the more reactive alkenes possible in the presence of less reactive one. Disubstituted internal (*Z*)-alkenes are more reactive to disiamylborane than their (*E*)-isomers (Eq. 125). Recently, it has been observed that the selectivity is much higher with ThxBHCl : SMe_2

$$\underset{18\%}{(CH_3CH_2)(H)C{=}C(CH_3)(H)} + \underset{82\%}{(CH_3CH_2)(H)C{=}C(H)(CH_3)}$$

$$\xrightarrow{Sia_2BH} \underset{2\%}{(CH_3CH_2)(H)C{=}C(CH_3)(H)} + \underset{98\%}{(CH_3CH_2)(H)C{=}C(H)(CH_3)} \qquad (125)$$

as compared to Sia_2BH in the hydroboration of *cis*/*trans* pairs (Chart 7)[19] in favor of more reactive *cis* olefin.

$$CH_3CH_2CH_2CH{=}CH_2 + CH_3CH_2CH{=}CHCH_2CH_3$$

$$\xrightarrow{Sia_2BH} CH_3CH_2CH{=}CHCH_2CH_3 + CH_3CH_2CH_2CH_2CH_2{-}BSia_2 \qquad (126)$$

Terminal alkenes can be removed from mixtures containing internal alkenes [43], and trisubstituted alkenes from tetrasubstituted isomers (Eqs 126 and 127) [198].

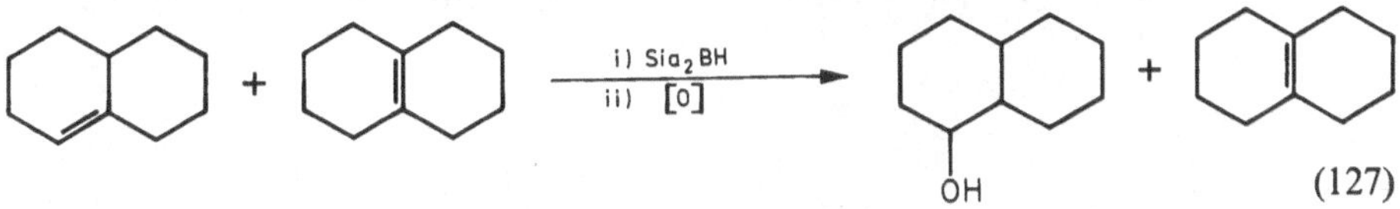

(127)

The separation of racemic mixture is the most difficult process and involves several steps. The selective hydroboration sequence provides a convenient procedure [199 200] for the resolution of 1,3-disubstituted allenes, while the process is tedious and lengthy by other methods [201]. Thus, the treatment of (±)-1,3-dimethyl allene with 50 mol percent of (+)-IPC_2BH provided unreacted (R)-(–)-1,3-dimethylallene (Eq. 128). The configuration of allenes is consistently R when resolved by (+)-IPC_2BH. As both forms of IPC_2BH are available, it should be possible to obtain the allenes of opposite configuration.

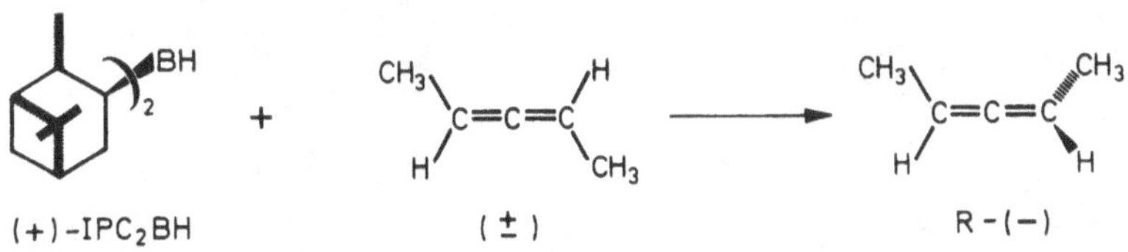

(128)

The reaction appears to be general as many allenes of different structures have been resolved by this method [202–204]. The enantiomeric purities increase in the following order: 1,3-dimethylallene < 1,2-cyclononadiene < 1,3-di-tert-butylallene < 1,3-diethylallene < 1,3-di-n-propylallene.

The procedure has also been nicely employed for the resolution of dienes [205], spirodienes [206], trienes [207], and α-pinene [208].

3.22 Asymmetric Synthesis

The synthesis of enantiomerically pure compounds is the challenging problem for organic chemists. The synthesis becomes obsolete if the intermediates produce racemic mixtures. The problem is particularly acute when the asymmetric centers do not reside in a rigid cyclic or polycyclic framework. To be able to carry out efficient syntheses of complex molecules, chemists have to control the sense of chirality at each chiral center as it is introduced in the course of synthesis. Monoalkyl- or dialkylboranes exhibit a remarkable chemo-, stereo-, and regioselectivity for the hydroboration of unsaturated compounds. This property, coupled with the capability for asymmetric creation of chiral centers with chiral hydroboration agents, makes the reaction most valuable for asymmetric organic synthesis. In some of the cases, however, this has been achieved by diborane itself as shown in the synthesis of monensin by Kishi et al. A stereospecific synthesis of its seven carbon component has been accomplished by two hydroboration reactions (Eq. 129) [209].

The stereospecific addition of borane is explained by the model (below) which depicts borane approaching the double bond from the least hindered side of the most likely conformation of the molecule.

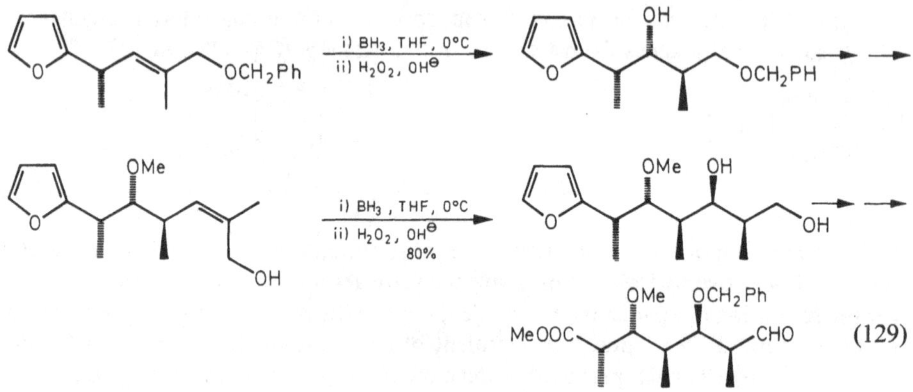

(129)

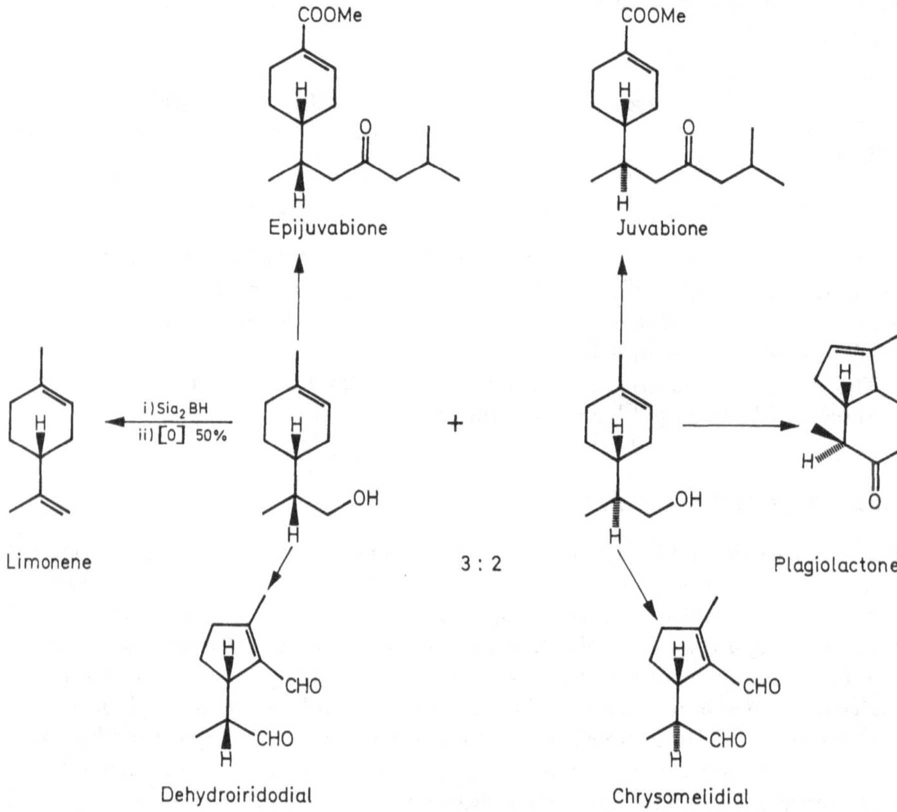

Scheme 6

Hydroboration of limonene with disiamylborane followed by oxidation afforded the isomeric alcohols as 3:2 mixture. These starting materials are utilized in the synthesis of juvabione isomers [210] and in the synthesis of beetle defense substances, chrysomelidial [211], plagiolactone [212] and dehydroiridodial [213] (Scheme 6).

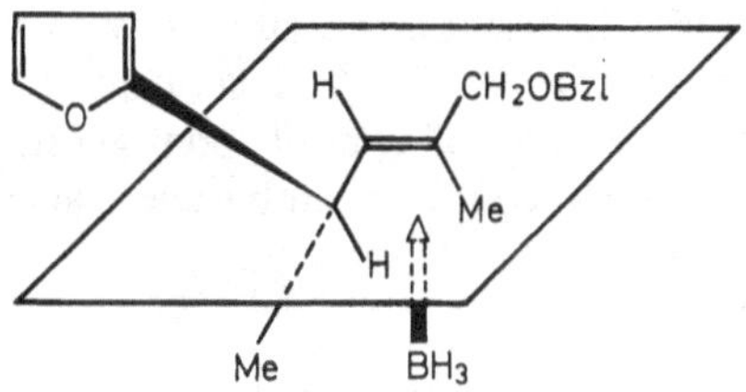

Both antipodes of α-pinene are available, which upon hydroboration with borane leads to both forms of diisopinocampheylborane (IPC_2BH) depicted below:

(+)-IPC_2BH (−)-IPC_2BH

These are excellent chiral hydroborating agents and have been utilized in the synthesis of many natural products. (+)-IPC_2BH has been satisfactorily employed to obtain intermediates required in prostaglandin $F_{2\alpha}$ synthesis [214] (Scheme 7) and the Fried prostaglandin intermediate [215]. Similar treatment of **35** with (–)-IPC_2BH affords **37**, an antipode of **36**.

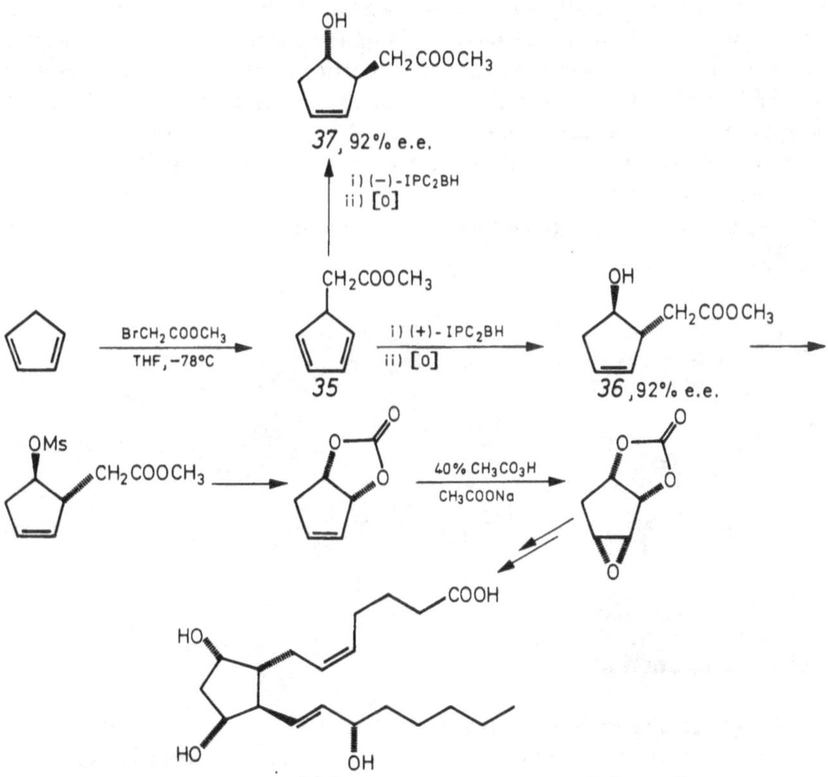

Scheme 7

Another useful application of (+)-IPC_2BH is the synthesis of loganin, an important building block in much of the plant world (Scheme 8) [216]. Treatment of the diene (**38**) with (–)-IPC_2BH again affords the alcohol (**40**) of opposite configuration with high and equivalent optical purity. See ref. 4 for chiral organoborane reagents.

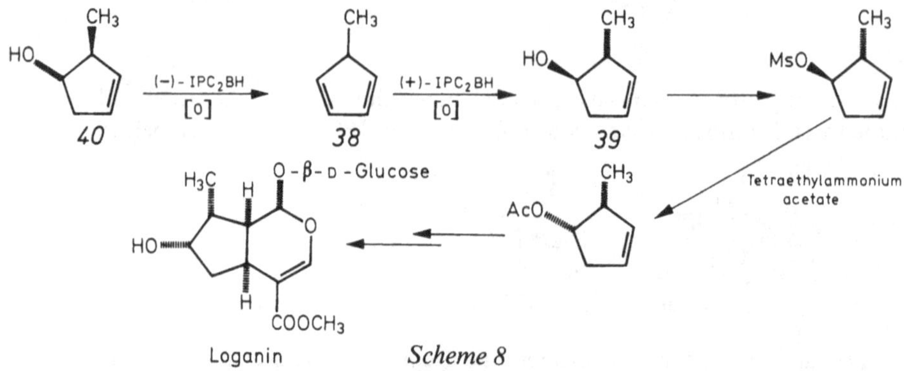

Scheme 8

Asymmetric hydroboration [217] of prochiral alkenes with monoisopinocampheylborane in the molar ratio of 1:1, followed by a second hydroboration of non-prochiral alkenes with the intermediate dialkylboranes, provides the chiral mixed trialkylboranes. Treatment of these trialkylboranes with acetaldehyde results in the selective, facile elimination of the 3-pinanyl group, providing the corresponding chiral borinic acid esters with high enantiomeric purities. The reaction of these intermediates with base and dichloromethyl methyl ether provides the chiral ketones (Eq. 130) [218]. A simple synthesis of secondary homoallylic alcohols with excellent enantiomeric purities via B-allyldiisopinocampheylborane has been also reported [219].

Recently, Matteson and his co-workers have demonstrated interesting boronic ester homologation with very high chiral selectivity and its use in syntheses of natural products [220].

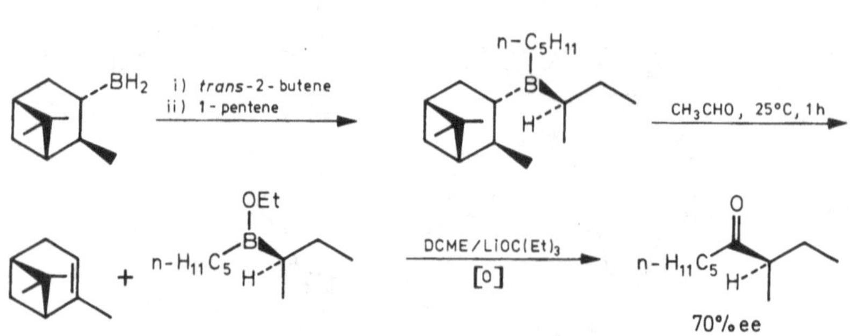

(130)

3.23 Selective Reduction

The sterically demanding organoboranes provide a efficient tool for non-catalytic *cis*-hydrogenation of carbon-carbon triple bonds. Coupled with protonolysis, this sequence gives an elegant method for the synthesis of corresponding *cis*-enynes or *cis, cis*-dienes from conjugated diynes [79]. Monohydroboration of diynes with Sia_2BH places the boron exclusively at the internal position, which followed by protonolysis affords the corresponding *cis*-enynes in good yields (Eq. 131).

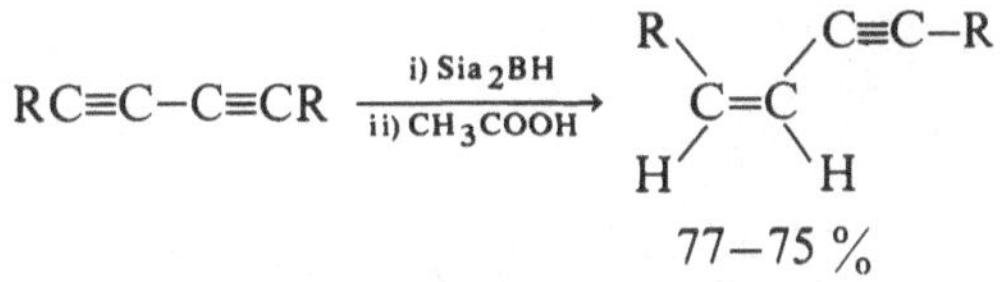

(131)

As the second addition of disiamylborane is very solw, the dihydroboration can be carried out with sterically less hindered dicyclohexylborane. Protonolysis of the intermediate organoboranes with acetic acid affords the corresponding *cis, cis*-dienes (Eq. 132).

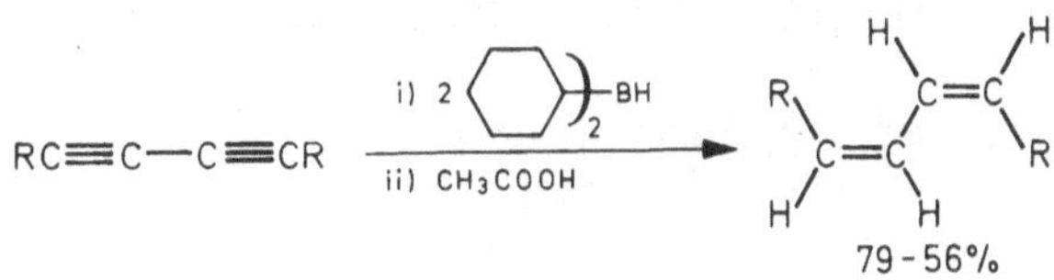

(132)

Organoboranes have proven to be excellent reducing agents where the functional group is preferentially reduced, without the concomitant hydroboration of carbon-carbon multiple bonds. For instance, citronellal is reduced to the unsaturated alcohol with catecholborane (Eq. 133) [221].

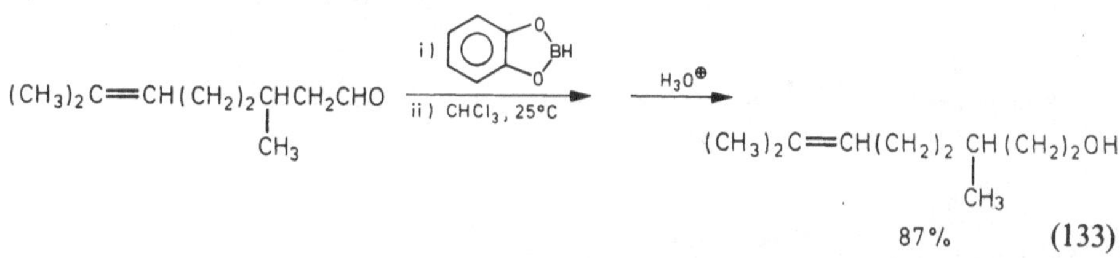

(133)

Catecholborane is a versatile reducing reagent, a mild and convenient alternative to the classical (Clemmensen, using acids and Wolff-Kishner, using bases) reduction procedure of tosylhydrazones. Regiospecific isomerization occurs during the reduction of α,β-unsaturated carbonyl derivatives often leading to unique alkenes (Eqs. 134 and 135) [222,223] and allenes (Eq. 136) [224].

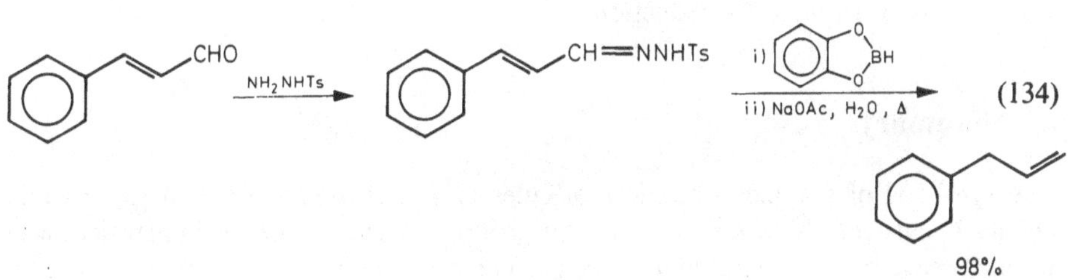

(134)

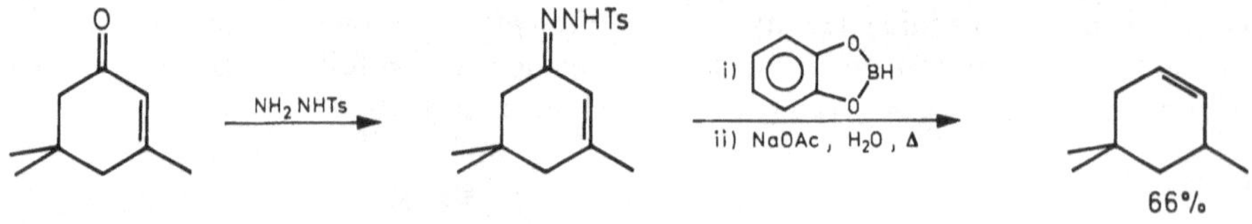

(135)

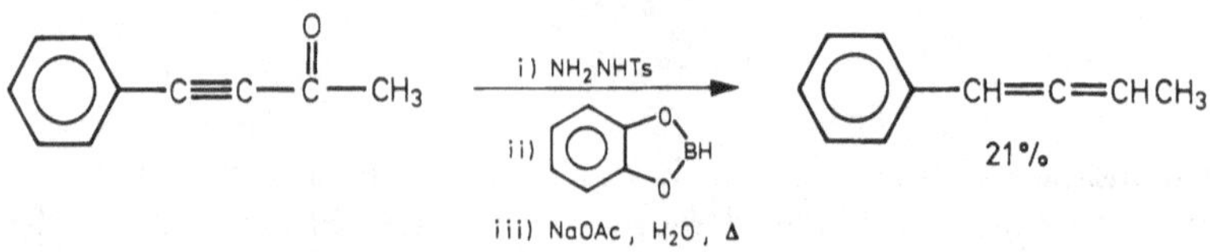

(136)

Similarly, bis(benzoyloxy)borane prepared easily from benzoic acid and diborane (Eq. 137) reduces the tosylhydrazones in excellent yield, e.g., the tosylhydrazone of 6-oxo-15-hexadecenoic acid affords 15-hexadecenoic acid (Eq. 138) [225].

$$2\ C_6H_5COOH + BH_3 \xrightarrow[0\ ^\circ C]{THF} (C_6H_5COO)_2BH + 2\ H_2\uparrow \tag{137}$$

$$CH_2{=}CH(CH_2)_8{-}\overset{\overset{\displaystyle O}{\|}}{C}{-}(CH_2)_4COOH \xrightarrow[\substack{ii)\ (C_6H_5COO)_2BH \\ iii)\ NaOAc,\ H_2O}]{i)\ NH_2NHTs} \underset{96\ \%}{CH_2{=}CH(CH_2)_{13}COOH} \tag{138}$$

Conjugated aldehydes and ketones can be selectively reduced in quantitative yields to the corresponding allylic alcohols by 9-BBN in THF (Eq. 139) [226].

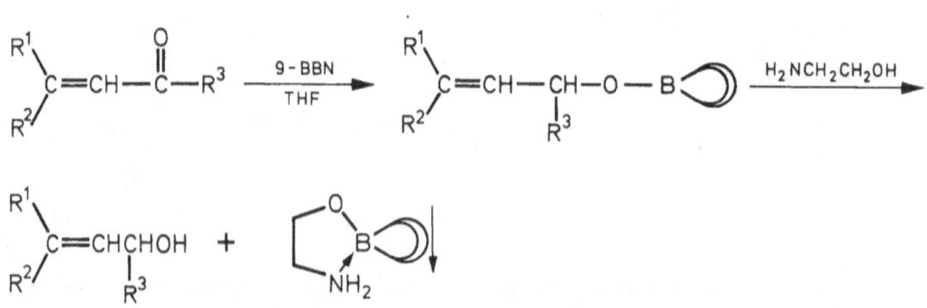

(139)

The mildness of the procedure provides the selective reduction of unsaturated carbonyl groups in the presence of almost any other functional groups such as ester, amide, carboxylic acid, nitro, halogen, nitrile, etc. It is evident that this reagent is far superior in purity and yield of products as compared to the conventional reagents, without any observable 1,4-reduction.

4 Summary

The synthesis of complex organic molecules is a challenging problem for organic chemists, particularly, when many similar groups which also behave in identical way to same reagents are present in a molecule. The problem becomes more acute when a particular form of chiral carbon is required in multi-chiral compounds. These problems then make the synthetic process a multistep, tedious and time consuming one. The resolution of racemic mixtures often results in loss of one half of the valuable materials. In recent times the organoboranes have emerged like a meteor to circumvent these difficulties. The developments of many mono- and dialkylboranes and haloboranes with varying steric and electronic requirements have made them possible to react only one of the carbon-carbon double or triple bonds in the presence of others. These reaction sequences become plentiful, particularly, as many sensitive groups can tolerate the reaction conditions. With the recent development of hydridation procedure, it is now possible to introduce selectively any of the alkyl groups and in any form of sequence to synthesize mixed organoboranes. The versatility of these reagents is further magnified in the reduction of carbonyl groups where the carbon-carbon double bonds remain unchanged. Moreover, the organic chemists for the first time have chiral hydroborating agents available on the shelf. The ease of the reaction have made them invaluable. No effort has been made to compile a comprehensive list of recent publications in this review article. It is the main goal of this review to demonstrate representative and useful examples for obtaining various organic compounds by employing hydroboration and organoboron compounds thus obtained, in which organic chemists, especially in a field of natural product chemistry, may be interested.

5 References

1. Gerrard, W.: The Organic Chemistry of Boron, Academic Press, New York 1961
2. Lappert, M. F.: Chem. Rev. *56*, 959 (1956)
3. Brown, H. C. in: Organic Synthesis Today and Tomorrow, Proceedings of the 3rd IUPAC Symp. on Organic Synthesis, Madison, Wisconsin, U.S.A. (eds. Trost, B. M., Hutchinson, C. R.) Pergamon Press, Oxford 1981
4. Brown, H. C., Jadhav, P. K., Mandal, A. K.: Tetrahedron *37*, 3547 (1981)
5. Brown, H. C., Jadhav, P. K., Mandal, A. K.: J. Org. Chem. *47*, 5074 (1982)
6. Brown, H. C., Kulkarni, S. U.: J. Organomet. Chem. *218*, 299 (1981)
7. Cragg, G. L. M.: Organoboranes in Organic Synthesis, Dekker, New York 1973;
   Brown, H. C.: Organic Synthesis via Boranes, Wiley, New York 1975;
   Pelter, A., Smith, K. in: Comprehensive Organic Chemistry, (ed. Barton, D. H. R., Ollis, W. D.) Vol. 3, p. 683, Pergamon, Oxford 1979;
   Negishi, E.: Organometallics in Organic Synthesis, Vol. 1, p. 283, Wiley, New York 1980;
   Brown, H. C.: Hydroboration, Benjamin/Cummings, Reading, Massachusetts 1980[2];
   Brown, H. C.; Zaidlewicz, M. in: Comprehensive Organometallic Chemistry, (ed. Wilkinson, G.) Vol. 7, Pergamon, Oxford 1982;
   Mikhailov, B. M., Bubnov, Yu. N.: Organoboron Compounds in Organic Synthesis, Harwood, Chur, Switzerland 1984;
   Köster, R.: Houben-Weyl Methoden der Organischen Chemie, Vol. 13, Thieme, Stuttgart 1984
8. For recent reviews, see: Negishi, E.: J. Organomet. Chem. *108*, 281 (1976);
   Weill-Raynal, J.: Synthesis 633 (1976);
   Cragg, G. M. L., Koch, K. R.: Chem. Soc. Rev. *6*, 393 (1977);
   Avasthi, K., Devaprabhakara, D., Suzuki, A. in: Organometallic Chemistry Reviews, p. 1, Elsevier, Amsterdam 1979;
   Brown, H. C., Campbell, J. B.: Aldrichimica Acta *14*, 3 (1981);
   Pelter, A.: Chem. Soc. Rev. *11*, 191 (1982);
   Suzuki, A.: Acc. Chem. Res. *15*, 178 (1982);
   Suzuki, A.: Topics Current Chem. *112*, 69 (1983)

9. Uzarewics, I., Zaidlewicz, A., Uzarewicz, A.: Rocz. Chem. *44*, 1403 (1970)
10. Ohloff, G., Uhde, G.: Helv. Chim. Acta. *48*, 10 (1965)
11. Zweifel, G., Whitney, C. C.: J. Org. Chem. *31*, 4178 (1966)
12. Nussim, M., Nazur, Y., Sondheimer, F.: ibid. *29*, 1120 (1964)
13. Kienzle, F., Minder, R. E.: Helv. Chim. Acta *58*, 27 (1975)
14. Mauldin, C. H., Biehl, E. R., Reeves, P. C.: Tetrahedron Lett. 2955 (1972)
15. Evans, G., Johnson, B. F. G., Lewis, J.: J. Organometal. Chem. *102*, 507 (1975)
16. Gunatilaka, A. A. L., Mateos, A. F.: J. Chem. Soc. Perkin Trans. I 935 (1979)
17. Brown, H. C., Liotta, R., Kramer, G. W.: J. Org. Chem. *43*, 1058 (1978)
18. Liotta, R., Brown, H. C.: ibid. *42*, 2836 1977)
19. Marfat, A., McGuirk, P. R., Helquist, P.: ibid. *44*, 3888 (1979)
20. Bestmann, H. J., Vostrowsky, O., Koschatzky, K. H., Platz, H., Brosche, T., Kantardjiew, I., Reinwald, M., Knauf, W.: Angew. Chem. Int. Ed. Èng. *17*, 768 (1978)
21. Köster, R., Griansnow, G., Larbig, W., Binger, P.: Liebigs Ann. Chem. *672*, 1 (1964)
22. Marshall, J. A., Greene, A. S., Ruden, R.: Tetrahedron Lett. 855 (1971)
23. Pacquette, L. A., Lang, S. A. Jr., Short, M. R., Parkinson, B., Clardy, J.: Tetrahedron Lett. 3141 (1972)
24. Larson, G. L., Fuentes, L. M.: Synth. Commun. *9*, 841 (1979)
25. Brown, H. C., Zweifel, G.: J. Am. Chem. Soc. *83*, 3834 (1961)
26. Brown, H. C., Moirikofer, A. W.: ibid. *85*, 2063 (1963)
27. Plamndon, J., Snow, J. T., Zweifel, G.: Organomet. Chem. Syn. *1*, 249 (1971)
28. Zweifel, G., Clark, G. M., Polston, N. L.: J. Am. Chem. Soc. *93*, 3395 (1971)
29. Zweifel, G., Brown, H. C.: ibid. *85*, 2066 (1963)
30. Negishi, E., Katz, J. J., Brown, H. C.: Synthesis 555 (1972)
31. Negishi, E., Yoshida, T.: J. Chem. Soc. Chem. Commun. 606 (1973)
32a. Knights, E. F., Brown, H. C.: J. Am. Chem. Soc. *90*, 5280 (1968)
32b. Scouten, C. G., Brown, H. C.: J. Org. Chem. *38*, 4092 (1973)
33. Brown, H. C., Scouten, C. G., Liotta, R.: J. Am. Chem. Soc. *101*, 96 (1979)
34. Brown, C. A., Coleman, R. A.: J. Org. Chem. *44*, 2328 (1979)
35. Brown, H. C., Gupta, S. K.: J. Am. Chem. Soc. *94*, 4370 (1972); Idem, ibid., *97*, 5249 (1975)
36a. Brown, H. C., Campbell, J. B. Jr.: J. Org. Chem. *45*, 389 (1980)
36b. Brown, H. C., Ravindra, N.: J. Am. Chem. Soc. *99*, 7097 (1977)
36c. Brown, H. C., Chandrasekharan, J.: J. Org. Chem. *48*, 644 (1983)
37. Brown, H. C., Ravindra, N.: J. Am. Chem. Soc. *98*, 1785 (1976)
38. Lane, C. F., Brown, H. C.: ibid. *92*, 7212 (1970)
39. Lane, C. F., Brown, H. C.: ibid. *93*, 1025 (1971)
40. Brown, H. C., Lane, C. F.: Synthesis 303 (1972)
41. Brown, H. C., Negishi, E.: J. Am. Chem. Soc. *89*, 5285 (1967)
42. Brown, H. C., Yamamoto, Y., Lane, C. F.: Synthesis 304 (1972)
43. Brown, H. C., Zweifel, G.: J. Am. Chem. Soc. *83*, 1241 (1961)
44. Zweifel, G., Nagase, K., Brown, H. C.: ibid. *84*, 190 (1962)
45. Brown, H. C., Singh, K. P., Garner, B. J.: J. Organometal. Chem. *1*, 2 (1962)
46. Mousseron, M., Chamayou, P.: Bull. Soc. Chim. France *3*, 403 (1962)
47. Henrick, C. A.: Tetrahedron *33*, 1845 (1977)
48. Brown, H. C., Liotta, R., Scouten, C. G.: J. Am. Chem. Soc. *98*, 5297 (1976)
49. Zweifel, G., Horng, A., Snow, J. T.: ibid. *92*, 1427 (1970)
50. Rao Ramana, V. V., Devaprabhakra, D.: Tetrahedron *34*, 2223 (1978)
51. Zweifel, G., Pearson, N. R.: J. Org. Chem. *46*, 829 (1981)
52. Jacob, P. III, Brown, H. C.: ibid. *42*, 579 (1977)
53. Lewis, R. G., Gustafson, D. H., Erman, W. F.: Tetrahedron Lett. 401 (1967)
54. Kretschmar, H. C., Erman, W. F.: ibid. 41 (1970)
55. Mikhailov, B. M., Bubnov, Yu. N., Korobeinikova, S. A., Frolov, S. I.: J. Organometal. Chem. *27*, 165 (1971) and references cited therein
56. Brown, H. C., Imai, T.: J. Am. Chem. Soc. *105*, 6285 (1983)
57. Miyaura, N., Maeda, K., Suginome, H., Suzuki, A.: J. Org. Chem. *47*, 2117 (1982)
58. Negishi, E., Brown, H. C.: Synthesis 196 (1972)
59. Negishi, E., Sabansky, M., Kats, J. J., Brown, H. C.: Tetrahedron *32*, 925 (1976)

60. Pelter, A., Hutchings, M. G., Smith, K.: J. Chem. Soc. Chem. Commun. 1048 (1971)
61. Pelter, A., Hutchings, M. G., Smith, K.: ibid. 1529 (1970)
62. For a review on the application of thexylborane see, Negishi, E., Brown, H. C.: Synthesis 77 (1974)
63. Katz, J. J.: Ph. D. Thesis, Purdue University 1974
64. Zweifel, G., Pearson, N. R.: J. Am. Chem. Soc. *102*, 5919 (1980) and references cited therein
65. Pelter, A., Smith, K., Hutchings, M. G., Rowe, K.: J. Chem. Soc. Perkin Trans. I 129 (1975)
66. Brown, H. C., Sikorski, J. A., Kulkarni, S. U., Lee, H. D.: J. Org. Chem. *45*, 4540 (1980)
67. Brown, H. C., Sikorski, J. A.: Organometallics *1*, 28 (1982)
68. Brown, H. C., Sikorski, J. A., Kulkarni, S. U., Lee, H. D.: J. Org. Chem. *47*, 863 (1982)
69. Sikorski, J. A., Brown, H. C.: ibid. *47*, 872 (1982)
70. Brown, C. A., Krishnamurthy, S., Kim, S. C.: J. Chem. Soc. Chem. Commun. 391 (1973)
71. Brown, C. A.: J. Am. Chem. Soc. *95*, 4100 (1973)
72. KIPBH is available as a 1.0 M solution in THF from Aldrich Chemical Co.
73. Kulkarni, S. U., Lee, H. D., Brown, H. C.: J. Org. Chem. *45*, 4542 (1980)
74. Brown, H. C., Basavaiah, D., Racherla, U. S.: Synthesis 886 (1983)
75. Brown, H. C., Basavaiah, D.: J. Org. Chem. *47*, 754 (1982); Brown, H. C., Basavaiah, D., Kulkarni, S. U.: ibid., *47*, 3808 (1982)
76. Midland, M. M., McDowell, D. C., Hatch, R. L., Tramontano, A.: J. Am. Chem. Soc. *102*, 867 (1980)
77. Brown, C. A., Yamashita, A.: J. Chem. Soc. Chem. Commun. 959 (1976)
78. Midland, M. M., Lee, P. E.: J. Org. Chem. *46*, 3933 (1981)
79. Zweifel, G., Polston, N. L.: J. Am. Chem. Soc. *92*, 4068 (1970)
80. Brown, H. C., Bhat, N. G., Basavaiah, D.: Synthesis 885 (1983)
81. Sinclair, J. A., Molander, G. A., Brown, H. C.: J. Am. Chem. Soc. *99*, 954 (1977)
82. Miyaura, N., Itoh, M., Suzuki, A.: Tetrahedron Lett. 255 (1976)
83. Molander, G. A., Singaram, B., Brown, H. C.: J. Org. Chem. *49*, 5024 (1984)
84. Hara, S., Dojo, H., Suzuki, A.: Chem. Lett. 285 (1983)
85. Hara, S., Kishimura, K., Suzuki, A.: ibid. 221 (1980)
86. Tahara, S., Hosokawa, K., Mizutani, J.: Agric. Biol. Chem. *44*, 193 (1980)
87. Hara, S., Kishimura, K., Suzuki, A.: Tetrahedron Lett. 2891 (1978)
88. Dhillon, R. S., Hara, S., Suzuki, A.: to be published
89. Brown, H. C., Imai, T.: J. Org. Chem. *49*, 892 (1984)
90. Onak, T.: Organoborane Chemistry, Academic Press, New York, N.Y. 1975
91. Mikhailov, B. M.: Intra-Sci. Chem. Rep. *7*, 191 (1973)
92. Mikhailov, B. M.: Organometal. Chem. Rev. Sect. A *8*, 1 (1972)
93. Mikhailov, B. M.: Pure Appl. Chem. *39*, 505 (1974)
94. Mikhailov, B. M.: Ups. Khim. *45*, 1102 (1976)
95. Zweifel, G., Nagase, K., Brown, H. C.: J. Am. Chem. Soc. *84*, 190 (1962)
96. Kramer, G. W., Brown, H. C.: J. Organometal. Chem. *132*, 9 (1977)
97. Negishi, E., Yoshida, T., Silveira, A. Jr., Chiou, B. L.: J. Org. Chem. *40*, 814 (1975)
98. Kramer, G. W., Brown, H. C.: ibid. *42*, 2292 (1977)
99. Benkeser, R. A.: Synthesis 347 (1971);
Courtois, G., Miginiac, L.: J. Organometal. Chem. *69*, 1 (1974)
100. Suzuki, A., Arase, A., Matsumoto, H., Itoh, M., Brown, H. C., Rogic, M. M., Rathke, M. W.: J. Am. Chem. Soc. *89*, 5708 (1967);
Brown, H. C., Rogic, M. N., Rathke, M. W., Kabalka, G. W.: ibid. *89*, 5709 (1967)
101. Midland, M. M., Preston, S. B.: ibid. *104*, 2330 (1982)
102. Yamamoto, Y., Komatsu, T., Maruyama, K.: J. Chem. Soc. Chem. Commun. 191 (1983) and references cited therein
103. Lukina, M. Yu., Rudashevskaya, T. Yu., Nesmeyanova, O. A.: Dokl. Akad. Nauk. SSSR *190*, 1109 (1970)
104. Bubnov, Yu., N., Nesmeyanova, O. A., Rudashevaskaya, T. Yu., Mikhailov, B. M., Kazansky, B. A.: Tetrahedron Lett. 2153 (1971)
105. Brown, H. C., Basavaiah, D., Kulkarni, S. U.: J. Organometal. Chem. *225*, 63 (1982)
106. Campbell, J. B. Jr., Molander, G. A.: ibid. *156*, 71 (1980)
107. Brown, H. C., Ravindran, N.: J. Org. Chem. *38*, 1617 (1973)

108. Zweifel, G., Polston, N. L., Whitney, C. C.: J. Am. Chem. Soc. *90*, 6243 (1968)
109. Yamamoto, Y., Yatagai, H., Moritani, I.: ibid. *97*, 5606 (1975)
110. Yamamoto, Y., Yatagai, H., Maruyama, K., Sonoda, A., Murahashi, S.: ibid. *99*, 5652 (1977)
111. Yamamoto, Y., Yatagai, H., Sonoda, A., Murahashi, S.: J. Chem. Soc. Chem. Commun. 452 (1976)
112. Bartocha, B., Douglas, C. M., Gray, Y. M.: Z. Naturforsch. B14, 809 (1959)
113. Brinckman, F. E., Stone, F. G. A.: J. Am. Chem. Soc. *82*, 6218 (1960)
114. Brown, H. C., Ravindran, N.: ibid. *98*, 1298 (1976)
115. Pasto, D. J., Balasubramaniyan, P.: ibid. *89*, 295 (1967)
116. Brown, H. C., Holmes, R. R.: ibid. *78*, 2173 (1956)
117. Zweifel, G., Arzoumanian, H., Whitney, C. C.: ibid. *89*, 3652 (1967)
118. Zweifel, G., Fisher, R. P., Snow, J. T., Whitney, C. C.: ibid. *93*, 6309 (1971)
119. Brown, H. C., Rathke, M. W., Rogic, M. M.: ibid. *90*, 5038 (1968)
120. Brown, H. C., Lane, C. F.: ibid. *92*, 6660 (1970)
121. Kulkarni, S. U., Basavaiah, D., Brown, H. C.: J. Organometal. Chem. *225*, C1 (1982)
122. Brown, H. C., Basavaiah, D.: J. Org. Chem. *47*, 3806 (1982)
123. Brown, H. C., Basavaiah, D.: Synthesis 283 (1983)
124. Basavaiah, D., Brown, H. C.: J. Org. Chem. *47*, 1792 (1982)
125. Wang, K. K., Chu, K.: ibid. *49*, 5175 (1984)
126. Zweifel, G., Arzoumanian, H.: J. Am. Chem. Soc. *89*, 5086 (1967)
127. Köbrich, G., Merkle, H. R.: Angew. Chem. *79*, 50 (1967); Angew. Chem. Int. Ed. Engl. *6*, 74 (1967)
128. Corey, E. J., Ravindranathan, T.: J. Am. Chem. Soc. *94*, 4013 (1972)
129. House, H. O.: Modern Synthetic Reactions, W. Benjamin, Menlo Park, California, p. 205 (1972)
130. Kulkarni, S. U., Lee, H. D., Brown, H. C.: Synthesis, 193 (1982)
131. Brown, H. C., Lee, H. D., Kulkarni, S. U.: ibid., 195 (1982)
132. Basavaiah, D., Brown, H. C.: J. Org. Chem. *47*, 1792 (1982)
133. Basavaiah, D.: Heterocycles *18*, 153 (1982)
134. Jacob, P. III, Brown, H. C.: J. Am. Chem. Soc. *98*, 7832 (1976)
135. Zweifel, G., Fisher, R. P.: Synthesis, 376 (1975)
136a. Pelter, A., Bentley, T. W., Harrison, C. R., Subrahmanyam, C., Laub, R. J.: J. Chem. Soc., Perkin Trans. I 2419 (1976)
136b. Pelter, A., Gould, K. J., Harrison, C. R.: ibid. 2428 (1976)
136c. Pelter. A., Harrison, C. R., Subrahmanyam, C., Kirkpatrick, D.: ibid. 2435 (1976)
136d. Pelter, A., Subrahmanyam, C., Laub, R. J., Harrison, C. R.: Tetrahedron Lett. 1633 (1975)
137a. Lalima, N. J. Jr., Levy, A. B.: J. Org. Chem. *43*, 1279 (1978)
137b. Levy, A. B., Angelastro, R., Marinello, E. R.: Synthesis 945 (1980)
138. Brown, H. C., Basavaiah, D., Kulkarni, S. U.: J. Org. Chem. *47*, 171 (1982)
139. Brown, H. C., Basavaiah, D.: ibid. *47*, 5407 (1982)
140. Zweifel, G., Snow, J. T., Whitney, C. C.: J. Am. Chem. Soc. *90*, 7139 (1968)
141. Zweifel, G., Backlund, S. J.: J. Organometal. Chem. *156*, 159 (1978)
142. Miyaura, N., Yamada, K., Suginome, H., Suzuki, A.: J. Am. Chem. Soc. *107*, 972 (1985)
143. Yamamoto, Y., Yatagai, H., Moritani, I.: ibid. *97*, 5606 (1975); Yamamoto, Y., Yatagai, H., Maruyama, K., Sonoda, A., Murahashi, S.: ibid. *99*, 5652 (1977)
144. Campbell, J. B. Jr., Brown, H. C.: J. Org. Chem. *45*, 549 (1980)
145. Miyaura, N., Yamada, K., Suzuki, A.: Tetrahedron Lett. 3437 (1979)
146. Dieck, H. A., Heck, R. F.: J. Org. Chem. *40*, 1083 (1975)
147. Miyaura, N., Suginome, H., Suzuki, A.: Tetrahedron Lett. 127 (1981)
148. Brown, H. C., Campbell, J. B. Jr.: J. Org. Chem. *45*, 550 (1980)
149. Miyaura, N., Yano, T., Suzuki, A.: Bull. Chem. Soc. Jpn. *53*, 1471 (1980)
150. Yano, T., Miyaura, N., Suzuki, A.: Tetrahedron Lett. *21*, 2865 (1980)
151. Normant, J. F., Commercon, A., Villiers, J.: ibid. 1465 (1975)
152. Brown, H. C., Molander, G. A.: J. Org. Chem. *46*, 645 (1981)
153. Negishi, E., Lew, G., Yoshida, T.: J. Chem. Soc. Chem. Commun. 874 (1973)
154. Utimoto, K., Kitai, N., Naruse, K., Nozaki, H.: Tetrahedron Lett. 4233 (1975)
155. Negishi, E., Abramovitch, A.: ibid. 411 (1977)

156. Rossi, R., Carpita, A., Cuirici, M. G.: Tetrahedron *37*, 2617 (1981)
157. Yoshida, T., Williams, R. M., Negishi, E.: J. Am. Chem. Soc. *96*, 3688 (1974)
158. Miyaura, N., Suginome, H., Suzuki, A.: Bull. Chem. Soc. Jpn. *55*, 2221 (1982)
159. Miyaura, N., Suginome, H., Suzuki, A.: Tetrahedron Lett. *24*, 1527 (1983); Idem: Tetrahedron *39*, 3271 (1983)
160. Miyaura, N., Suginome, H., Suzuki, A.: Tetrahedron Lett. *25*, 761 (1984)
161. Zweifel, G., Fisher, R. P.: Synthesis 557 (1972)
162. Miyaura, N., Suzuki, A.: J. Organometal. Chem. *213*, C53 (1981)
163. Miyaura, N., Suzuki, A.: Unpublished work
164. Lambrecht, R. M., Wolf, A. P. in: Radiopharmaceuticals (ed. Subramanian, G., Rhodes, B. A., Cooper, J. F., Sodd, V. J.) Soc. of Nuclear Medicine, New York, pp 109–145 (1975)
165. Heindel, N. D., Burns, H. D., Honda, T., Brady, L. W.: Chemistry of Radiopharmaceuticals, Masson, New Yotk 1977
166. Mazaitis, J. K., Gibson, R. E., Komai, T., Eckelman, W. C., Francis, B., Reba, R. C.: J. Nucl. Med. *21*, 142 (1980)
167. Brown, H. C., De Lue, N. R., Kabalka, G. W., Hedgecock, H. C. Jr.: J. Am. Chem. Soc. *98*, 1290 (1976)
168. Brown, H. C., De Lue, N. R.: Synthesis, 114 (1976)
169. Kabalka, G. W., Gooch, E. E.: J. Org. Chem. *45*, 3578 (1980)
170. Kabalka, G. W., Sastry, K. A. R., Hsu, H. C., Hylarides, M. D.: ibid. *46*, 3113 (1981)
171a. Suzuki, A.: Acc. Chem. Res. *15*, 182 (1982)
171b. Negishi, E., Matsushita, H., Okukado, N.: Tetrahedron Lett. 2715 (1981)
171c. Rand, C. L., Van Horn, D. E., Moore, M. W., Negishi, E.: J. Org. Chem. *46*, 4093 (1981)
171d. Normant, J. F., Alexakis, A.: Synthesis 841 (1981)
172. For a review on catecholborane see, Lane, C. F., Kabalka, G. W.: Tetrahedron *32*, 981 (1976)
173. Brown, H. C., Hamaoka, T., Ravindran, N.: J. Am. Chem. Soc. *95*, 5786 (1973)
174. Brown, H. C., Hamaoka, T., Ravindran, N.: ibid. *95*, 6456 (1973)
175. Rossi, R., Carpita, A.: Tetrahedron *39*, 287 (1983)
176. Gooch, E. E., Kabalka, G. W.: Synth. Commun. *11*, 521 (1981)
177. Kabalka, G. W., Gooch, E. E.: J. Org. Chem. *46*, 2582 (1981)
178. Kabalka, G. W., Sastry, K. A. R., Somayaji, V.: Heterocycles *18*, 157 (1982)
179. Kabalka, G. W., Gooch, E. E.: J. Chem. Soc. Chem. Commun. 1011 (1981)
180. For the preparation of homo 1,1-dihalo-1-alkenes, see: Corey, E. J., Fuchs, P. L.: Tetrahedron Lett. 3769 (1972);
Salmond, W. G.: Tetrahedron Lett. 1239 (1977);
Mendoza, A., Matteson, D. S.: J. Organometal. Chem. *152*, 1 (1978);
Yoshida, T., Negishi, E.: J. Am. Chem. Soc. *103*, 1276 (1981)
181. For stereodefined synthesis of mixed 1,1-dihaloalkenes, see: Schlosser, M., Ladenberger, V.: Chem. Ber. *100*, 3893 (1967);
Schlosser, M., Heinz, G.: Chem. Ber. *102*, 1944 (1969).
For a non-stereospecific method, see: Köbrich, G., Trapp, H.: Chem. Ber. *99*, 670 (1966)
182. Fisher, R. P., On, H. P., Snow, J. T., Zweifel, G.: Synthesis 127 (1982)
183. Negishi, E.: Organometallics in Organic Synthesis, Vol. 1, J. Wiley & Sons, New York 1980
184. Brown, H. C., Kulkarni, S. U.: J. Org. Chem. *44*, 2422 (1979); Idem: J. Organometal. Chem. *168*, 281 (1979)
185. Hara, S., Dojo, H., Takinami, S., Suzuki, A.: Tetrahedron Lett. 731 (1983)
186. Hara, S., Satoh, Y., Ishiguro, H., Suzuki, A.: ibid. 735 (1983)
187. Hara, S., Kato, T., Shimizu, H., Suzuki, A.: ibid. *26*, 1065 (1985)
188. Satoh, Y., Serizawa, H., Hara, S., Suzuki, A.: J. Am. Chem. Soc. *107*, 5225 (1985)
189. Satoh, Y., Serizawa, H., Hara, S., Suzuki, A.: Synth. Commun. *14*, 313 (1984)
190. Satoh, Y., Tayano, T., Koshino, H., Hara, S., Suzuki, A.: Synthesis, 406 (1985)
191. Honeycutt, J. B. Jr., Riddle, J. M.: J. Am. Chem. Soc. *82*, 3051 (1960)
192. Larock, R. C., Brown, H. C.: ibid. *92*, 2467 (1970)
193. Larock, R. C., Brown, H. C.: J. Organometal. Chem. *36*, 1 (1972)
194. Larock, R. C.: Tetrahedron *38*, 1713 (1982); Larock, R. C.: Angew. Chem. Int. Ed. Engl. *17*, 27 (1978)
195. Brown, H. C., Rei, M. H.: J. Chem. Soc. D 1296 (1969)

196. Larock, R. C., Gupta, S. K., Brown, H. C.: J. Am. Chem. Soc. *94*, 4371 (1972)
197. Pappo, R., Collins, P. W.: Tetrahedron Lett. 2627 (1972)
198. Benkeser, R. A., Kaiser, E. M.: J. Org. Chem. *29*, 955 (1964)
199. Waters, W. L., Caserio, M. C.: Tetrahedron Lett. 5233 (1968)
200. Waters, W. L., Linn, W. S., Caserio, M. C.: J. Am. Chem. Soc. *90*, 6741 (1968)
201. Walbrick, J. M., Wilson, J. W. Jr., Jones, W. M.: ibid. *90*, 2895 (1968)
202. Moore, W. R., Anderson, H. W., Clark, S. D.: ibid. *95*, 835 (1973)
203. Byrd, L. R. P., Caserio, M. C.: ibid. *93*, 5758 (1971)
204. Pasto, D. J., Borchardt, J. K.: Tetrahedron Lett. 2517 (1973)
205. Wharton, P. S., Kretchner, R. A.: J. Org. Chem. *33*, 4258 (1968)
206. Hulshof, L. A., McKervey, M. A., Wynberg, H.: J. Am. Chem. Soc. *96*, 3906 (1974)
207. Furukawa, J., Kakuzen, T., Morikawa, H., Yamamoto, R., Okuno, O.: Bull. Chem. Soc. Jpn. *41*, 155 (1966)
208. Brown, H. C., Jadhay, P. K., Desai, M. C.: J. Org. Chem. *47*, 4583 (1982)
209. Schmid, G., Fukuyama, T., Akasaka, K., Kishi, Y.: J. Am. Chem. Soc. *101*, 259 (1979)
210. Pawson, B. A., Cheung, H. C., Gurbaxani, S., Saucy, G.: ibid. *92*, 336 (1970)
211. Pawson, B. A., Fales, H. M., Thompson, M. J., Vebel, E. C.: Science *154*, 1020 (1966)
212. Meinwald, J., Jones, T. H.: J. Am. Chem. Soc. *100*, 1883 (1978)
213. Yoshihara, K., Sakai, T., Sakan, T.: Chem. Lett. 433 (1978)
214. Partridge, J. J., Chadha, N. K., Uskokovic, M. R.: J. Am. Chem. Soc. *95*, 7171 (1973)
215. Fried, J., Sih, J. C., Lin, C. H., Dalven, P.: ibid. *94*, 4343 (1972); Fried, J., Lin, C. H.: J. Med. Chem. *16*, 429 (1973)
216. Partridge, J. J., Chadha, N. K., Uskokovic, M. R.: J. Am. Chem. Soc. *95*, 532 (1973)
217. Brown, H. C., Singaram, B.: ibid. *106*, 1797 (1984)
218. Brown, H. C., Jadhav, P. K., Desai, M. C.: ibid. *104*, 6844 (1982)
219. Brown, H. C., Jadhav, P. K.: ibid. *105*, 2092 (1983)
220. Matteson, D. S., Sadhu, K. M.: ibid. *105*, 2077 (1983), and references cited therein
221. Kabalka, G. W., Baker, J. D.: Abstracts of papers, 170th National Meet. of the American Chemical Soc., Chicago, III, Sept. 1975, ORGN No. 77
222. Kabalka, G. W., Yang, D. T. C., Baker, J. D.: J. Org. Chem. *41*, 574 (1976)
223. Kabalka, G. W., Hutchins, R. O., Natale, N. R., Yang, D. T. C., Broach, V.: Org. Synth. *59*, 42 (1979)
224. Kabalka, G. W., Newton, R. J., Chandler, J. H., Yang, D. T. C.: J. Chem. Soc. Chem. Commun. 727 (1978)
225. Kabalka, G. W., Summers, S. T.: J. Org. Chem. *46*, 1217 (1981)
226. Krishnamurthy, S., Brown, H. C.: ibid. *42*, 1197 (1977)

Synthesis of Ynamines

Juliette Collard-Motte and Zdenek Janousek

Laboratoire de Chimie Organique Place L. Pasteur, 1
B-1348 Louvain-la-Neuve, Belgium

Table of Contents

1 Introduction

The direct linking of a secondary amino nitrogen with a carbon-carbon triple bond to create yne-amines (ynamines) had sporadically been attempted since the end of the past century.

While this was a noble and worthwile goal, it was not at all easy. Thus where planned strategies ended in frustration, serendipity suddenly intervened. Propargylic bromide was found to react with phenothiazine leading to the first ynamine via an unexpected propargylic rearrangement (Eq. 62) [109]. This was in 1958 and the structure proof based only on IR evidence and acid hydrolysis has elicited a certain incredulity.

The first direct synthesis of diethylamino-phenylacetylene two years later [51] was certainly encouraging for those willing to tackle such a difficult task. From the synthetic point of view, however, the yield of 1.7% could be considered disastrous.

Nevertheless the last citadel has fallen and ynamines were commonplace in 1965 following the busy activity of three major research groups: H. G. Viehe's group in Brussels and those of J. Ficini in Paris and J. F. Arens group in Utrecht. Each of them developed an original approach. They were closely followed by others who are duly quoted throughout the text.

The field expanded so rapidly that the first review published in 1969, contains more than fifty original ynamines and 95 references [14]. Now in 1985, ynamines are about 20 years *young* and are doing well. Their number is close to 500 and every year there are about 30–40 papers concerning the use and/or preparation of these versatile compounds. This interest is not likely to subside.

Since then, many highly adequate and selective approaches have been designed and it is the purpose of this review to present a comprehensive picture of them.

2 Ynamines from other Heterosubstituted Acetylenes via Nucleophilic Substitutions

Such substitutions using lithium amides, secondary and in some cases tertiary amines as nucleophiles, have been introduced in early sixties as the first expedient method for this unique class of compounds. It is relevant that β,β-difluoro- and chlorofluoroolefins readily available through modified Wittig reaction from aldehydes constitute also good ynamine precursors. In the past decade, however, the more versatile lithium aminoacetylide method has gained more prominence. Substitution reactions are still used, among others, for phenyl, tert.-butyl, cinnamyl and cyclopropyl ynamines.

2.1 Ynamines from 1-Halo-substituted Acetylenes

First attemps to achieve nucleophilic substitutions of halogene in chloro- and bromoalkynes using protic solvents met invariably with failure. Consequently such reactions were considered as hopeless [1,2]. Yet evidence gradually accumulated that 1-haloalkynes are not inherently inert and the year 1962 was a real breakthrough [1,3,4]. Nucleophilic substitutions on an sp carbon became routine with the advent of dipolar aprotic solvents such as DMF, DMSO and HMPT, but their use is not always mandatory. Moreover first representatives of long-sought 1-fluoroacetylenes were prepared (mostly in situ) and provided the first practicable synthesis of ynamines [5-9] (Eqs. 1–4):

$$PhCH{=}CClF \xrightarrow{LiNEt_2} Ph{-}C{\equiv}C{-}F \xrightarrow{LiNEt_2} Ph{-}C{\equiv}C{-}NEt_2 \quad (1)$$

$$t{-}Bu{-}CHBrCHBrF \xrightarrow{LiNMe_2} t{-}Bu{-}C{\equiv}C{-}F \xrightarrow{LiNMe_2} t{-}Bu{-}C{\equiv}C{-}NMe_2 \quad (2)$$

$$CF_2{=}CH_2 \xrightarrow{sec\text{-}BuLi} HC{\equiv}CF \xrightarrow{2LiNEt_2} LiC{\equiv}C{-}NEt_2 \xrightarrow{BuBr} \underset{36\,\%}{Bu{-}C{\equiv}C{-}NEt_2} \quad (3)$$

$$Me_3Si{-}C{\equiv}C{-}F + \xrightarrow[\text{or } 2HNEt_2]{LiNEt_2} Me_3Si{-}C{\equiv}C{-}NEt_2 \quad (4)$$

This is in harmony with the usual leaving group order for S_N at unsaturated carbon, i.e. $k(F) > k(Cl) > k(Br) > k(I)$ [10,11] since the rate-determining association is facilitated by electron-withdrawing substituents.

Similarly 1,1-difluoroalkenes [12] lead via fluoroacetylenes to the corresponding ynamines in 30–40% overall yields based on the starting aldehydes [7,13] and the whole process amounts to a versatile one-carbon homologation. Even vinyl ynamines are available by this sequence (5, 6):

$$RCHO \xrightarrow[PPh_3]{F_2ClCCO_2Na} RCH{=}CF_2 \xrightarrow{LiNEt_2} R{-}C{\equiv}C{-}NEt_2 \quad R = C_5H_{11}, C_6H_{13}, \text{subst. aryl groups} \quad (5)$$

$$PhCH{=}CH{-}CH{=}CF_2 \xrightarrow{LiNEt_2} PhCH{=}CH{-}C{\equiv}C{-}NEt_2 \quad (6)$$

It was soon realized that the far more readily available 1-chloro and bromo acetylenes [14-16] could be used too and in fact phenylchloroacetylene gives with lithium dimethylamide in ether, 87% of the corresponding ynamine [6]. Similarly, N-lithium pyrroline affords 67% of the corresponding ynamine [17].

In the following example (Eq. 7) $P{-}C_6H_4$ means styrene polymer where a diethylaminoethynyl moiety is built up via a formylation, Wittig reaction with dichloromethylene phosphorane and elimination-substitution steps. This polymer-bound ynamine is till now unique [18].

$$P{-}C_6H_4{-}CH{=}CCl_2 \xrightarrow{LiNEt_2} P{-}C_6H_4{-}C{\equiv}C{-}Cl \rightarrow P{-}C_6H_4{-}C{\equiv}C{-}NEt_2 \quad (7)$$

It is important to note that tertiary amines can be sometimes advantageously used instead of metallated secondary amines [19,20]. This obviously necessitates more vigorous reaction conditions (autoclave) and longer reaction times but bulk quantities of ynamines can be obtained at lower cost (8) [20-22].

$$\text{Ph-C}\equiv\text{C-X} + \text{NMe}_3 \xrightarrow[40-60\,\text{hr}]{55\,^\circ\text{C}} \text{Ph-C}\equiv\text{C-NMe}_2$$

$$\text{X=Cl, Br} \quad 54-61\,\% \tag{8}$$

Reactions with tertiary amines proceed via ethynylammonium salts which in many cases can be isolated [23-26]. Their degradation to ynamines can be brought about thermally. This process becomes more easy in the presence of an excess of tertiary amine upon which an alkyl group is transfered (9, 10) [24,25].

$$\text{R-C}\equiv\text{C-X} + \text{NR}'_3 \rightarrow \text{R-C}\equiv\text{C-NR}'_3\text{X}^-$$

$$\xrightarrow{\text{NR}'_3} \text{R-C}\equiv\text{C-NR}'_2 + \text{NR}'_4\text{X}^- \tag{9}$$

$$\text{Ph-C}\equiv\text{C-Br} + \text{NMe}_3 \xrightarrow[\text{a few days}]{0^\circ} \text{Ph-C}\equiv\text{C-NMe}_2 \quad 50\,\% \tag{10}$$

Even secondary amines occasionally give very good results (11) [26].

$$\text{R-C}\equiv\text{C-C}\equiv\text{C-Br} + \text{HNMe}_2 \rightarrow \text{R-C}\equiv\text{C-C}\equiv\text{C-NMe}_2 \tag{11}$$

$$\text{R = Me, 91\%; Me}_2\text{CHOH, 69\%}$$

It is therefore rather surprising that this useful variation has been used scarcely during the past decade.

Any substitution reaction (i.e. by tertiary or secondary amines or by metallated secondary amines) is greatly facilitated when the β-carbon carries an electron-withdrawing or simply a conjugative group. Accordingly, chlorocyanoacetylene reacts smoothly at $-50°$ with free secondary amines to cyanoynamines, an example of push-pull substituted acetylenes (12) [27].

$$\text{NC-C}\equiv\text{C-Cl} \xrightarrow{\text{HNR}_2} \text{NC-C}\equiv\text{C-NR}_2 \qquad 20-55\,\% \tag{12}$$

Even less activated phosphonyl chloroacetylenes react still very smoothly with triethylamine [28] or secondary amines (13) [28,29].

$$(\text{EtO})_2\text{P(=O)-C}\equiv\text{C-Cl} + \text{Et}_3\text{N} \xrightarrow[10\,\text{min.}]{\text{Ph-H, reflux}} (\text{EtO})_2\text{P(=O)-C}\equiv\text{C-NEt}_2 \quad 44\,\% \tag{13}$$

$$(\text{n-PrO})_2\text{P(=O)-C}\equiv\text{C-Cl} + \text{R}_2\text{NH}$$

$$\xrightarrow{0^\circ,\ 30\,\text{min.}} (\text{n-PrO})_2\text{P(=O)-C}\equiv\text{C-NR}_2$$

$$\text{R = Me; R = Et, 85\%}$$

As already shown, phenyl and ethynyl groups favorize the substitution. The same is apparently also true for a conjugated vinyl group (14) [30, 31] and even the cyclopropane moiety (15) [32].

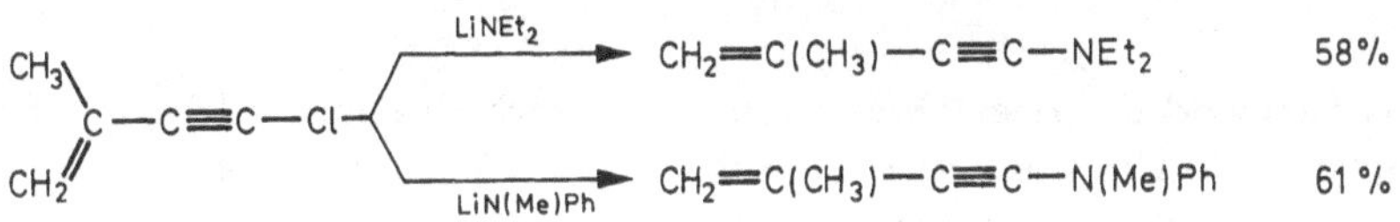

(14)

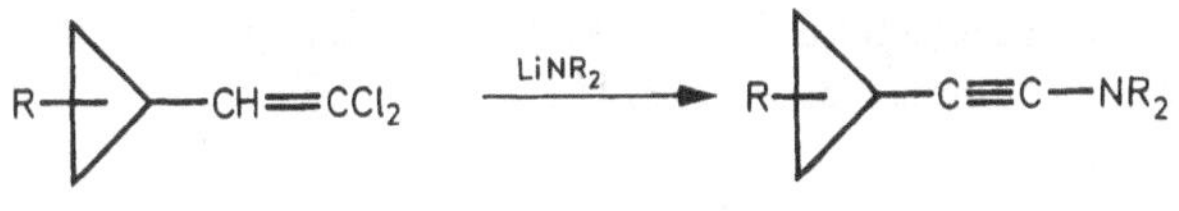

(15)

A number of trichlorovinyl ynamines 2 were prepared from tetrachloroenyne 1 and secondary amines. The yields are almost quantitative but such ynamines are too fragile to allow a purification by distillation or even TLC on Alumina (16) [33, 34].

$$\underset{1}{CCl_2{=}CCl{-}C{\equiv}C{-}Cl} + HNR_2 \xrightarrow[20^\circ]{Ether} \underset{2}{CCl_2{=}CCl{-}C{\equiv}C{-}NR_2} \qquad (16)$$

In view of the fact that the trichlorovinyl group represents a masked lithoethynyl group, (Sect. 5.3) ynamines 2 have become key intermediates in the synthesis of push-pull di- and polyacetylenes [34–38] and diynediamines 4 (17) [39].

$$2 + 2\,BuLi \quad Li{-}C{\equiv}C{-}C{\equiv}C{-}NR_2 \xrightarrow{E^+} E{-}C{\equiv}C{-}C{\equiv}C{-}NR_2 \qquad (17)$$

$$\downarrow BrCN$$

$$\underset{3}{Br{-}C{\equiv}C{-}C{\equiv}C{-}NR_2} \xrightarrow{LiNR^1R^2} \underset{4}{R^1R^2N{-}C{\equiv}C{-}C{\equiv}C{-}NR_2}$$

It is enlightening to compare 3 with 2-chloro- and bromoynamines which in addition of being highly unstable, react with amines only by addition to 2-haloketeneN,N-acetals (Sect. 6) Aminoethynyl moiety in 3 apparently does still activate the nucleophilic substitution. Let us also note that this method of preparation of diynediamines 4 is greatly superior to the Glaser coupling of ynamines. Consequently, 4 where R are alkyl groups are a very little known class of ynamines.

The most strenuous conditions are required for tert-butyl and generally all tert-alkyl haloacetylenes, but this problem can be alleviated by the use of HMPT [40]. On the other hand, this solvent may cause dificulties during the work-up stage. Both the trialkylamine and lithium amide methods are very general but they may be complementary in certain cases. Thus with triethylsilylbromoacetylene as a sluggish substrate, (although the silicon atom acts as a weak π-acceptor toward a triple bond) triethylamine gives the expected ynamine 5 (18) [41], whereas lithium diethylamide reacts reportedly only at 110–20° to give a mixture containing among others bis-(triethylsilyl) acetylene. In the presence of HMPT, however, the reaction proceeds smoothly but the fractionation is impracticable[41].

$$Et_3Si{-}C{\equiv}C{-}Br + Et_3N \xrightarrow[60\,hrs]{90-95^\circ} Et_3Si{-}C{\equiv}C{-}NEt_2 \qquad 5,\ 80\,\% \quad (18)$$

Nevertheless, later work describes the preparation of a series of triorganosilylynamines from silylethynyl chlorides and lithium dialkylamides in ether at room temperature. The yields are good, ranging between 50 and 80%[42].

Sulphur atom in chloroethynyl sulphides 6 renders the chlorine sufficiently reactive so that the reaction proceeds at low temperature (19)[43].

$$\underset{6}{MeS{-}C{\equiv}C{-}Cl} + LiNEt_2 \xrightarrow[-40^\circ]{Ether} MeS{-}C{\equiv}C{-}NEt_2 \qquad 50\,\% \qquad (19)$$

Again, it is more practical to use formal hydrogen chloride adducts of 6 together with an excess of lithium dialkylamide which acts both as base in the elimination step and reagent as in the substitution step (20)[44].

$$HClC{=}CClSR \xrightarrow{3\,eq \cdot LiNR_2} RS{-}C{\equiv}C{-}NR_2 \xleftarrow{3\,eq \cdot LiNR_2} Cl_2C{=}CHSR \qquad (20)$$

This valuable class of ynamines is accessible even from β-trifluoroethyl sulphides which are made in turn from trifluoroethanol, a common solvent (21)[45,46].

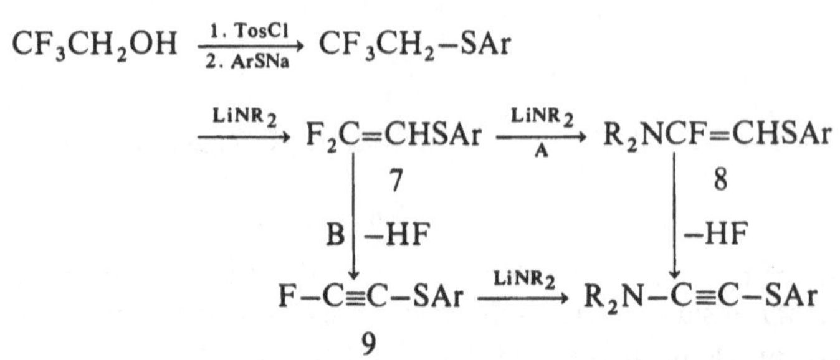

(21)

The authors favor the nucleophilic addition of lithium amide to arylthiodifluoroolefin 7 leading to α-fluoroenamine 8 and finally to 9 (Path A). Nevertheless, intermediacy of 9 cannot be entirely discarded (Path B).

β-Trifluomethyl ethers react only by elimination followed by metallation to 10. Alkoxy or aryloxy-ynamines remain therefore as elusive as ever (22)[46].

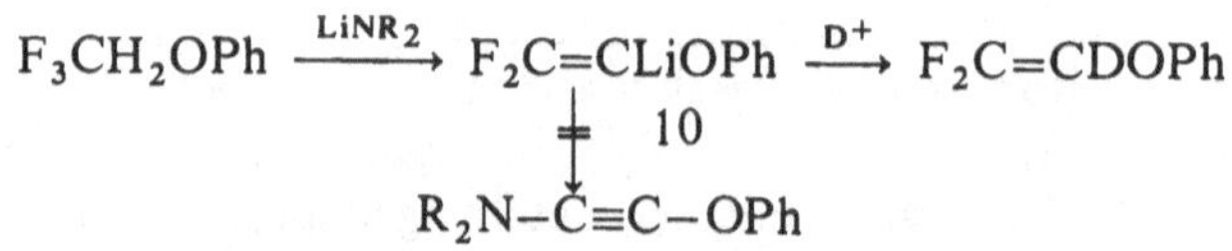

(22)

Phenylselenenyl ynamines are obtained analogously from the corresponding selenides (23)[47]:

$$CF_3CH_2SePh \xrightarrow{LiNR_2} PhSe{-}C{\equiv}C{-}NR_2 \qquad R{=}Me,\ 95\%;\ Et\ 85\% \tag{23}$$

This method is quite general and yields are good. It should be noted however, that there exists no example where the nitrogen bears one (or two) aryl groups due to the lower reactivity of lithium aralkyl and especially diarylamides.

Strongly nucleophilic secondary amines may give to undesired side-reactions. Thus haloacetylenes bearing carbonyl groups react eagerly with secondary amines but the reaction cannot be arrested at the ynamine stage; ketene N,N-acetals are formed instead (24)[48].

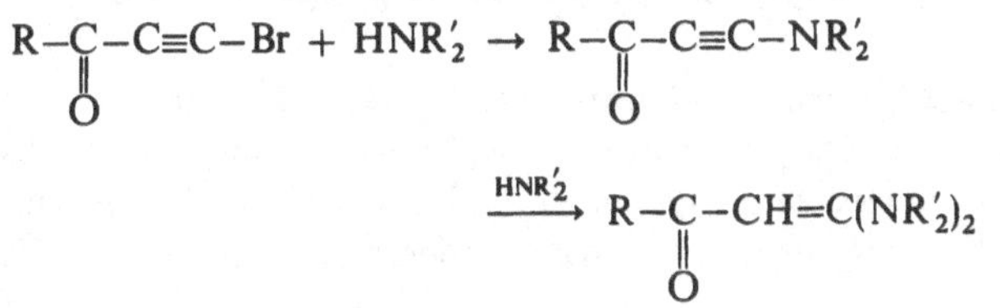
(24)

R = Ph, OEt

More conveniently accessible, β,β-dichloroketones react with secondary amines at $-50°$ to give α-chloro-β-chlorocarbonylenamines 11, whereas at room temperature in protic solvents, the corresponding aminals 12 are obtained (25)[49].

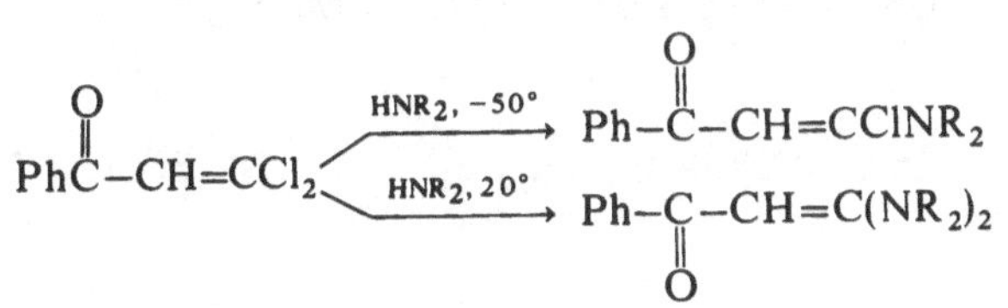
11
12
(25)

11 undergo base promoted elimination, but the corresponding ynamines can only be detected but not isolated. The sequences (24, 25) are not suitable as synthetic methods for 2-acylated (push-pull) ynamines. An earlier report on the isolation of the 2-benzoyl-N,N-dibenzylynamine following the sequence (25) might therefore be erroneous[50].

The ease of addition of diethylamine even to diethylamino phenylacetylene was noticed as early as in 1960 during the first successfull although unproductive (yield 1.7%) synthesis of phenylethynyldiethylamine[51].

The rather basic lithiated secondary amines may give rise to another kind of side-reactions e.g. isomerisation of ynamines into aminoallenes (see Propargylic Rearrangements)

2.2 Ynamines from Acetylenic Ethers

Stable and readily available alkyn-1-yl ethers [15,16,52] react with lithium dialkylamides at 110–120° to give ynamines in good yields (26) [53–56].

$$R{-}C{\equiv}C{-}OEt + Li{-}NR'_2 \rightarrow R{-}C{\equiv}C{-}NR'_2 + EtOLi \quad (26)$$

The yield is low when R = CH_3 (14%) but such ynamines are easily available through propargylic rearrangements. No substitution at all can be achieved when R = H. Also, the much less reactive sodium and lithium aralkyl and diarylamides react too slowly and yields are low [57].

On the other hand, this method is particulary suitable for ynamines where R = Et and higher alkyl groups, where the haloacetylene method is unpractical. This method could be extended even to the preparation of silylated ynamines (27) [41,58].

$$R_3Si{-}C{\equiv}C{-}OEt + LiNEt_2 \xrightarrow[120°]{30\ min.} R_3Si{-}C{\equiv}C{-}NEt_2 \quad (27)$$

Silylated amines are formed as side-products. Silylated ynamines are thermally very stable and can be distilled at normal pressure without decomposition [41]. As a corollary, their reactivity is reduced as compared to 2-alkyl ynamines [59]. The same procedure was applied to the synthesis of a few triorganogermyl ynamines (28) [58].

$$R_3Ge{-}C{\equiv}C{-}OEt + LiNR'_2 \rightarrow R_3Ge{-}C{\equiv}C{-}NR'_2 \quad (28)$$

50–80%

In sharp contrast, stannylated acetylenes are substituted at the "wrong" end and a metal-metal exchange takes place (29, 30) [58].

$$R_3Sn{-}C{\equiv}C{-}OEt + LiNR'_2 \rightarrow Li{-}C{\equiv}C{-}OEt + R_3SnNR'_2 \quad (29)$$

$$Me_3Sn{-}C{\equiv}C{-}OMe + Et_2NSnMe_3 \rightarrow \underset{13}{Me_3Sn{-}C{\equiv}C{-}NEt_2}\ 34\% \quad (30)$$

$$Me_3Si{-}C{\equiv}C{-}OEt + Et_2NSnMe_3 \rightarrow \underset{39\%}{13} + \underset{17\%}{Me_3Si{-}C{\equiv}C{-}NEt_2} \quad (31)$$

As shown, this problem could be circumvented in one case by using trimethylstannyl diethylamine (30) but the sequence given is obviously impractical. As shown in the Section 5.3 stannylated ynamines are most conveniently prepared from aminoethynyl lithium and triorganostannyl halides.

2.3 Mechanisms

As some rewiews [1,4] deal with this challenging and as it happens not yet completely settled problem, we shall only briefly resume some facts. In fluoroacetylenes, the nucleophile attacks at C-1 tends to localize the negative charge at C-2, away from fluorine. This process is favourable both by the kinetic (low energy of activation due to easy rate-determining association) and thermodynamic criterions (high energy content due to the presence of fluorine on a sp carbon). The fluorine is expelled in the next stage and the whole process is reminiscent of S_N in 2,4-dinitrofluorobenzene (32):

$$R{-}C{\equiv}C{-}F \quad :Nu \rightarrow R{-}\overset{\ominus}{C}{=}C\begin{smallmatrix}Nu\\F\end{smallmatrix} \rightarrow R{-}C{\equiv}C{-}Nu + F^{\ominus} \qquad (32)$$

Acetylenic ethers follow the same pathway, but the less favorable energetical aspects dictate the use of harsher reaction conditions.

In a striking contrast, chloroalkynes are attacked at C-2 since chlorine stabilizes a negative charge. Not only the initial attack but also the remaining steps are rather difficult since the chloride is expelled via an α-elimination with concomittant migration of either the group R or the nucleophile (33):

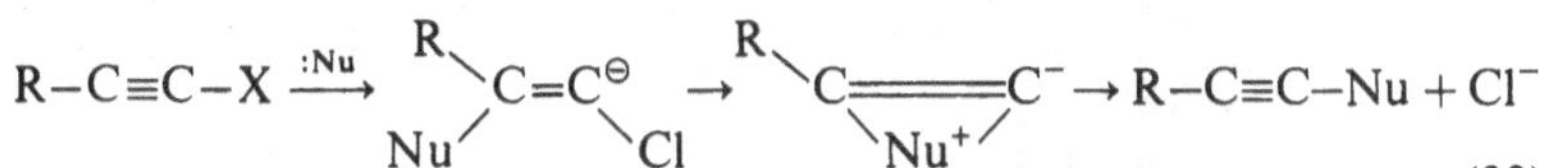

(33)

$X = Cl, Br, I \quad Nu = R_2N^-$ etc.

This process which has been called onium rearrangement [60–62] provides an excellent method for the synthesis of ynediamine. On the other hand if the group R in the chloroacetylene stabilizes the negative charge to a greater extent than does a chlorine atom the nucleophile shows again a marked bias for an attack at C-1 and substitution proceds smoothly. This is probably the case when R is a phenyl group or a silicon atom certainly with thioalkyl, phosphoryl, cyano groups, and the like. Non activated chloro- and bromoalkynes might also react through radical ion pairs [1]. In the past years no new examples have been added to the existing well established procedures.

3 Ynamines Via Elimination Sequences from Various Amino Compounds

In the preceding paragraphs we dealt with procedures where amino groups are introduced in the last step into halo- and alkoxyacetylenes which can eventually be prepared in situ by elimination from the corresponding haloalkanes and alkenes.

In this chapter we will deal with eliminations from substrates already possessing the requisite secondary amino group.

3.1 Ynamines from Thioamides

Thioamides arise from hydrogen sulphide addition to ynamines and it is clear that the reverse process would be much more interesting owing to the easy availability of thioamides.

Thus far only phenylthioacetamides could be converted to ynamines upon refluxing with sodium amide in xylene (34)[63-65].

$$\mathrm{Ph{-}CH_2{-}\underset{\underset{S}{\|}}{C}{-}NR_2 \xrightarrow{NaNH_2,\ 130^\circ} Ph{-}C{\equiv}C{-}NR_2} \qquad (34)$$

$$\mathrm{R{=}CH_3,\ \ 60\%; \qquad R{=}Et,\ \ 90\%}$$

Unexplicably, the morpholino ynamine was obtained in only 10% yield and aliphatic thioamides form only thiolate salts[63].

These problems can be obviated by using alkylated hiolates-ketene-S,N-acetals (35)[66-68].

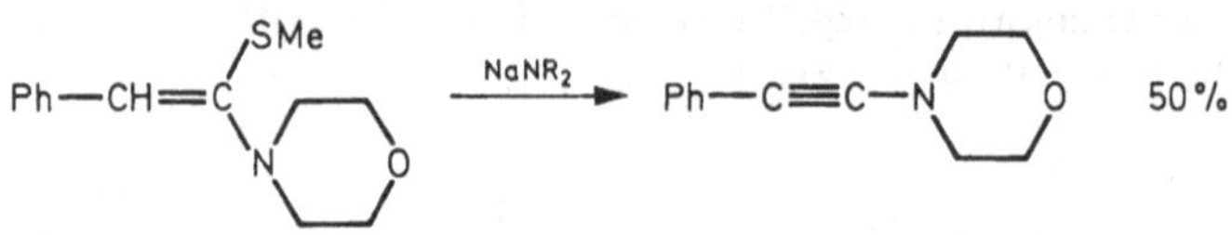

(35)

An illustrating example is the synthesis of the most simple ynamine, 15 which is also extremely reactive, by this method (36)[65,67].

$$\mathrm{H_2C{=}C\begin{smallmatrix}S{-}Bu\\NMe_2\end{smallmatrix} \xrightarrow{KNH_2,\ HMPT} \underset{14}{K{-}C{\equiv}C{-}NMe_2}} \qquad (36)$$

$$\mathrm{14 \xrightarrow{C_2H_4Br_2} \underset{15}{H{-}C{\equiv}C{-}NMe_2} \qquad 55\%}$$

The starting materials are cheap but the scope of this method remains rather limited.

3.2 Ynamines from α-Chloroenamines and Amide Chlorides

Amides being formally hydrated ynamines would constitute cheap starting materials for the latter and many stratagems have been devised to achieve this goal. It must be kept in mind, however, that direct dehydration of amides is not feasible. One indirect approach has already been exemplified, namely the prior conversion of amides into thioamides. Another classical method involves halogen chloride elimination from amide chlorides. These versatile salts have received only scant attention prior to 1960.

Amide chlorides are readily available from amides via phosgenation but other reactive acid halides are equally used (thionyl chloride, oxalyl chloride, phosphorus pentachloride and oxychloride, pyrocatechol phosphorus trichloride) (37)[69].

$$\mathrm{RCH_2{-}\overset{\overset{O}{\|}}{C}{-}NR'_2 \xrightarrow{COCl_2} R{-}CH_2{-}\overset{\overset{Cl}{|}}{C}{=}NR'_2 \quad Cl^- \ \text{or}\ HCl_2^-} \qquad (37)$$

$$\mathrm{\xrightarrow[-HCl]{Base} R{-}CH{=}CClNR'_2 \xrightarrow[-HCl]{Base} R{-}C{\equiv}C{-}NR'_2}$$

Obviously, the elimination goes through α-chloroenamines, which have themselves become useful synthones [70, 71]. Using two or more equivalents of a base, ynamines are formed in a straightforward manner (38, 39) [66, 72].

$$PhCH_2CONMe_2 \xrightarrow[3.\,Et_3N]{1.\,HCl,\ 2.\,COCl_2} \underset{32\,\%}{Ph{-}C{\equiv}C{-}NMe_2} \qquad (38)$$

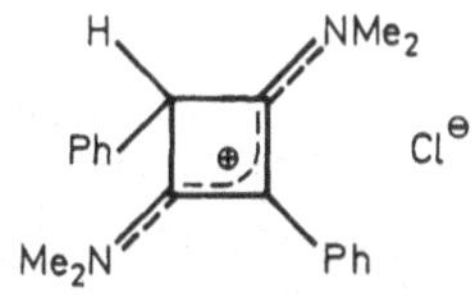

32 %

$$EtCH_2CCl{=}\overset{\oplus}{N}Et_2\ Cl^{\ominus} \xrightarrow[\text{or } LiNEt_2]{LiN(C_6H_{11})_2} \underset{38-77\,\%}{Et{-}C{\equiv}C{-}NEt_2} \qquad (39)$$

Amide should be first transformed to its hydrochloride in order to cut down the formation of α-chloro-β-chlorocarbonyl enamines [72, 73]. The method is very general but the choice of experimental conditions and of the base is critical and yields may vary strongly depending on the case (39). There is also one side-reaction which can hardly be avoided: the ynamine already formed reacts with the yet unreacted amide chloride, the α-chloroenamine [70] or even with the tert-amine hydrochloride to form the very stable cyclobutene cyanines and/or allenamidinium salts. These by-products have an interest of their own and will be discussed later. Formation of these salts may strikingly lower the yield of ynamines, but because of their salt character they are not harmfull during the work-up [72].

Further, many amide chlorides having an α-methylene group undergo a selfcondensation to trimethine cyanines. This undesired side-reaction could be exploited synthetically to prepare rare push-pull ynamine salts (40, 41) [75−77] ynamine amidines (41) [76, 78] and amides (41) [76, 79].

$$MeCONMe_2 \xrightarrow[3.\,NaClO_4]{1.\,COCl_2,\ 2.\,\Delta} \underset{16}{Me_2N{-}CCl{=}CH{-}CMe{=}\overset{+}{N}Me_2ClO_4^-} \qquad (40)$$

$$16 \xrightarrow{Et_3N} Me_2N{-}C{\equiv}C{-}\underset{\displaystyle Me}{\underset{|}{C}}{=}\overset{+}{N}Me_2 \quad ClO_4^- \quad ^{77)}$$

$$MeCONMe_2 + 2\,\underset{PI}{Cl_2C{=}\overset{+}{N}Me_2Cl^-} \rightarrow \underset{17}{Me_2N{-}\underset{\displaystyle Cl}{\underset{|}{C}}{=}CH{-}\underset{\displaystyle Cl}{\underset{|}{C}}{=}\overset{+}{N}Me_2 \quad Cl^-} \qquad (41)$$

$$17 \xrightarrow{Et_3N,\ 0^\circ} Me_2N{-}C{\equiv}C{-}CCl{=}\overset{+}{N}Me_2 \quad Cl^- \quad \text{unstable}^{76)}$$

$$17 \xrightarrow[2.\,Base]{1.\,HI} Me_2N{-}C{\equiv}C{-}CI{=}\overset{+}{N}Me_2 \quad I^- \quad 83\,\%,\ \text{stable}^{75)}$$

$$17 \xrightarrow{RNH_2} Me_2N{-}C{\equiv}C{-}\dot{C}({=}NR)NMe_2 \qquad R{=}H,\ 53\,\%^{78)},\ {=}\text{alkyl},\ 50{-}63\,\%$$

$$17 \xrightarrow{K_2CO_3,\ H_2O} Me_2N{-}C{\equiv}C{-}C({=}O)NMe_2 \qquad 70\,\%$$

Dichloromethine cyanine 17 which can be obtained in high yield from phosgeniminium chloride (PI) and N,N-dimethylacetamide has proven to be a particularly versatile source of push-pull ynamines. Arylamines, however cyclise on the aryl group and 2,4-bis(dimethylamino)quinolines are the final products [80]. The intermediary ynamines can only be detected by the IR. The process is catalysed by traces of hydrogen chloride see also (5.3).

Secondary amides react with two equivalents of phosgene or PI to N-acylated chloroenamines 18, 19 respectively (42) [81,82].

$$R{-}CH_2{-}\overset{O}{\overset{\|}{C}}{-}NHMe + COCl_2 \rightarrow \underset{18}{R{-}CH{=}CCl{-}NMeCOCl} \xrightarrow{MeOH} \underset{20}{R{-}CH{=}CCl{-}NMeCO_2Me} \qquad (42)$$

$$R{-}CH_2{-}\overset{O}{\overset{\|}{C}}{-}NHMe + PI \rightarrow \underset{19}{RCH{=}CCl{-}NMeCCl{=}\overset{+}{N}Me_2\ Cl^-} \xrightarrow{H_2O} \underset{21}{RCH{=}CCl{-}NMe{-}CONMe_2}$$

$$20 \xrightarrow{t\text{-}BuOK} R{-}C{\equiv}C{-}NMeCO_2Me$$

$$21 \xrightarrow{t\text{-}BuOK} R{-}C{\equiv}C{-}NMeCONMe_2$$

These products can be selectively hydrolysed, alcoholysed and aminolysed to furnish 20 and 21 which give the corresponding N-acylated ynamines in high yields [81,82]. Such an N-acylation results in a marked decrease of reactivity of the ynamine triple bond.

The amide chloride method has not found wide-spread use except for phenylethynylamines [72].

3.3 Ynamines from Enamines and β-Haloenamines

Halogen additions to olefins followed by double dehydrohalogenation is a classical sequence which has been widely used for making carbon-carbon triple bonds [2].

Thus in principle, aldehydes can be converted to ynamines through the corresponding enamines. In practice however, this is described only in the case of phenylacetaldehyde (43) [83].

$$PhCH_2CHO \rightarrow PhCH{=}CHNEt_2 \xrightarrow{Br_2} PhCHBrCHBrNEt_2 \qquad (43)$$

$$\xrightarrow{t\text{-}BuOK} PhCBr{=}CHNEt_2 \xrightarrow{t\text{-}BuOK} PhC{\equiv}C{-}NEt_2 \quad 55\%$$

Dehydrohalogenation of β-haloenamines having an alkyl group in beta is very difficult. When β-bromoenamines are reacted with organo-lithium reagents only halogen-metal exchange is observed (44) [84].

$$t{-}BuCBr{=}CHNR_2 \xrightarrow{n\text{-}BuLi} t{-}BuCLi{=}CHNR_2 \qquad (44)$$

β-Chloroenamines [85] under the same conditions are deprotonated in α to the chlorine atom. Chlorine is then substituted by alkyl from alkyllithium which leads again to β-lithiated enamines [84].

β,β-Dihaloamines as formal precursors of β-haloenamines cannot serve as ynamine precursors, either [86]. Only reduction or substitution reaction take place (45).

$$t\text{-}BuCCl_2CH_2NEt_2 \xrightarrow{LiNEt_2} t\text{-}BuCHClCH_2NEt_2 \qquad (45)$$

$$HCCl_2CH_2NR^1_2 \xrightarrow{LiNR^2_2} (R^2_2N)_2CHCH_2NR^1_2 \rightarrow R^2_2NCH{=}CHNR^1_2$$

$$MeCCl_2CH_2NEt_2 \xrightarrow{LiNEt_2} \text{no reaction}$$

Recent work shows that 2,2-dichloroenamines are available from 2,2,2-trichloroethylamines and t-butoxide. Using two equivalents of base, the elimination apparently proceeds further to give 2-chloroethynyl amines 22. These compounds cannot be isolated but their formation is suggested by a strong IR band at $\nu = 2200\ cm^{-1}$. Only 1-(t-butoxy)-2-chlorovinylamines 23 are isolated with excess tert-butoxide (46) [87].

$$Cl_3CCH_2NR_2 \xrightarrow[-HCl]{t\text{-}BuOK} Cl_2C{=}CHNR_2 \xrightarrow{-HCl} \underset{22}{Cl{-}C{\equiv}C{-}NR_2} \qquad (46)$$

$$22 \xrightarrow{t\text{-}BuO^-} ClHC{=}C(OBu{-}t)NR_2 \quad 23 \qquad 44{-}58\ \%$$

Olefins 23 are hold interest as potential precursors of the still unknown 2-alkoxy-ynamines via onium rearrangement.

Enediamines 24 undergo bromination to diiminium salts 25 which give ynediamines in moderate yields upon treatment with base (47) [88].

$$\underset{24}{R_2NCH{=}CHNR_2} \xrightarrow{Br_2} \underset{25}{R_2\overset{+}{N}{=}CHCH{=}\overset{+}{N}R_2} \xrightarrow{NEt_3} \underset{26}{R_2N{-}C{\equiv}C{-}NR_2} \qquad (47)$$

26a, NR_2 = piperidino 32%
b, = morpholino 44%

Enamides 27 can be transformed to ynamides without any difficulty due to the increased acidity of the hydrogen atom (48) [82,89].

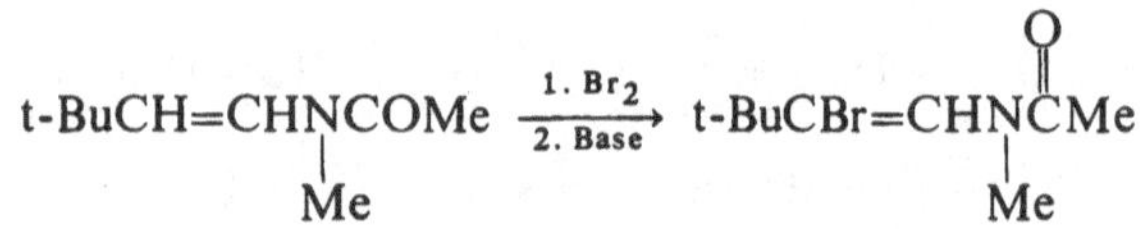

(48)

27 28

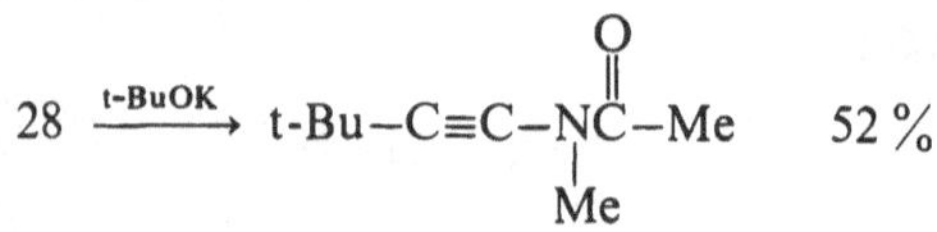

The same is true when an electron-withdrawing group is present in beta to the amino group and great numbers of vinylogous amides, carbamates and ureas have been transformed to their ethynylogs 29 for which the term "push-pull acetylenes" is used (49) [90,91]. Their chemistry has been thoroughly studied and they represent the best known class of ynamines.

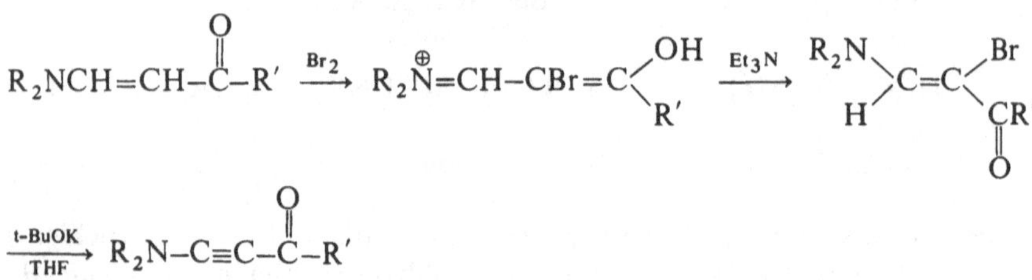

(49)

29

Despite the extended conjugation, dimethylaminopropynal 29 R′ = H polymerizes completely during a few minutes at 20° and 4-dimethylamino-3-butyne-2-one (29, R′ = CH_3) polymerizes completely after 24 hours. Higher homologs and aroyl derivatives are much easier to handle and can be stored for a long time. These ynamines are reagents of choice among others for activating the carboxyl group of amino acids and many of them have been developed to be used in peptide synthesis [92-94].

In addition to the widely used N,N-diethylamino-1-propyne, 1-(4-chlorophenyl)-3-(4-methyl-1-piperazinyl)-2-propyne-1-one 30 has become the second commercially available ynamine:

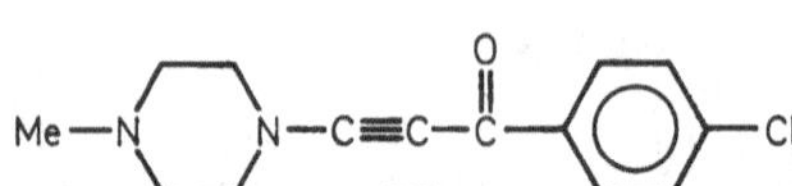

30

This second main method is especially suitable for relatively unstable push-pull acetylenes and is regularly applied in cases where the corresponding aminoethynyl lithium compounds are not available [95]. This addition-elimination approach was abortive during attempts to prepare the corresponding push-pull diacetylenes. Nor did the Cadiot-Chodkiewitz coupling furnish the expected acetal (50, 51) [96].

$$R_2NCH{=}CBrCH{=}CBrC({=}O){-}H \not\longrightarrow R_2N{-}C{\equiv}C{-}C{\equiv}C{-}C({=}O){-}H \quad (50)$$

$$PhN(Me){-}C{\equiv}C{-}H + Br{-}C{\equiv}C{-}CH(OEt)_2 \overset{Cu^+}{\not\longrightarrow} PhN(Me){-}C{\equiv}C{-}C{\equiv}C{-}CH(OEt)_2 \quad (51)$$

3.4 Ynamines from α,β-Di and Trichloroenamines

Multiple bonds are created not only through eliminations of hydrogen halides but also via removal of two atoms of halogene from 1,2-dihalo-compounds. This principle has also been applied to the ynamine synthesis.

Equation 52 illustrates a very straightforward if not general approach starting with α,β-dichloroenamines which are transformed to push-pull ynamines by the action of lithium amalgame (52) [97].

$$Et_2NCCl{=}CCl{-}C({=}O){-}X + Li/Hg \rightarrow Et_2N{-}C{\equiv}C{-}C({=}O){-}X \qquad X{=}OMe\ 45\%;\ NEt_2\ 66\% \quad (52)$$

But the most versatile entry into the whole ynamine chemistry starts with cheap and readily available trichloroacetamides. Phosphines and phosphites are capable of abstracting a positive chlorine atom thereby leading to trichloroenamines. The conversion is smooth regardless of substituents on the amide nitrogen (53).

$$Cl_3C{-}C({=}O){-}NR_2 \xrightarrow[P(OEt)_3]{PPh_3,\ P(Bu)_3} Cl_2C{=}C(Cl)NR_2 \quad (53)$$

R = Ph, 55%; Et, 69–74% [98]; Me, Ph, 85% [99]; Et, Ph, 85% [99]; i-Pr, Ph, 72% [99]; Bz, Ph, 77% [99]; $-(CH_2)_5-$, 61% [95]

Dichloroacetamides react only when the nitrogen bears one aryl or better two aryl groups, but α,α-dichlorophenylacetamides react readily whatever are the substituents on nitrogen (54, 55). This reminds that the chlorine must be sufficiently activated by electronwithdrawing groups.

$$Cl_2CH{-}C({=}O){-}N(Ph)R \xrightarrow{PR''_3} ClCH{=}CClN(Ph)R \qquad R = Me,\ 11\%\ [100];\ Ph\ 84\%\ [101] \quad (54)$$

$$PhCCl_2{-}C({=}O){-}NR_2 \xrightarrow{PR'_3} PhCCl{=}CClNR_2 \qquad R = Et, Et\ 70\%\ [101,102];\ Ph, Me\ 95\%\ [103];\ 77\%\ [104] \quad (55)$$

These thermally stable α,β-dichloroenamines eliminate both chlorine atoms when exposed to one equivalent of butyl lithium [105,106]. Neither magnesium nor sodium work properly and Grignard reagents substitute the chlorine in alpha to the amino group by an aryl or alkyl group [107].

$$PhCCl{=}CClNEt_2 \xrightarrow{BuLi} Ph{-}C{\equiv}C{-}NEt_2 \quad 85\,\%^{106)} \tag{56}$$

$$PhCCl{=}CClN(Me)Ph \xrightarrow{BuLi} Ph{-}C{\equiv}C{-}N(Me)Ph \quad 73\,\%^{104)} \tag{57}$$

Ynamines can be thereby obtained in global yields of 40–50 % from the corresponding arylacetic acids [106].

Di- and especially trichloroenamines which derive from acetic acid have become the most versatile precursors of ynamines, known to date (58–60) [106].

$$ClCH{=}CClNPh_2 \xrightarrow{2\text{ eq. BuLi}} Li{-}C{\equiv}C{-}NPh_2 \xrightarrow{H_2O} \underset{\textbf{31}}{H{-}C{\equiv}C{-}NPh_2} \quad 65\,\% \tag{58}$$

$$Cl_2C{=}CClN(R)Ph \xrightarrow{2\text{ eq. BuLi}} Li{-}C{\equiv}C{-}N(R)Ph \xrightarrow[NH_4OH,\ 0^\circ]{t\text{-BuBr or}} \underset{\textbf{32}}{H{-}C{\equiv}C{-}N(R)Ph} \tag{59}$$

32, R = Me, 72 % [106]; Et, 51 % [99]; Bz, 34 % [99]; i-Pr, 77 % [99]

$$Cl_2C{=}CClNEt_2 \xrightarrow{2\text{ eq. BuLi}} Li{-}C{\equiv}C{-}NEt_2 \xrightarrow{Ph_3CH} \underset{\textbf{33}}{H{-}C{\equiv}C{-}NEt_2} \quad 32\,\% \tag{60}$$

The haloenamines are first metallated in alpha to the amino group followed by an expulsion of the second chlorine atom. The intermediate chloroynamines could be trapped when using only one mole of butyllithium, but these ethynylogs of chloramines are too unstable when both groups R are alkyl [60,87,107]. Chloroynamines then undergo a rapid halogen metal exchange to give lithioynamines. Only the less reactive N-phenyl derivatives 31 and 32 can be liberated with water whereas 33 is neutralized with triphenylmethane [106]. The above sequence is the best method for unsubstituted, "free" ynamines. However, the utility of lithiated ynamines goes far beyond their protonation and they represent an extremely valuable building block for construction of other ynamines via alkylation, acylation (aminoethynylation) and aminoethynyl metallation. These methods will be dealt with separately in (Chapt. 5.3).

4 Ynamines by Isomerization of Propargylamines

It has been known for a century that an acetylenic bond can migrate either to the middle of a hydrocarbon chain or to the end giving metal akynydes [108].

More specifically, propargylic substrates may rearrange upon treatment with bases to give the corresponding heterosubstituted acetylenes via allene intermediates (61).

$$\mathrm{R{-}C{\equiv}C{-}CH_2{-}X} \gtrless \mathrm{RHC{=}C{=}CHX} \gtrless \mathrm{R{-}CH_2{-}C{\equiv}C{-}X} \quad (61)$$
$$\mathrm{X = NR_2, OR, SR, SeR} \ldots$$

Actually, an attempt to introduce a propargyl group on phenothiazine in the presence of sodium hydride led fortuitously to the first ynamine synthesis (62) [109].

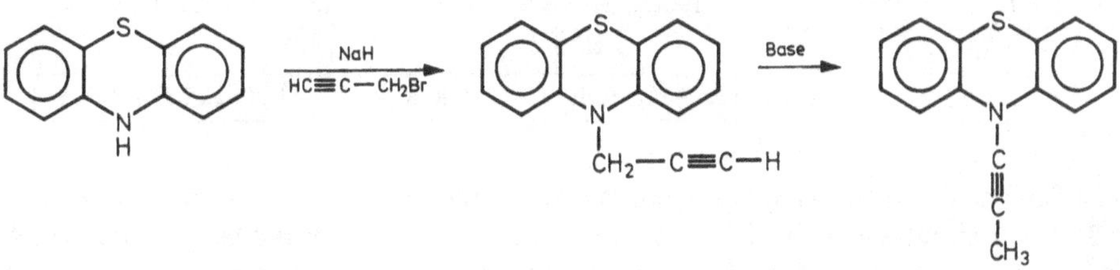

(62)

More recent work illustrates how important the reaction conditions are. Either aminoallene 34 or ynamine 35 can be obtained (63) [110].

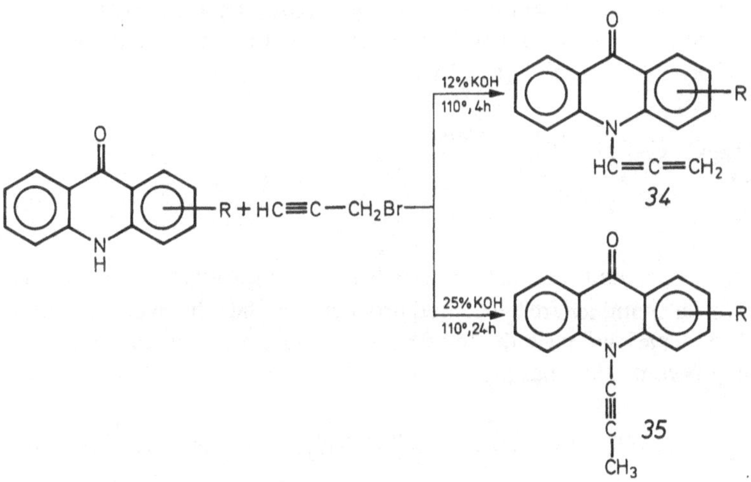

(63)

This case is rather exceptional and even under favorable circumstances only one compound i.e. ynamine or aminoallene can be obtained. Worse, equilibrium mixtures of both compounds can be formed which makes such procedures completely unatractive as synthetic tool.

Let us a consider a general case of propargylic amine:

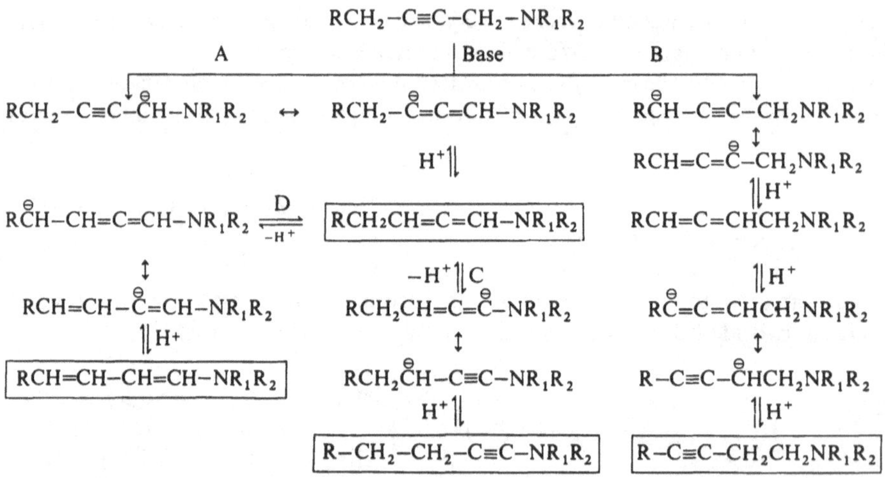

The Scheme illustrates vividly the complexity of the process. Two or more pathways may occur simultaneously. The strength and concentration of the base, character of the solvent, reaction time and temperature are of crucial importance and vary strongly from one case to another.

All four pathways A–D are documented. 1-Aminoallenes and ynamines being more or less conjugated are thermodynamically more stable than their propargylic precursors. Terminal acetylenes are final products if an excess of a strong base is used and the acetylide irreversibly precipitates out of the reaction mixture (R = H; path B). Such bases are sodium and potassium amides, excess butyllithium and especially the potassium salt of propylenediamine (KAPA) (64) [108].

$$CH_3{-}C{\equiv}C{-}CH_2{-}NEt_2 \xrightarrow[2.\ H_2O]{1.\ KNH_2,\ Na\ or\ KAPA} H{-}C{\equiv}C{-}CH_2CH_2NEt_2 \qquad (64)$$

This means that such conditions are to be avoided or the base must be present only in catalytic amounts if ynamines are the desired products. On the other hand, 1-amino-1,3-dienes (Path A) have obviously the lowest energy content due to extended conjugation and may become the final products (65) [111].

$$Me_2NCH_2{-}C{\equiv}C{-}CH_2NMe_2 \xrightarrow[xylene]{Na} Me_2NCH{=}CH{-}CH{=}CHNMe_2 \qquad (65)$$

Nevertheless, the deprotonation steps in paths C and D are very difficult and when R is an alkyl or phenyl group the reaction may stop at the aminoallene stage [112–115]. This lack of acidity can not always be invoked inasmuch as the obviously more acidic pyrrolyl, pyrazolyl, and imidazolyl compounds give only the corresponding 1-aminoallenes [24, 113]. In such cases the thermodynamic and not the kinetic factors, are responsible for the outcomme of the isomerisation.

In practice, ynamines are formed in good yields only from non-substituted propargylamines. The following reaction (66) can be run on a large scale [108, 111, 112].

$$HC{\equiv}C{-}CH_2NEt_2 \xrightarrow[\text{or t-BuOK, DMSO}]{KNH_2/Al_2O_3} CH_3{-}C{\equiv}C{-}NEt_2 \quad (66)$$

$$65{-}93\ \%$$

Diethylaminoprop-1-yne has been commercial for a long time thanks to this efficient and straightforward procedure [112, 115, 116].

N-Propargyldimethylamine and morpholine give lower yields of ynamines (25 %) [112] but the corresponding 1-propynyl diisopropylamine and 2,2,6,6-tetramethylpiperidine can be obtained in a 75 % yield [117]. The allenic amines are sometimes still present in crude ynamines but they form high-boiling dimers and oligomers above 80°. This difference in thermostability is exploited to prepare pure ynamines. Thus when the equilibrium mixture of diethylaminoallene and the ynamine (22:78) is heated for 30 minutes at 125°, the allene is decomposed whereas the ynamine is obtained thereupon by distillation [115]. This updated study also shows that the equilibrium is reached very quickly in DMSO but it contains 80 % of aminoallene when R = Me and R_2 = morpholino. This explains the low yields of ynamines in such cases. In the absence of DMSO, when using for example HMPT-potassium tertbutanolate, the conversion of allenes to ynamines becomes very slow [115]. This allowed the first efficient synthesis of the elusive dimethylaminoallene. Aminoallenes are more reactive than ynamines but their chemistry is not yet sufficiently explored [118].

N,N-Diaryl ynamines are formed already with weakly basic systems like potassium hydroxide (67) [24, 25].

$$HC{\equiv}C{-}CH_2{-}NPh_2 \xrightarrow[KOH]{DMSO} Me{-}C{\equiv}C{-}NPh_2 \quad 50\ \% \quad (67)$$

The same pattern is observed when the hydrogen atom in propargylamines is replaced by ethynyl, vinyl or isopropenyl groups (68) [111].

$$HC{\equiv}C{-}C{\equiv}C{-}CH_2NEt_2 \xrightarrow[HMPT]{\text{T-BuOK}} Me{-}C{\equiv}C{-}C{\equiv}C{-}NEt_2 \quad 78\ \% \quad (68)$$

Instead of using diacetylenes as such, the appropriate carbinols may also lead to diynamines in a one-pot reaction (69) [24–26].

$$R_2N{-}CH_2{-}C{\equiv}C{-}H + BrC{\equiv}C{-}C(CH_3)_2OH \xrightarrow{Cu^+} R_2N{-}CH_2{-}C{\equiv}C{-}C{\equiv}C{-}C(CH_3)_2OH \quad (69)$$

$$\xrightarrow[-\text{Acetone}]{KOH} R_2NCH_2(C{\equiv}C)_2H \xrightarrow{KOH} R_2N{-}(C{\equiv}C)_2{-}CH_3$$

R = Ph, 38 %

= Me, Ph 50 %

The last method was originally developped to synthesize N,N-diaryl and arylalkyl diynamines but it is sufficiently general as illustrated by the succesfull preparation of dimethylaminobutadiyne in 73% yield [26].

Another ingenious variation starts with 4-methoxypropargylamines 36 which are readily accessible via copper-catalyzed Mannich reaction of propargyl ethers [119]. The group R can again only be a hydrogen atom or methyl group.

$$CH_3O-\underset{\large R}{\underset{|}{C}}H-C{\equiv}C-H \xrightarrow[Cu^{2+}]{R'_2N-CH_2-OH} CH_3O-\underset{\large R}{\underset{|}{C}}H-C{\equiv}C-CH_2NR'_2$$

36

$$\xrightarrow[\text{or } NaNH_2]{\text{t-BuOK}} RCH{=}CH-C{\equiv}C-NR'_2 \qquad (70)$$

37

R = H, R′ = Me 70%; R = H, R′ = Et 70–86%; R = H, R′ = iPr 70%; R = Me, R′ = Et 80%

Yields of very sensitive vinyl and propenyl ynamines 37 bearing methyl, ethyl and isopropyl groups on the nitrogen atom are very good (68) [119].

Aminodiynes having a straight chain of six carbon atoms and more are better prepared from the appropriate bromoacetylenes by substitution with secondary or tertiary aliphatic amines [26], because strong bases used to induce the propargylic rearrangement give mixtures of various conjugated dienynes. Analogously, propargylic enynes having at least six carbon atoms isomerize to conjugated 1-aminohexatrienes rather than to ynamines (71) [111,120].

$$RCH_2-CH{=}CH-C{\equiv}C-CH_2NR'_2 \xrightarrow{\text{Base}} R(CH{=}CH)_3NR'_2 \qquad (71)$$

Vinyl propargylamines work also very well (72–73) [120–122] provided the longest straight chain has no more than five carbon atoms or it is sufficienlly ramified as to preclude the formation of aminohexatrienes.

$$CH_2{=}\underset{\large CH_3}{\underset{|}{C}}-C{\equiv}C-CH_2NR_2 \xrightarrow[\text{DMSO}]{\text{t-BuOK}} CH_3-\underset{\large CH_3}{\underset{|}{C}}{=}CH-C{\equiv}C-NR_2 \quad 84-89\% \qquad (72)$$

$$+\; CH_2{=}\underset{\large CH_3}{\underset{|}{C}}-CH{=}C{=}CHNR_2 \quad 10\%$$

$$CH_2{=}CH-C{\equiv}C-CH_2NR_2 \xrightarrow{\text{t-BuOK}} CH_3-CH{=}CH-C{\equiv}C-NR_2 \qquad (73)$$

38, 65–80%

$$+\; CH_3-C{\equiv}C-CH{=}CH-NR_2$$

39, 10%

It has been demonstrated that the isomeric enamine 39 arise from 3-ene 1-ynamines 38 irreversibly by the action of very strong bases as they seem to be more stable thermodynamically [119,121,123].

This conclusion is based upon the observation that the double and the triple bond exchange their positions not only upon treatment with potassium amide (this might be due to the insolubility of 40 in ammonia) but also with catalytic amounts of potassium tert. butanolate in DMSO or HMPT (74) [123].

$$\underset{38}{H_2C{=}CH{-}C{\equiv}C{-}NR_2} \xrightarrow[NH_3]{KNH_2} \underset{40}{KC{\equiv}C{-}CH{=}CHNR_2}$$

$$\xrightarrow{NH_4Cl} HC{\equiv}C{-}CH{=}CHNR_2 \qquad (74)$$

$$R = Me,\ 80\%; \qquad R = Et,\ 92\%$$

Butatrienylamines 41 must occur as intermediates but they isomerise too fast to give 39 in t-BuOK/DMSO or KNH_2/NH_3 mixtures. They are indeed isolated if BuLi/t-BuOK is used to achieve a rapid deprotonation of 38 followed by kinetic quenching with tert. butanol at −100°. This sequence not only clarifies the reaction mechanismus but provides also a new class of very reactive intermediates (75) [124,125].

$$38 \xrightarrow{-H^+} CH_2{=}\overset{\ominus}{C}{-}C{\equiv}C{-}NR_2 \leftrightarrow H_2C{=}C{\cdots}C{=}C{-}NR_2$$

$$\xrightarrow{H^+} \underset{41}{H_2C{=}C{=}C{=}CHNR_2} \qquad R = Me,\ 78\%; \quad i-Pr,\ 89\% \qquad (75)$$

It can be concluded that propargylic rearrangements make an excellent and expeditious method of access to various 2-methyl ynamines, their vinylogs and ethynylogs.

5 Ynamine Interconversion

Although the aminoethynyl moiety in ynamines is very reactive, it permits certain selective transformations which lead to other usually more complex ynamines. Protic acids and frequently protic solvents must be avoided. Basic reagents on the other hand, are well tolerated since they add either slowly or not at all to the electron-rich triple bond in ynamines.

5.1 Metallation at Methyl Group in γ-Position

Diethylaminopropyne can be metallated yielding lithoynamine 42 which gives good yields of γ-silylated ynamines upon silylation with trimethylsilyl chloride (76) [117,126].

$$Me{-}C{\equiv}C{-}NEt_2 \xrightarrow[TMEDA]{BuLi} \underset{42}{[CH_2C{\equiv}C{-}NEt_2 \leftrightarrow CH_2{=}C{=}C{-}NEt_2]^{\ominus}Li^+}$$

$$\xrightarrow{Me_3SiCl} Me_3SiCH_2{-}C{\equiv}C{-}NEt_2 \qquad 74{-}86\% \qquad (76)$$

Less hindered methyl and alkyl halides react at the α-carbon of 42 but the resulting allenic amines are too unstable. By chosing bulkier diisopropylamino and especially tetramethylpiperidino groups, selective γ-alkylations can be effected (77) [117].

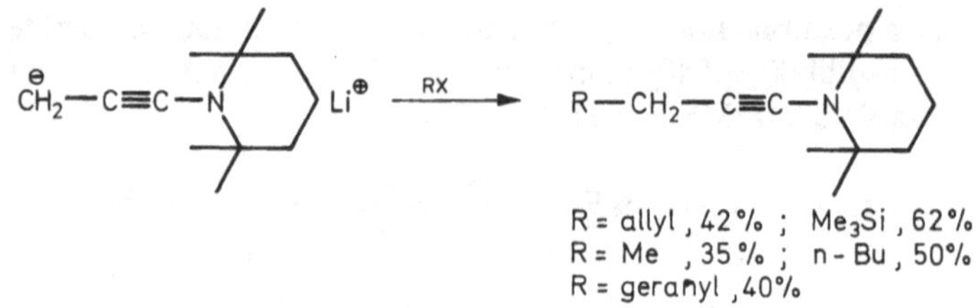

(77)

This process represents an interesting three carbon atom extension which upon hydrolysis leads to various carboxylic acid amides.

5.2 Oxidative Dimerization of Ynamines

Two examples of the still very little known diynediamines have been obtained by cuprous salt catalyzed oxydative dimerisation of diphenylamino and N-methylanilinoacetylene (78) [127].

$$\begin{matrix}Ph \\ R\end{matrix}\!\!>N-C\equiv C-H \xrightarrow[O_2,\ NH_4OH]{Cu_2Cl_2,\ MeOH} \begin{matrix}Ph \\ R\end{matrix}\!\!>N-C\equiv C-C\equiv C-N<\!\!\begin{matrix}Ph \\ R\end{matrix} \qquad (78)$$

43

R = Ph, 55 % R = Me, 60 %

This valuable duplication is a particular case of the Glaser coupling [128,129]. It should be noted that ynamines carrying one and especially two aryls on the nitrogen are able to resist to water in the absence of an acid. Ynamines deriving from purely aliphatic amines cannot be duplicated using this sequence owing to the extended hydrolysis. The scope of this reaction has therefore remained confined to the examples given in (78). Recent work shows that diynediamines are readily obtained from 1-bromo-4-amino-1,3-diynes [39]. Mixed couplings are achieved through Cadiot-Chodkiewitz coupling [128,129] of an acetylene with a l-haloacetylene, but haloynamines being ethynylogous haloamines are too unstable to be employed [130]. An unsuccesfull case of a mixed coupling has already been mentioned in (Eq. 49).

5.3 Ynamines from Aminoethynyllithium Compounds

Alkylation. As shown in Section 3.4, trichloroenamines react with two equivalents of butyl lithium to afford lithiated ynamines 44 (79)

$$Cl_2C{=}CClNR_2 \xrightarrow{n\text{-}BuLi} Li{-}C\equiv C{-}NR_2 \xrightarrow{E^+} E{-}C\equiv C{-}NR_2 \qquad (79)$$

44 aminoethynyl group

These compounds have proven to be the most versatile building blocks for construction of other ynamines mainly through protonation, alkylation and acylation. Such reactions are called aminoethynylation [131]. More recent development involves the use of stable stannyl ynamines. Lithio ynamines 44 are prepared in situ and alkylation is achieved by adding the corresponding alkyl halide or tosylate. The D_3-ynamine 45 was prepared via the latter method (80) [132]. Two selected examples (81, 82) where the acetylene chain carries a latent functionality follow [133,134].

$$\text{Tos–O–CD}_3 + \underset{44a}{\text{Li–C}\equiv\text{C–NEt}_2} \longrightarrow \text{CD}_3\text{–C}\equiv\text{C–NEt}_2 \quad 45\text{-}50\% \tag{80}$$

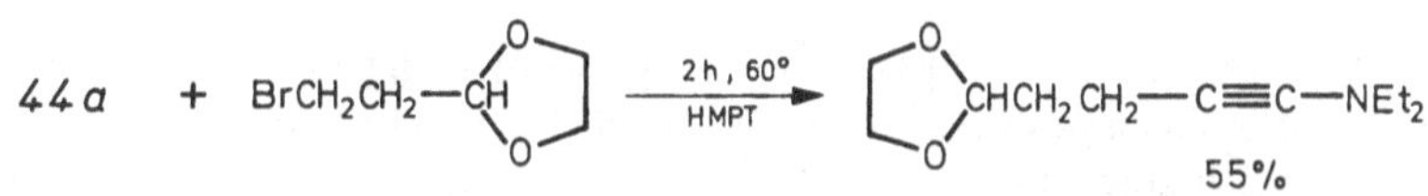

(81)

$$44a + \text{BrCH}_2\text{CH}_2\text{OBu-t} \xrightarrow[\text{HMPT}]{2\text{h}, 60^\circ} \text{t-BuOCH}_2\text{CH}_2\text{–C}\equiv\text{C–NEt}_2 \quad 60\% \tag{82}$$

A small amount (up to one equivalent) of HMPT is necessary in order to accelerate the alkylation step. Sec.-butyl bromide gives 38% of the expected ynamine together with 25% butene and 22% of diethylaminoacetylene. Expectedly, t-butyl bromide undergoes only elimination and it serves as a convenient proton source to liberate N-ethynyl-methylaniline [106]. Primary substrates react smoothly, however, even in complex cases and with this in mind, the scope of this reaction is almost unlimited. One more example illustrates this technique (83) [135].

$$\text{CH}_2\text{=CBrCH}_2\text{O(CH}_2)_{12}\text{–I} + \text{Li–C}\equiv\text{C–N(Me)Ph}$$

$$\xrightarrow[\text{HMPT}]{3\text{hr}, 70^\circ} \text{CH}_2\text{=CBrCH}_2\text{O(CH}_2)_{12}\text{–C}\equiv\text{C–N(Me)Ph} \quad 78\% \tag{83}$$

$$\xrightarrow[2.\ \text{H}_2\text{O}]{1.\ \text{t-BuLi}} \text{HO–(CH}_2)_{12}\text{–C}\equiv\text{C–N(Me)Ph} \quad \sim 100\%$$

Metal exchange. Lithiated ynamines 44 are smoothly silylated to the corresponding silyl ynamines (84) [42,59,136]. Such ynamines formally replace the very sensitive non-substituted ynamines but the increase in stability entails a loss of reactivity toward electrophilic reagents because silicon acts as a weak electron acceptor.

$$44\text{a} \xrightarrow{\text{Me}_3\text{SiCl}} \text{Me}_3\text{SiC}\equiv\text{C–NEt}_2 \quad 65\% \tag{84}$$

Silylated ynamines react only with very reactive acid chlorides, but the analogous stannyl ynamines 46 are much more versatile and have recently become available in fifty gram lots (85) [95,131,137–139].

$$\text{Cl}_2\text{C=CCl–NR}_2 \xrightarrow[2.\ \text{R}'_3\text{SnX}]{1.\ n\text{-BuLi}} \underset{46}{\text{R}'_3\text{Sn–C}\equiv\text{C–NR}_2} \tag{85}$$

Germylated ynamines 47 can be obtained in the same fashion from lithium ynamines (86) [58, 140, 141].

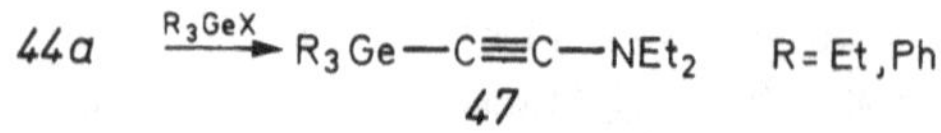

(86)

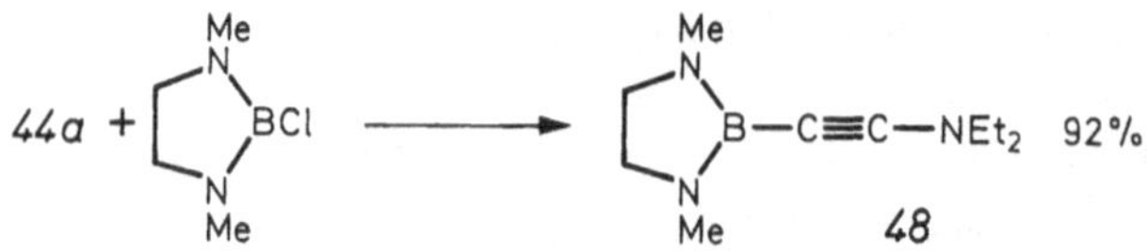

(87)

The compound 48 is conceptually very intersting. Boron being a strong electronaccepting group, 48 is in fact a push-pull ynamine. ^{13}C NMR data fully support the view [142].

Aminoethynyl arsines 49 are usually prepared by metal exchange from stannyl ynamines and diphenylarsenyl chloride (88) [141].

$$(n\text{-}Bu)_3Sn{-}C{\equiv}C{-}N(Me)Ph \xrightarrow{Ph_2AsCl} \underset{49}{Ph_2As{-}C{\equiv}C{-}N(Me)Ph} \quad 87\,\% \tag{88}$$

An analogous sequence leads to ynamino phosphines 50 (89) which are particularly versatile precursors of push-pull phosphorylated ynamines 51–54 through subsequent oxydation with hydrogen peroxide, elemental sulphur, selenium and sulphonyl azides [143].

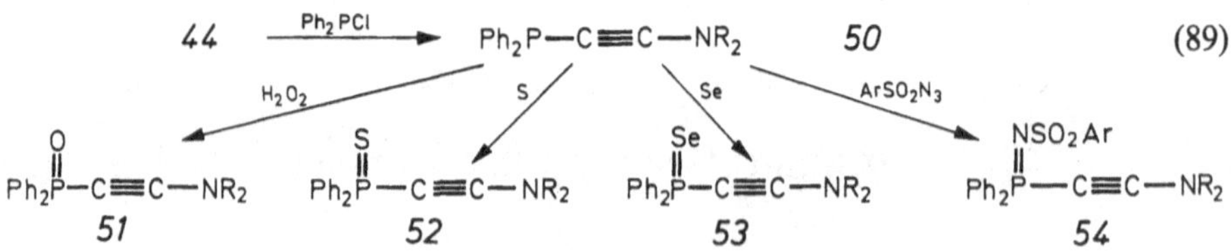

(89)

Sulphur can be dissolved in 44 thereby giving ynethiolate anion, which might become an expeditious precursor of thioynamines (90) [144].

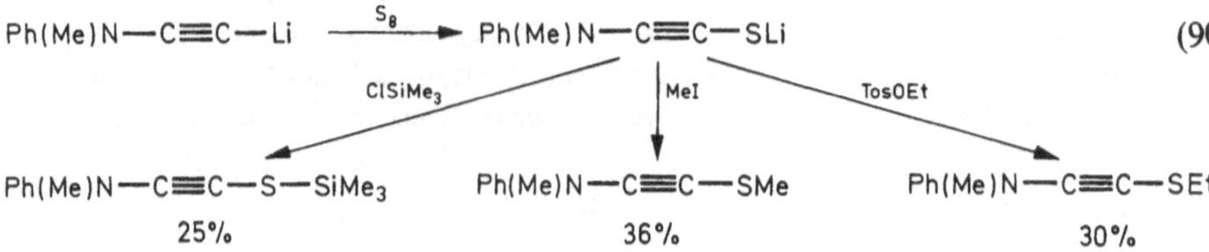

(90)

S-allylthioynamine, however, cannot be obtained at all and the yields generally ought to be improved.

Acylation. Ynamine synthons 44 and 46 can be readily acylated. This approach has led to an amazing number of push-pull ynamines which illustrates the usefulness of this approach (91–94)

$$44\,a \xrightarrow{ClCO_2Me} Et_2N{-}C{\equiv}C{-}CO_2Me \quad 70\,\%^{54)} \tag{91}$$

$$44\,a \xrightarrow{ClCN} Et_2N{-}C{\equiv}C{-}CN \quad 63\,\%^{145)} \tag{92}$$

Cyanogen bromide, however, effects bromination giving rather unstable bromoynamines [130].

$$Li{-}C{\equiv}C{-}N(Me)Ph + Me_2N{-}\underset{\displaystyle Cl}{C}{=}NTos \rightarrow Me_2N{-}\underset{\displaystyle \| \atop NTos}{C}{-}C{\equiv}C{-}N(Me)Ph \quad 40\,\% \tag{93}$$

$$44 \text{ or } 46 \quad + R'{-}\underset{\displaystyle \| \atop O}{C}{-}Cl \rightarrow R'{-}\underset{\displaystyle \| \atop O}{C}{-}C{\equiv}C{-}NR_2 + LiCl \text{ or } R_3''SnCl \tag{94}$$

Tributyltin chloride may cause problems of separation during the work-up stage and for sensitive dialkylaminoacetylene derivatives 55 the use of trimethyltin residue is advisable (95) [139].

$$Et_2N{-}C{\equiv}C{-}SnMe_3 + R{-}\underset{\displaystyle \| \atop O}{C}{-}Cl \rightarrow \underset{55}{Et_2N{-}C{\equiv}C{-}\underset{\displaystyle \| \atop O}{C}{-}R} \tag{95}$$

55, R = Pr, 36%; t-Bu, 55%; C_5H_{11}, 20%; Ph, 40%; OMe, 40%

The yields are lower because of competing hydrolysis during the aqueous work-up. Lithioynamines 44 must be used with unreactive halides such as carbamoyl chlorides and imidoyl halides (93) [95]. The scope of this reaction seems to be limited only by the availability of the requisite acyl chlorides. It is less surprising, therefore, that a single paper reports the preparation of about sixty push-pull acetylenes [95]. In contrast, silylated ynamines, react only with very strong acylation agents, e.g. trichloroacetyl and nitrobenzoyl chloride [95] as well as with phosgeniminium chloride.

A perusal of the literature discloses that ketones and aldehydes have rarely been reacted with 44. In view of the tendency of ynamines to add water and alkohols, it is obvious that ynamines 56 undergo intramolecular addition of water (Meyer-Schuster rearrangement) quite readily (96). This process is provoked already by silicagel, alumina or it even occurs spontaneously [146].

$$CH_3COCH_3 + \underset{44\,a}{Li{-}C{\equiv}C{-}NEt_2} \rightarrow \underset{56}{(Me_2\overset{\displaystyle HO \atop |}{C}{-}C{\equiv}C{-}NEt_2)}$$

$$\rightarrow \underset{50\,\%}{Me_2C{=}CH{-}CONEt_2} \tag{96}$$

Even camphor reacts very well [146] and 44 is therefore the synthetic equivalent of the N,N-dialkylacetamide anion.

Weakly electrophilic aliphatic isocyanates and carbodiimides are good reaction partners toward 44. The overall reaction is an acylation which leds to an easy short-cut to push-pull secondary amides 57 and amidines 58. A few examples follow (97) [138], (98) [147,148].

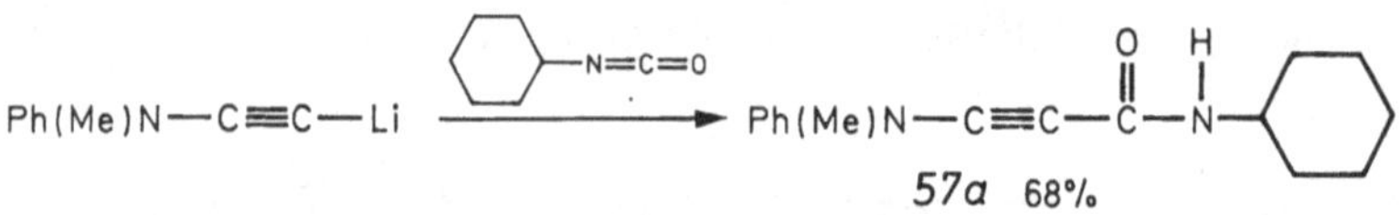

(97)

$$\text{Ph(Me)N–C}\equiv\text{C–Li} \xrightarrow{\text{t-BuN=CO}} \text{Ph(Me)N–C}\equiv\text{C–C(=O)–N(H)–t–Bu}$$

57b, 75%

$$44 \xrightarrow{\text{R}'\text{N=C=NR}'} \text{R}_2\text{N–C}\equiv\text{C–C}\begin{matrix}\text{NHR} \\ \text{NR}'\end{matrix} \qquad 58 \qquad (98)$$

$$\text{Ph(Me)N–C}\equiv\text{C–C(=N–C}_6\text{H}_{11}\text{–c)–N(H)–C}_6\text{H}_{11}\text{–c} \xrightarrow{\text{MeI}} \text{Ph(Me)N–C}\equiv\text{C–C(=N–C}_6\text{H}_{11}\text{)–N(Me)–C}_6\text{H}_{11}$$

37% (99)

$$\text{Ph(Me)N–C}\equiv\text{C–C(=O)–N(H)–C}_6\text{H}_4\text{–Me–4}$$

$$\xrightarrow[\text{MeI}]{\text{NaNH}_2} \text{Ph(Me)N–C}\equiv\text{C–C(=O)–N(Me)–C}_6\text{H}_4\text{–Me–4}$$

39% (100)

As the (Eqs. 99, 100) demonstrate, the secondary amides and amidines can be alkylated at the nitrogen [95,148]. Push-pull amidines 58 can in certain cases be N-acylated with diphenylketene whereas methylene sulfone introduces a methylsulphonyl group [148].

With arylsubstituted carbodiimides. however, mixtures of ynamines and 2,4-diaminoquinolines are obtained which reduces the scope of such acylations at least as far as the ynamine synthesis is concerned [147].

Aminoethynylmetallation.

Electron-poor multiple bonds in acetylene dicarboxylates [42,137,149], ketenes [150–152] isocyanates, isothiocyanates, carbodiimides [138,153] and arynes [42] add metallated ynamines to give adducts that still carry the ynamine moiety. This behaviour contrast with that of other ynamines which mostly undergo various cycloadditions but it must be stressed that even here cycloadducts may be formed as by-products.

$$\text{R}^3_n\text{M–C}\equiv\text{C–NR}^1\text{R}^2 + \text{X=Y} \rightarrow \text{R}^3_n\text{M–X–Y–C}\equiv\text{C–NR}^1\text{R}^2$$

Acetylene dicarboxylates were studied independently by two groups (101) [42,137,154].

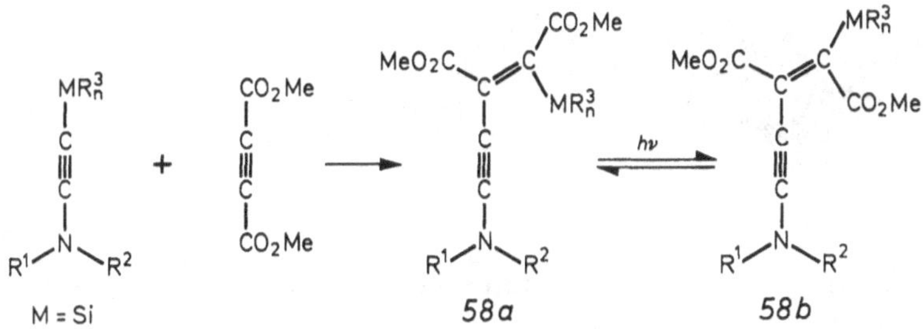

(101)

The addition takes place exclusively in a cis fashion when silylated ynamines are used [42,137] to give 58a having maleic acid configuration. The isomers with fumaric acid configurations are formed in 40–50% conversions upon sensitized photoisomerization in benzene solutions [137]. Methyl propiolate behaves in a like manner [42].

Diazomethane reacts with 58a to give homologous vinyl ynamines with unchanged configuration at the carbon carbon double bond (102) [149]. The reaction can be explained in terms of a regiospecific 1-3dipolar addition to give a pyrazoline which loses nitrogen with a concomittant silyl group migration. Partial isomerisation to the trans form can also here be brought about by UV light in 30–50% conversions.

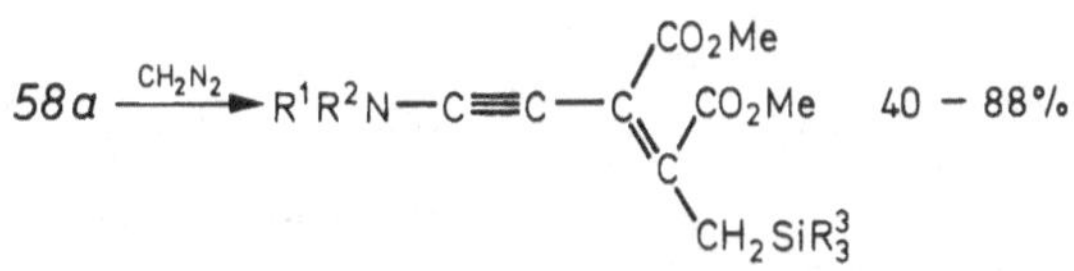

(102)

Aminoethynylstannylation proceeds in a less satisfactory manner. Not only the yields are lower but the arising products are mixtures of cis and trans isomers (103) [137,154].

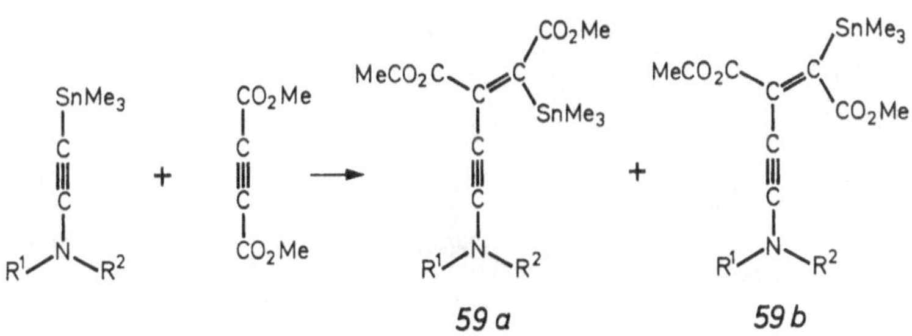

(103)

Cis-addition is favoured by alkyl groups on the ynamine nitrogen, high solvent polarity (acetonitrile) low temperature and low concentration. Thus when R^1, R^2 = Me, Ph the ratio 59a, 59b in acetonitrile is 78/22, but in benzene it becomes 30/70. A 40/60 ratio in benzene at 20° becomes 25/75 when the reaction is run at 80°. A fivefold dilution changes the ratio from 30/70 to 40/60.

Diazomethane brings about the insertion of one methylene group as happens with the silylated counterparts, and the yields are good [154].

In one case, diazomethane cycloadds upon the homologation product forming the ynamine 60 which carries a heterocyclic ring (104) [154].

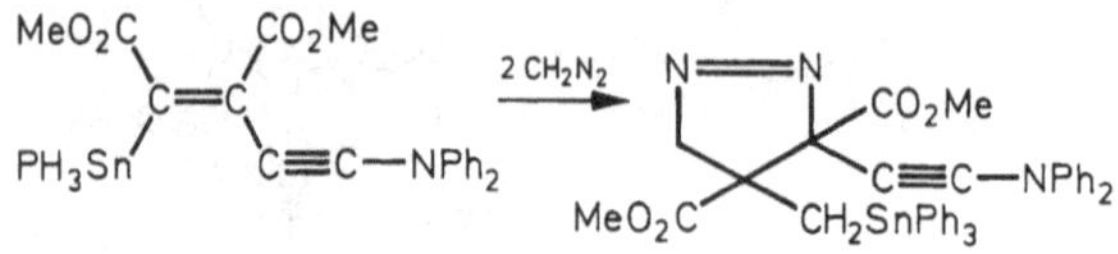

60 36% (104)

(Methylanilinoethynyl) triphenylgermane forms the germylated cis vinyl ynamine with acetylene dicarboxylate in 47% yield [137].

Arylisocyanates and isothiocyanates form adducts 61 which belong to push-pull acetylenes (105) [138,153].

$$R^1R^2N-C\equiv C-SnR^3_3 + RNCX \rightarrow R^1R^2N-C\equiv C-C(=NR)XSnR^3_3$$

61 $X = O, S$ (105)

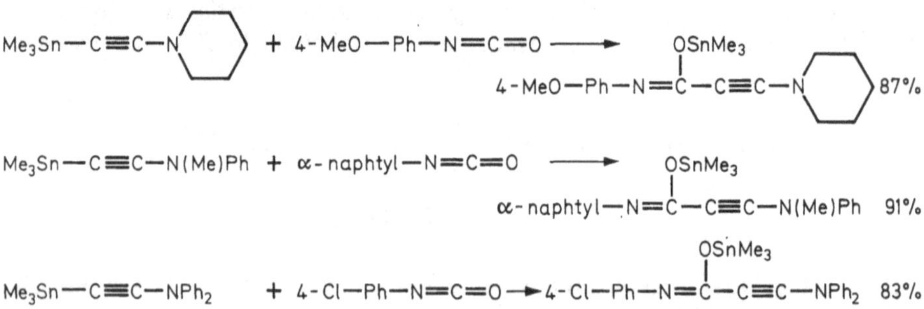

The arising 3-aminopropiolic imidates can be selectively N-acylated or alkylated (106) [138].

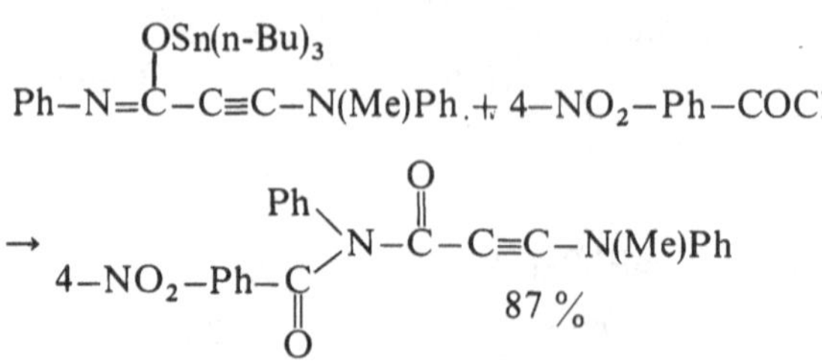

(106)

Mild hydrolysis of O-stannylated imidates leads to a whole series of secondary propiolic amides, inacessible by other methods (107) [95,138].

$$PhN=C(OSnMe_3)-C\equiv C-N(Me)Ph \xrightarrow{H_2O} Ph-NH-C(=O)-C\equiv C-N(Me)Ph$$

88% (107)

Only the more strongly electrophilic diarylcarbodiimides react with stannyl ynamines to give the transient dipolar adducts 62. They evolve further either by undergoing a 1,3-shift of the organyl group to give ynamines 63 or they cyclise to diaminoquinolins 64 (108) [148, 153].

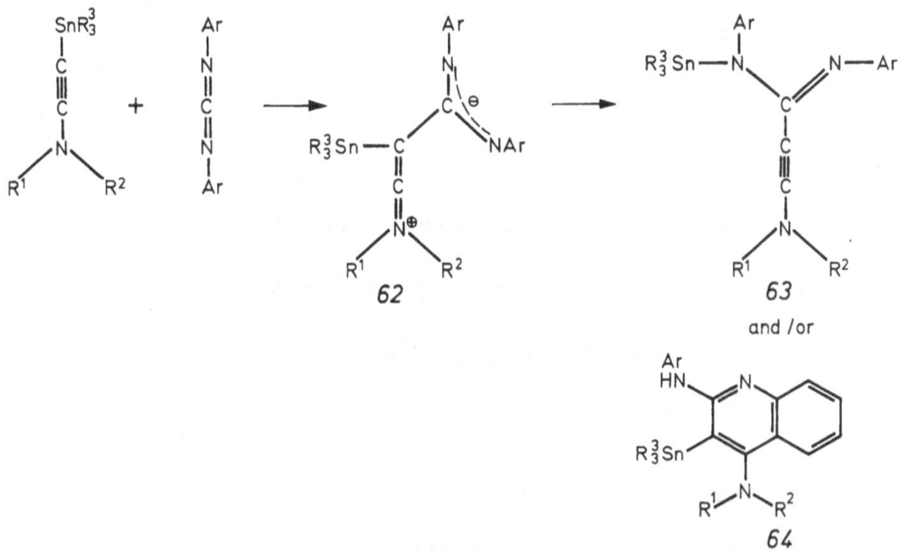

(108)

In practice, 64 are the predominant products and ynamines 63 are only by-products.

Isothiocyanates react with stannyl ynamines to give adducts 65 which upon careful hydrolysis yield the still rare ynamines thioamides 66 (109) [153].

$$R_2N{-}C{\equiv}C{-}SnPh_3 + PhN{=}C{=}S \rightarrow \underset{65}{R_2N{-}C{\equiv}C{-}\overset{\overset{\displaystyle NPh}{\|}}{C}{-}S{-}SnPh_3}$$

$$\xrightarrow{H_2O} \underset{66}{R_2N{-}C{\equiv}C{-}C({=}S){-}NHPh} \qquad (109)$$

Diphenylketene [131, 150, 151, 155, 156] 9-fluorenylidene ketene [156] and cyanoketenes (112) [153] have been thoroughly studied. In all cases, the silylated, germylated and stannylated ynamines add across the carbonyl group of ketenes to furnish a whole series of new vinyl ynamines 67 and of ynamine ketones 68 (110) upon mild hydrolysis (110–113).

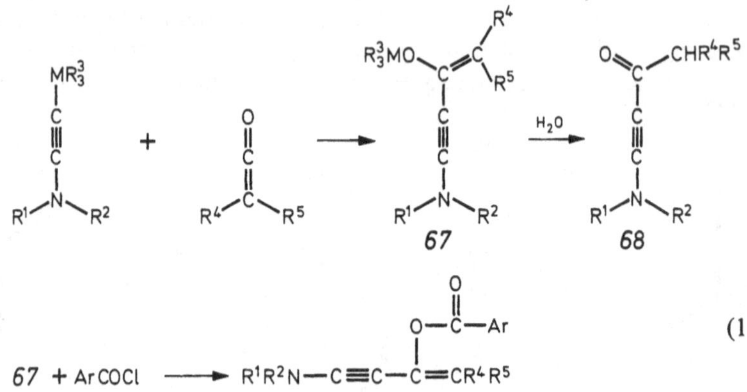

(110)

As shown, besides hydrolysis, selective O-acylations are feasible (113) [131, 152].

$$Me_3Sn{-}C{\equiv}C{-}N(CH_3)C_6H_5 + Ph_2C{=}C{=}O \rightarrow Ph_2C{=}C(OSnMe_3){-}C{\equiv}C{-}N(Me)Ph \quad \textbf{69} \qquad (111)$$

$$Ph_3Si{-}C{\equiv}C{-}NPh_2 + t\text{-}Bu(CN)C{=}C{=}O \rightarrow t\text{-}Bu(CN)C{=}C(OSiPh_3){-}C{\equiv}C{-}NPh_2 \qquad (112)$$

$$Ph_3Si{-}C{\equiv}C{-}N(Me)Ph + Ph(CN)C{=}C{=}O \rightarrow Ph(CN)C{=}C(OSiPh_3){-}C{\equiv}C{-}N(Me)Ph$$

$$\textbf{69} + p\text{-}NO_2{-}PhCOCl \rightarrow Ph_2C{=}C(OPhNO_2\text{-}p){-}C{\equiv}C{-}N(Me)Ph \qquad (113)$$

$$\textbf{69} + Ph{-}CH{=}CHCOCl \rightarrow Ph_2C{=}C(OCOCH{=}CHPh){-}C{\equiv}C{-}N(Me)Ph$$

Aminoethynylmetallation with reactive ketenes usually leads to mixtures of products because of competing [2 + 2]cycloadditions. Thus when only 68 are desired one will use the very common acid chlorides instead of ketenes. Stable silylated and germylated ketenes react only by [2 + 2]cycloaddition [157]. Aliphatic ketenes, isopropylidene and cyclopentylidene ketenes react more slowly and yields are low (114) [156].

$$Me_3Si{-}C{\equiv}C{-}NEt_2 + R_1R_2C{=}C{=}O \rightarrow R_1R_2C{=}C(OSiMe_3){-}C{\equiv}C{-}NEt_2$$

$$R_1, R_2 = CH_3, 39\%; -(CH_2)_4\ 15\% \qquad (114)$$

Benzyne has also been reported[42] to add ynamines but the yields are low (115):

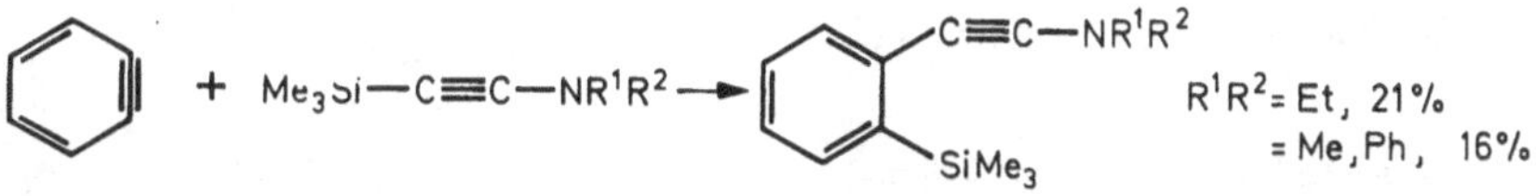

(115)

Amino di- and polyacetylenes. Trichlorovinyl group is an excellent precursor of the lithium acetylide moiety (116). R can be almost any group compatible with butyl lithium (116)[158].

$$\mathrm{R{-}CCl{=}CCl_2} \xrightarrow{\text{2 BuLi}} \mathrm{R{-}C{\equiv}C{-}Li} \qquad (116)$$

This approach has been very fruitful in the construction of hitherto inaccessible aminopolyacetylenes. The little stable trichlorovinylynamines 70 or perchloroenamine 71 can be put to good use by reacting them with butyl lithium (117)[37–39].

It should be noted here that the analogous sodium salt 72a has previously been prepared from dimethylamino diacetylene and sodium amide (118)[26].

$$\mathrm{R_2N{-}C{\equiv}C{-}CCl{=}CCl_2}\ (\mathbf{70}) \xrightarrow{\text{2 BuLi}} [\mathrm{R_2N{-}C{\equiv}C{-}C{\equiv}C{-}Li}]\ (\mathbf{72})$$

or

$$\mathrm{R_2N{-}CCl{=}CCl{-}CCl{=}CCl_2}\ (\mathbf{71}) \xrightarrow{\text{3–4 BuLi}} [\mathrm{R_2N{-}C{\equiv}C{-}C{\equiv}C{-}Li}]\ (\mathbf{72}) \qquad (117)$$

$$\mathrm{Me_2N{-}C{\equiv}C{-}C{\equiv}CH} \xrightarrow{\mathrm{NaNH_2}} \underset{\mathbf{72\,a}}{\mathrm{Me_2N{-}C{\equiv}C{-}C{\equiv}C{-}Na}} \qquad (118)$$

72 seems to be as promising as is its lower ethynylog 44. Thus protonation, alkylation and organometallation of 72 gives the corresponding diynamines (119–121)[26, 34, 37, 38].

$$\mathbf{72} \xrightarrow{\mathrm{H^+ donor}} \mathrm{R_2N{-}C{\equiv}C{-}C{\equiv}C{-}H} \qquad (119)$$

$\mathrm{R_2}$ = morpholino, 60–87%; Me, Ph, 54%; Me, Me, 73%

$$\underset{\mathbf{72\,b}}{\mathrm{Ph(Me)N{-}C{\equiv}C{-}C{\equiv}C{-}Li}} \xrightarrow[\mathrm{HMPT}]{\mathrm{MeI}} \mathrm{Ph(Me)N{-}C{\equiv}C{-}C{\equiv}C{-}Me}\quad 56\,\% \qquad (120)$$

$$\mathbf{72\,a} \xrightarrow{\mathrm{MeI}} \mathrm{Me_2N{-}C{\equiv}C{-}C{\equiv}C{-}Me}\quad 89\,\%$$

$$\mathbf{72\,b} \xrightarrow{\mathrm{ClMPh_3}} \mathrm{Ph(Me)N{-}C{\equiv}C{-}C{\equiv}C{-}MPh_3} \qquad (121)$$

Diphenylarsinyldiynamine 74 is prepared via metal-metal exchange from the trimethylstannyl compound 73 (122) [37]. Diphenylphosphynyl compound 75 can again be selectively oxidized at phosphorus (122) [37].

$$\underset{73}{\text{Ph(Me)N}-\text{C}\equiv\text{C}-\text{C}\equiv\text{C}-\text{SnMe}_3} \xrightarrow{\text{Ph}_2\text{AsCl}} \underset{74,\ 75\%}{\text{Ph(Me)N}-\text{C}\equiv\text{C}-\text{C}\equiv\text{C}-\text{AsPh}_2} \qquad (122)$$

$$\text{72b} \xrightarrow{\text{Ph}_2\text{PCl}} \underset{75,\ 50\%}{\text{Ph(Me)N}-\text{C}\equiv\text{C}-\text{C}\equiv\text{C}-\text{PPh}_2}$$

$$\xrightarrow[\text{S}_8]{\text{H}_2\text{O}_2\text{ or}} \text{Ph(Me)N}-\text{C}\equiv\text{C}-\text{C}\equiv\text{C}-\overset{\overset{\text{X}}{\|}}{\text{P}}\text{Ph}_2 \qquad (123)$$

$$\text{X} = \text{O},\ 67\% \qquad \text{X} = \text{S},\ 74\%$$

72b reacts also with elemental sulphur to diynethiolate which can be converted by alkylation to the first thiodiynamine 76 (124) [37].

$$\text{72 b} \xrightarrow[\text{2. MeI}]{\text{1. S}_8} \underset{76,\ 85\%}{\text{Ph(Me)N}-\text{C}\equiv\text{C}-\text{C}\equiv\text{C}-\text{SMe}} \qquad (124)$$

As already mentioned in a previous section, cyanogen bromide effects bromination of 72 (125). The arising bromodiynamines cannot be isolated but they are reacted in situ to mixed diaminodiynes 77 (126) [39].

$$\text{Ph(R)N}-\text{C}\equiv\text{C}-\text{C}\equiv\text{C}-\text{Li} \xrightarrow{\text{BrCN}} [\text{Ph(R)N}-\text{C}\equiv\text{C}-\text{C}\equiv\text{C}-\text{Br}] \qquad (125)$$

$$\text{R} = \text{Me or Ph}$$

$$\text{73} \xrightarrow{\text{R}^1\text{R}^2\text{NH}} \underset{77}{\text{Ph(R)N}-\text{C}\equiv\text{C}-\text{C}\equiv\text{C}-\text{NR}^1\text{R}^2} \qquad (126)$$

R = Me; NR^1R^2 = piperidino 46%
morpholino 40%
R = Ph; piperidino, 54%; morpholino, 56%

Methyleneiminium salt (127) and a number of carbonyl compounds [26, 38] have been reacted with free aminodyne or its salts 72 (127, 128):

$$\text{Me}_2\text{N}-\text{C}\equiv\text{C}-\text{C}\equiv\text{C}-\text{H} \xrightarrow[\text{Cu}^{\oplus}]{\text{Et}_2\overset{\oplus}{\text{N}}=\text{CH}_2} \underset{57\%}{\text{Me}_2\text{N}-\text{C}\equiv\text{C}-\text{C}\equiv\text{C}-\text{CH}_2\text{NEt}_2} \qquad (127)$$

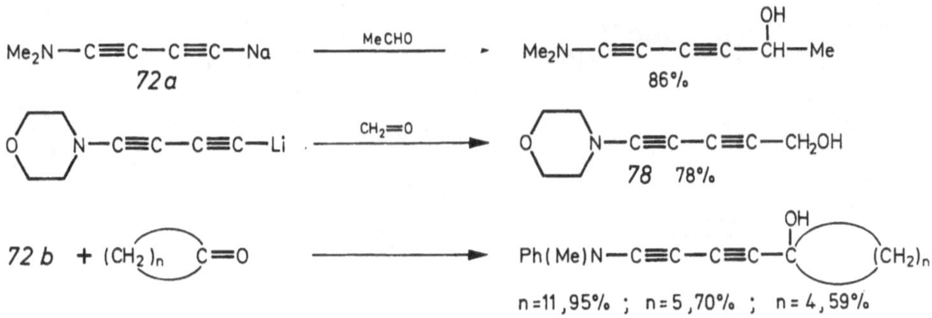

(128)

Ynamine alcohol 78 can be succesfully oxidized using manganese dioxide to give the first push-pull diyne aldehyde 79 (129) [38].

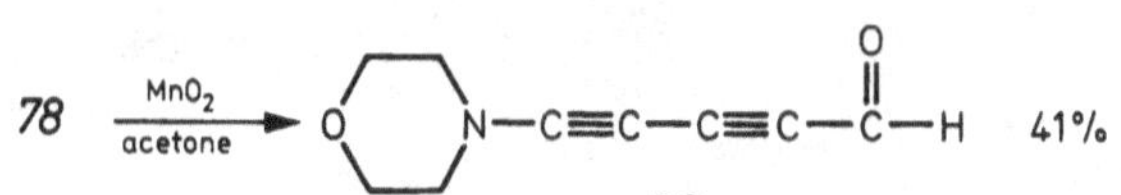

79

(129)

An attempt to effect acylation using acetyl chloride leads only to liberation of aminodiynes but acetylation does take place with acetic anhydride (130) [38].

$$72 \xrightarrow{Ac_2O} R_2N{-}C{\equiv}C{-}C{\equiv}C{-}\underset{\underset{O}{\|}}{C}{-}Me \qquad (130)$$

R = Me, 41 %; morpholino 63 %

Acetylene bonds carrying a diphenylamino group are only moderately reactive, which permits the synthesis of the first ynamine carboxylic acid 80 using carbon dioxide as reagent (131) [37].

$$[Ph_2N{-}C{\equiv}C{-}C{\equiv}C{-}Li] \xrightarrow{CO_2} Ph_2N{-}C{\equiv}C{-}C{\equiv}C{-}CO_2H \qquad (131)$$

80, 52 %

Expectedly, chloroformiate introduces an ester group [34,38] whereas isocyanates, isothiocyanates and carbodiimides lead to the respective push-pull aminodiyne amides, thioamides and amidines (132–135) [34,37].

$$72 \xrightarrow{ClCO_2Me} R_2N{-}C{\equiv}C{-}C{\equiv}C{-}CO_2Me \qquad (132)$$

R_2 = morpholino, 63 %; Et, 51 %; Me, Ph, 40 %

$$72\,b \xrightarrow{t\text{-}BuN=C=O} Ph(Me)N{-}C{\equiv}C{-}C{\equiv}C{-}\overset{\overset{O}{\|}}{C}{-}\underset{H}{N}{-}Bu{-}t \quad 44\,\% \qquad (133)$$

$$\text{72 b} \xrightarrow{C_6H_{11}N=C=S} \text{Ph(Me)N-C}{\equiv}\text{C-C}{\equiv}\text{C-}\overset{\overset{\displaystyle S}{\|}}{C}\text{-}\underset{\displaystyle H}{N}\text{-}C_6H_{11} \quad 46\% \qquad (134)$$

$$\text{72 b} \xrightarrow{(p\text{-MePh-N=})_2C} \text{Ph(Me)N-C}{\equiv}\text{C-C}{\equiv}\text{C-}\overset{\overset{\displaystyle N\text{-PhMe-p}}{\|}}{C}\text{-}\underset{\displaystyle H}{N}\text{-PhMe-p} \quad 30\% \qquad (135)$$

There has been reported a far-reaching extension, namely perchlorobutenyne 1 reacts with both 44 and 72 to give the corresponding trichlorovinyldiynamine 81 and triynamine 82 (136) [36]. Then the subsequent treatment with butyl lithium creates one more acetylide group. The interesting tri- and tetraacetylide intermediates 83, 84 could be converted to esters 85 and 86 (137).

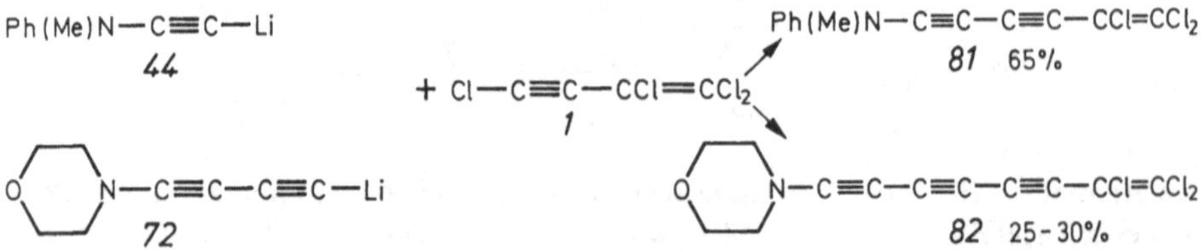

(136)

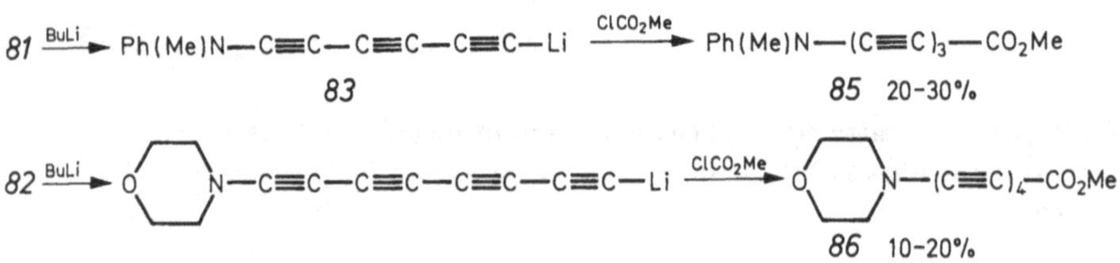

(137)

According to what has already been exposed, it is overly evident that this imaginative approach may have great impact on polyacetylene chemistry.

6 Ynamines by α-Elimination and Onium Rearrangement

An α-elimination of hydrogen halide or halogene from a sp^2 carbon leads formally to a vinyl carbene species which may eventually rearrange to the corresponding acetylene. This reaction was discoverd in 1894 and is called Fritsch-Buttenberg-Wiechell rearrangement [159].

$$RR'C{=}CXY \xrightarrow{-XY} [RR'C{=}Cl] \text{—} R\text{—}C{\equiv}C\text{—}R$$

One or both groups RR′ may be tertiary amino groups. This particular case has been termed Onium Rearrangement and it apparently proceeds via carbenoide species (138) [61].

$$R_2N(Ph)C{=}CClX \xrightarrow[-33^\circ]{NaNH_2,\,NH_3} Ph{-}C{\equiv}C{-}NR_2 \qquad R = Et,\ 70\,\% \quad (138)$$
$$87a\ X = H \qquad\qquad = Me,\ 80\,\%$$

$$(t\text{-}Bu)(Me_2N)C{=}CHCl \xrightarrow[\Delta]{NaH} t\text{-}Bu{-}C{\equiv}C{-}NMe \qquad 20\,\%$$

Organolithium reagents are not suitable as bases since they may lead to β-lithioenamines via metal-halogen exchange especially with bromo and iodo vinylic compounds [84,160]. Chloroolefins e.g. 87a react mostly by the desired hydrogen abstraction since chlorine acidifies the adjacent hydrogen. Alkyl lithium then substitutes the mobile chlorine atom leading to an alkylation of the vinylic moiety rather than to the onium rearrangement (139) [84,85].

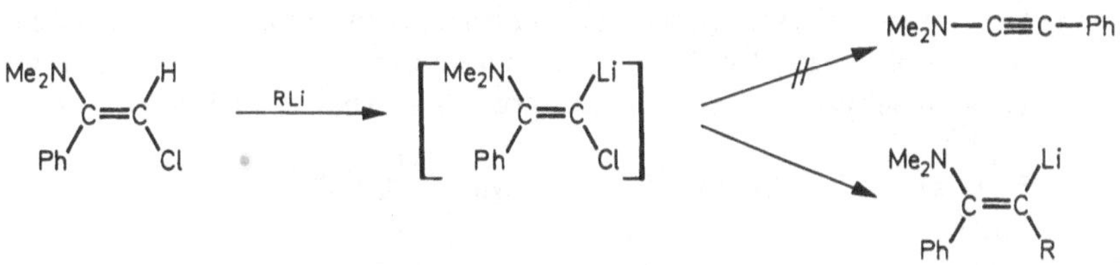

(139)

The synthesis of the requisite compounds 87a may be cumbersome since monochlorination of enamines is not an easy task. By the way of example, the β-chloroenamine 87a, R = Et was synthesized using the following sequence (140) [61,85].

$$Cl{-}C{\equiv}C{-}Cl\ .Et_2O \xrightarrow{HNEt_2} Et_2N{-}CCl{=}CHCl$$
$$\xrightarrow{PhMgBr} (Et_2N)(Ph)C{=}CHCl \qquad (140)$$

Grignard reagents are not enough basic to abstract a vinylic hydrogen, nor do they participate in halogen metal exchange. As shown, they instead substitute the reactive chlorine atom in α-chloroenamines. More recent work has shown that 87a, R = Me can be obtained from ω-chloroacetophenone and tris-dimethylamino) arsine in 86% yield [161].

β,β-Dihaloenamines 87b (X = Cl or Br) may also be anticipated to give a carbenoide species upon halogen metal exchange with organolithium reagents but dihaloenamine 87b, X = Cl; R = Et gives only 10% yield of the ynamine upon treatment with phenyl lithium and this is the only example known in the literature [162]. Moreover there are plenty of methods for the preparation of β-phenyl ynamines. Therefore, the use of onium rearrangement would be severely limited if it were not for the fact that it opens the access to bis-(dimethylamino) acetylenes (ynediamines) as a highly interesting class of electron-rich compounds (141) [6,61].

$$FCH{=}CCl_2 \xrightarrow{LiNR_2} FCLi{=}CCl_2 \xrightarrow[-LiCl]{} F{-}C{\equiv}C{-}Cl \xrightarrow{LiNR_2} \tag{141}$$

$$Cl{-}C{\equiv}C{-}NR_2 \xrightarrow{HNR_2} ClCH{=}C(NR_2)_2 \xrightarrow{LiNR_2} \underset{88}{R_2N{-}C{\equiv}C{-}NR_2}$$

The reaction sequence (141) is quite complicated but fluorodichloroethylene affords smoothly the corresponding ynediamines 88 when treated with three equivalents of lithium amide (142) [6,60,61].

$$FCH{=}CCl_2 \xrightarrow[3\,hr\ 20^\circ]{3\ eq.\ LiNEt_2} \underset{88\,a}{Et_2N{-}C{\equiv}C{-}NEt_2} \tag{142}$$

All intermediates shown in the sequence (141) have been isolated or identified including the explosive fluorochloroacetylene [60]. Γhis compound reacts with lithium methylanilide by *fluorine* substitution to give the unstable chloroynamine 89 in 20% yield.

$$F{-}C{\equiv}C{-}Cl + LiN(Me)Ph \rightarrow \underset{89}{Ph(Me)N{-}C{\equiv}C{-}Cl}$$

The chlorine atom in chloroynamines which are formed in situ cannot be displaced with secondary amines but an addition takes place instead, leading to β-chloroketeneaminals 91. The latter compounds undergo onium rearrangement to ynediamines on exposure to strong bases e.g. sodium amide, phenyllithium or lithium diethylamide. Cheap trichlorethylene is an attractive alternative and this method has been widely used [61,163]. Although it is possible to use isolated dichloroacetylene which is very toxic and explosive it is much more practical to generate it in situ using one equivalent of lithium amide in the presence of excess secondary amine. Dichloroacetylene then adds avidly one mole of amine present to give dichloroenamines 90. But the next step i.e. the chlorine substitution in 90 by excess amine is sluggish and the whole mixture must be heated in an autoclave or in a sealed tube to 100° for 8–15 hrs (143) [61].

$$Cl_2C{=}CHCl \xrightarrow{LiNR_2} Cl{-}C{\equiv}C{-}Cl \xrightarrow{HNR_2} \underset{90}{ClCH{=}CClNR_2} \tag{143}$$

$$90 \xrightarrow{HNR_2,\,100^\circ} \underset{91}{ClCH{=}C(NR_2)_2} \xrightarrow[or\ RLi]{NaNH_2} \underset{88}{R_2N{-}C{\equiv}C{-}NR_2}$$

This lack of reactivity can be exploited to form mixed chloroketeneaminals and subsequently ynediamines (144) [61].

$$\mathrm{Cl{-}C{\equiv}C{-}Cl + HNMe_2 \xrightarrow{0^\circ} \underset{90\,a}{ClCH{=}CClNMe_2} \xrightarrow{HNPr_2,\ 100^\circ}} \qquad (144)$$

$$\mathrm{ClCH{=}C\begin{matrix}NMe_2\\NPr_2\end{matrix} \xrightarrow[-35^\circ,\ 15\,h]{NaNH_2} \underset{89\ \%}{Me_2N{-}C{\equiv}C{-}NPr_2}}$$

Chloroketeneaminals 91 which are much more stable than α,β-dichloroenamines 90 are usually isolated and purified by distillation.

Difficulties arise when the most simple ynediamine bis(dimethylamino)acetylene is prepared starting with trichloroethylene since dimethylamine may react further giving tris(dimethylamino)ethylene [163]. A simple one-pot procedure for this ynamine has been elaborated in which trichloroethylene is treated with sodium amide and dimethylamine under normal pressure. The average yield is about 60% [163]. Dichloroacetylene ether complex can now be prepared under phase-transfer (PT) conditions from trichloroethylene. It reacts with dimethylamine under PT conditions to N,N,N'N'-tetramethylglycinamide apparently via the ynediamine intermediate [164].

Thioynamines are discussed as being formed via onium rearrangement from aryl and alkylthio dichlororethylenes [44], but alternatively chlorothioalkylacetylenes may be formed in a simple 1,2 elimination process. The ensuing chlorine substitution by lithium amides has been documented [43].

7 Miscellaneous Methods

The by now well-established flash vacuum pyrolysis methods (FVP) can also be applied in the field of ynamine synthesis. Thus norbornenyl ynamine 92 fragments at 500° to vinyl ynamine 93 in almost quantitative yield (145) [165]. FVP of diamino cyclopropenones gives ynediamines in very good yields (146) [166].

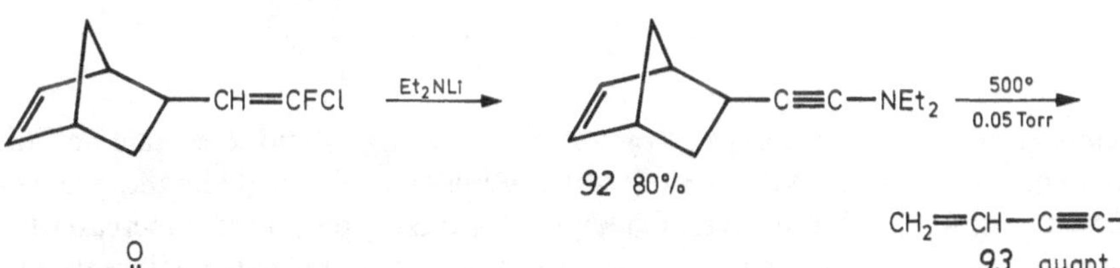

(145)

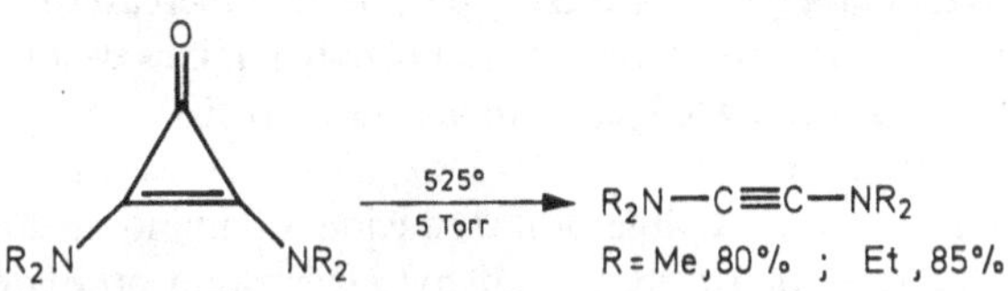

(146)

Perfluoromethyl ynediamine 95 can be obtained, among others, via FVP of the 1,2,4-triazine 94 (147) [167]. Pyrolysis of aminomethyleneisoxazolones 96 leads nonambigously to vinylidene carbenes 97 which rearrange to ynamines (148) [168]. Unfortunately, these FVP have not been run preparatively, but the process provides corroborating evidence for the onium rearrangement.

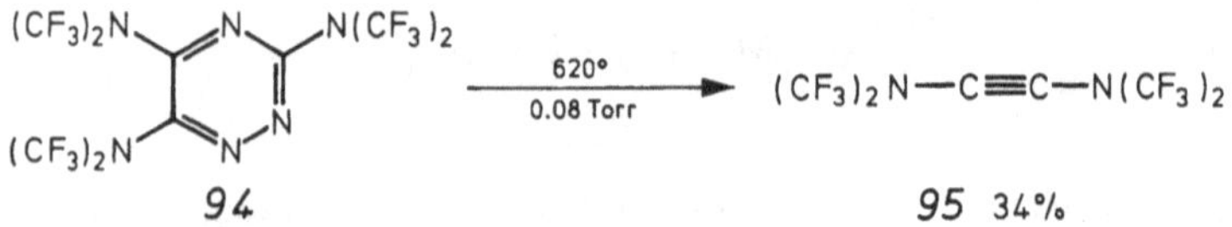

(147)

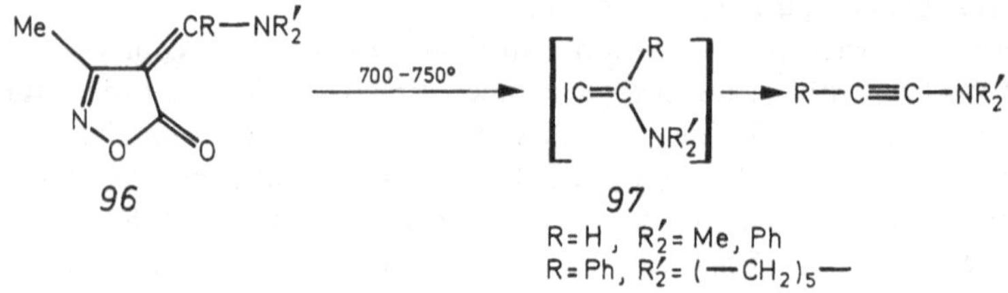

(148)

Diynediamines can enter into various cycloadditions leaving one ynamine moiety intact. The overall process amounts in fact, to ynamine interconversion as illustrated by the following examples (149) [39].

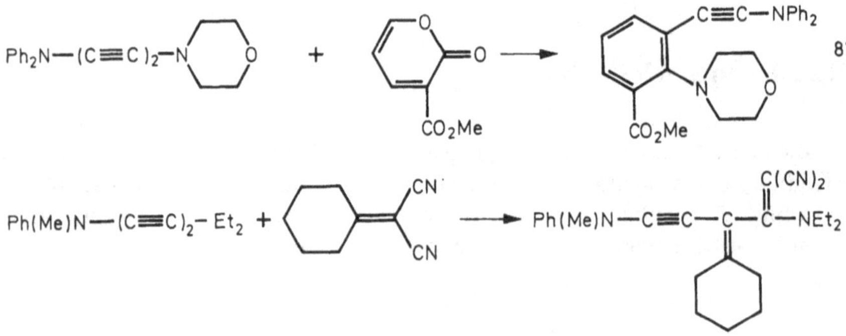

(149)

Oxidation of phenylacetylene with oxygen in the presence of cupric acetate and dimethylamine is claimed to give a mixture constituting of 58 % dimethylamino phenylacetylene and of 42 % 1,4 diphenylbutanediyne. Hydrazine purportedly increases the ynamine/diyne ratio [169]. The patent claims isolation of pure phenylynamines despite aqueous work-up. Unfortunately, to our knowledge, there are no reported attempts to duplicate this potentially useful approach.

Amines carrying good leaving groups are capable of introducing an amino group onto carbanionic centers. Thus, the reaction of phenyl ethynyl magnesium bromide with chlorodiethylamine has the merit of being the first successful ynamine synthesis. This original version can be, of course, hardly recommended because the yield does not reach even 2 % [51]. The recent state-of-art permits much more efficient ynamine synthesis, but more sophisticated reagents are to be employed. Three different procedures are used to generate acetylenic cuprates 98 which can be aminated with N,N-dimethylhydroxylamine derivatives 99 or 100 (150, 151)[170].

$$\underset{99}{Ph_2P(=O){-}O{-}NMe_2} \qquad \underset{100}{MeSO_2{-}O{-}NMe_2}$$

$$\underset{98,\ R=Ph}{(R{-}C{\equiv}C)_3CuLi_2} \xrightarrow[100]{99\ \text{or}} \underset{26\text{–}83\,\%}{Ph{-}C{\equiv}C{-}NMe_2} \tag{150}$$

$$98,\ R=n\text{-}Bu \xrightarrow{99} \underset{78\,\%,\ \text{isolated}}{n\text{-}Bu{-}C{\equiv}C{-}NMe_2} \tag{151}$$

Groups R studied were various alkyl groups, phenyl, trimethylsilyl and phenylthio groups. The yields depend strongly upon the method by which 98 is generated and on the reaction conditions in general. Only one detailed experimental procedure (151) is available and certain yields were assayed by the NMR or by hydrolysis[170]. It can be concluded that this approach is a valuable extension of electrophilic amination but thus far it is of only limited value for the preparation of ynamines.

Acetylenes can be prepared by intramolecular Wittig reaction but at least one of the groups R, R′ must be an electron-withdrawing group-aryl, ketone or ester function (152)[171]. Thus understably this approach could not be applied to the synthesis of ynamines starting with 101 (153).

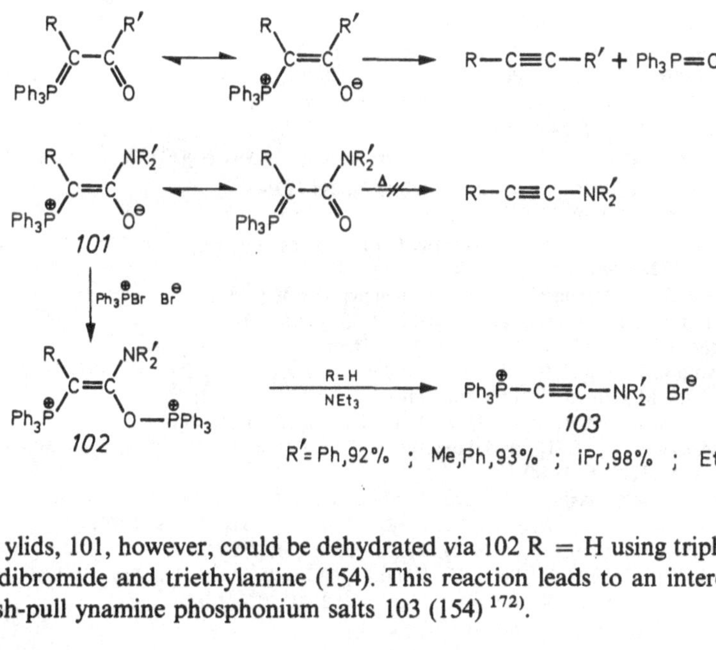

(152)

(153)

(154)

Carbamoyl ylids, 101, however, could be dehydrated via 102 R = H using triphenylphosphine dibromide and triethylamine (154). This reaction leads to an interesting class of push-pull ynamine phosphonium salts 103 (154)[172].

8 References

1. Miller, S. I., Dickstein, J. I.: Acc. Chem. Res. *9*, 358 (1976)
2. Jäger, V. in: Methoden der Organischen Chemie, (Houben Weyl), 4th Ed., Vol. V/2a, p. 622, Stuttgart, Georg Thieme Verlag 1977
3. Ziegler, G. R., Welch, C. A., Orzech, C. E., Kikkawa, S., Miller, S. I.: J. Am. Chem. Soc. *85*, 1648 (1963)
4. Dickstein, J. I., Miller, S. I. in: The Chemistry of Carbon-Carbon Triple Bond, S. Patai ed., p. 813, Wiley, J. 1978
5. Viehe, H. G.: Angew. Chem. *75*, 638 (1963)
6. Viehe, H. G., Reinstein, M.: Angew. Chem. *76*, 537 (1964); Angew. Chem. Intern. Ed. Engl. *3*, 506 (1964)
7. Strobach, D. R.: J. Org. Chem. *36*, 1438 (1971)
8. Houbion, J.: Dissertation, Université de Louvain 1973
9. Sauvêtre, R., Normant, J. F.: Tetrahedron Lett. *23*, 4325 (1982)
10. Rappoport, Z.: Adv. Phys. Org. Chem. *7*, 1 (1969)
11. Modena, G.: Acc. Chem. Res. *4*, 73 (1971)
12. Wheaton, G. A., Burton, D. J.: J. org. Chem. *48*, 917 (1983)
13. Hayashi, S., Nakai, T., Ishikawa, N.: Chem. Lett. *1980*, 935
14. Delavarenne, S. Y., Viehe, H. G.: Chemistry of Acetylenes M. Dekker, N.Y. 1969
15. Brandsma, L.: Preparative Acetylenic Chemistry, Elsevier 1971
16. Brandsma, L., Verkruijse, M. D.: Synthesis of Acetylenes, Allenes and Cumulenes, Elsevier 1981
17. Reinhoudt, D. N., Kouwenhoven, C. G.: Recl. Trav. Chim. Pays-Bas *95*, 67 (1976)
18. Moore, J. A., Kennedy, J. F.: J. Chem. Soc. Chem. Commun. *1978*, 1079
19. Viehe, H. G., Miller, S. I., Dickstein, J. I.: Angew. Chem. *76*, 537 (1964); Angew. Chem. Internat. Ed. Engl. *3*, 582 (1964)
20. Fuks, R., Viehe, H. G.: Chem. Ber. *103*, 564 (1970)
21. Harmon, R. E., Zenarosa, C. V., Gupta, S. K.: J. Org. Chem. *35*, 1936 (1970)
22. Eish, J. J., Gopal, H., Rhee, S. G.: J. Org. Chem. *40*, 2064 (1975)
23. Tanaka, R., Miller, S. I.: J. Org. Chem. *36*, 3856 (1971)
24. Dumont, J. L., Chodkiewicz, W., Cadiot, P.: Bull. Soc. Chim. France *1967*, 1197
25. Dumont, J. L., C. R. Acad. Sci. (Paris) *261*, 1710 (1965)
26. Gusev, B. P., Tsurgoven, L. A., Kucherov, V. F.: Izv. Akad. Nauk SSSR, Ser. Khim. *5*, 1098 (1972); C.A. *77*, 125815 (1972)
27. Sasaki, T., Kojima, A.: J. Chem. Soc. C *1970*, 476
28. Jonin, B. I., Petrov, A. A.: J. Gen. Chem. USSR *35*, 2247 (1965): C.A. *64*, 11240a (1965)
29. Garibina, V. A., Dogadina, A. V., Ionin, B. I., Petrov, A. A.: Zh. Obshch. Khim. *49*, 2385 (1979)
30. Viehe, H. G.: ref. 14 p. 861
31. Maretina, I. A., Petrov, A. A.: Zh. Org. Khim. *2*, 1994 (1966); Engl. Transl. *2*, 1912 (1966)
32. De Meyere, A.: Bull. Soc. Chim. Belges *93*, 241 (1984)
33. Roedig, A., Fouré, M.: Chem. Ber. *109*, 2159 (1976)
34. Himbert, G., Feustel, M.: Angew. Chem. Intern. Ed. Engl. *21*, 282 (1982); Angew. Chem. Suppl. 722 (1982)
35. Stämpfli, U., Neuenschwander, M.: Chimia *35*, 336 (1981)
36. Stämpfli, U., Neuenschwander, M.: ibid. *38*, 157 (1984)
37. Feustel, M., Himbert, G.: Liebigs Ann. Chem. *1984*, 586
38. Stämpfli, U., Galli, R., Neuenschwander, M.: Helv. Chem. Acta *66*, 1631 (1983)
39. Feustel, M., Himbert, G.: Tetrahedron Lett. *24*, 2165 (1983)
40. Cuvigny, T., Normant, H.: C. R. Acad. Sci. (Paris) *268*, 834 (1969)
41. Shchukovskaya, L. L., Budakova, L. D., Palchick, R. I.: Zh. Obshch. Khim. *43*, 1989 (1973); C. A. *80*, 15001b (1974)
42. Sato, Y., Kobayashi, Y., Sigura, M., Shirai, H.: J. Org. Chem. *43*, 199 (1978)
43. Verboom, W., Bos, H. J. T.: Recl. Trav. Chim. Pays-Bas *98*, 559 (1979)
44. Delavarenne, S. Y., Viehe, H. G.: Tetrahedron Lett. *1969*, 4761
45. Nakai, T., Tanaka, K., Setoi, H., Ishikawa, N.: Bull. Chem. Soc. Jpn. *50*, 3069 (1977)
46. Nakai, T., Tanaka, K., Ishikawa, N.: Chem. Lett. *1976*, 1263

47. Piettre, S., Janousek. Z., Viehe, H. G.: Synthesis *1982*, 1083
48. Vereshchagin, L. I., Gavrilov, L. D., Bolshedrorskaya, R. L., Kirillova, L. P., Okhapkina, L. L.: Zh. Org. Khim. *9*, 506 (1973); Engl. Transl. *9*, 516 (1973)
49. Schroth, W., Spitzner, R., Huggo, S.: Synthesis *1982*, 199
50. Aoyama, H., Hasegawa, T., Nishio, T., Omok, Y.: Bull. Soc. Chem. Jpn. *48*, 1671 (1975)
51. Wolf, V., Kowitz, F.: Liebigs Ann. Chem. *638*, 33 (1960)
52. Brandsma, L., Bos, H. Y. T., Arens, J. F.: ref. 14, p. 751
53. Montijn, P. P., Harryvan, E., Brandsma, L.: Recl. Trav. Chim. Pays-Bas, *83*, 1211 (1964)
54. Kuehne, M. E., Sheeran, P. J.: J. Org. Chem. *33*, 4406 (1968)
55. Pitacco, G., Valentin, E. in The chemistry of amino, nitroso and nitro compounds, S. Patai Editor, p. 623, Wiley 1982
56. ref. 2, p. 646
57. ref. 15, p. 85
58. Ponomarev, S. V., Zakharova, O. A., Lebedev, S. A., Lutsenkov, I. F.: Zh. Obshch. Khim. *45*, 2680 (1975); Engl. Transl. *45*, 2655 (1975)
59. Ficini, J., Dureault, A.: C. R. Acad. Sci. (Paris) *273*, 289 (1971)
60. Delavarenne, S. Y., Viehe, H. G.: Chem. Ber. *103*, 1198 (1970)
61. Delavarenne, S. Y., Viehe, H. G.: ibid. *103*, 1209 (1970)
62. ref. 2, p. 214
63. Halleux, A., Reimlinger, H., Viehe, H. G.: Tetrahedron Lett. *1970*, 3141
64. ref. 2, p. 172
65. ref. 15, p. 105
66. Buyle, R., Halleux, A., Viehe, H. G.: Angew. Chem. *78*, 593 (1966); Angew. Chem. Internat. Ed. Engl. *11*, 584 (1966)
67. Brandsma, L.: Recl. Trav. Chim. Pays-Bas, *111*, 265 (1971)
68. Eicher, T., Urban, M.: Chem. Ber. *113*, 408 (1980)
69. Kantlehner, W. in: Iminium Salts in Organic Chemistry Vol. 9, Part 2 p. 65; J. Wiley 1979
70. Ghosez, L., Haveaux, B., Viehe, H. G.: Angew. Chem. *81*, 468 (1969); Angew. Chem. Internat. Ed. Engl. *8*, 454 (1969)
71. Ghosez, L., Marchand-Brynaert, J.: in ref. 69, p. 421
72. Fuks, R., Viehe, H. G.: Bull. Soc. Chim. Belge *86*, 219 (1977)
73. Buyle, R., Viehe, H. G., Tetrahedron *1968*, 3987
74. Viehe, H. G., Buyle, R., Fuks, R., Merényi, R., Oth, J. M. F.: Angew. Chem. *79*, 53 (1967); Angew. Chem. Internat. Ed. Engl. *6*, 77 (1967)
75. Weiss, R., Wolf, H., Schubert, U., Clark, T.: J. Am. Chem. Soc. *103*, 6142 (1981)
76. Viehe, H. G., Janousek, Z.: Angew. Chem. Internat; Ed. Engl. *10*, 806 (1983)
77. Mee, J. D.: J. Am. Chem. Soc. *96*, 4712 (1974)
78. Caillaux, B., George, P., Tataruch, F., Janousek, Z., Viehe, H. G.: Chimia *30*, 387 (1976)
79. Viehe, H. G., Janousek, Z., Gompper, R., Lach, D.: Angew. Chem. *85*, 581 (1973); Angew. Chem. Internat. Ed. Engl. *12*, 566 (1973)
80. Viehe, H. G., de Voghel, G. J., Smets, F.: Chimia *30*, 189 (1976)
81. Janousek, Z., Collard, J., Viehe, H. G.: Angew. Chem. Internat. Ed. Engl. *11*, 917 (1972)
82. Goffin, E., Legrand, Y., Viehe, H. G.: J. Chem. Res. (S) 105 (1977)
83. Union Carbide Corp. Neth. Appl. 6,504,567; C. A.: *64*, 8032 (1966)
84. Duhamel, L., Poirier, J. M.: Bull. Soc. Chim. France, II-297 (1982)
85. For a review see: de Kimpe, N., Schamp, N.: Org. Prep. Proced. Int. *13*, 241 (1981); ibid. *15*, 71 (1983)
86. Halleux, A., Viehe, H. G.: J. Chem. Soc. C *1968*, 1726
87. Takamatsu, M., Sekiya, M.: Chem. Pharm. Bull. *28*, 3098 (1980)
88. Duhamel, L., Duhamel, P., Plé, G.: Tetrahedron Lett. *1972*, 85
89. ref. 2, p. 119
90. Lienhard, U., Fahrni, H. P., Neuenschwander, M.: Helv. Chim. Acta *61*, 1609 (1978)
91. Gais, H. J., Hafner, K., Neuenschwander, M.: ibid. *52*, 2641 (1969)
Gais, H.-J., Lied, T.: Angew. Chem. *20*, 283 (1978); Angew. Chem. Int. Ed., Engl. *17*, 267 (1978)
Gais, H.-J.: Angew. Chem. *90*, 625 (1983); Angew. Chem. Int. Ed. Engl. *17*, 597 (1983)
Gais, H.-J.: Tetrahedron Lett. *1984*, 273

92. Neuenschwander, M., Lienhard, U., Farhni, H. P., Hurni, B.: Chimia *32*, 212 (1978)
93. Neuenschwander, M., Farhni, H. P., Lienhard, U.: ibid. *32*, 214 (1978)
94. Neuenschwander, M., Farhni, H. P., Lienhard, U.: Helv. Chim. Acta *61*, 2437 (1978)
95. Himbert, G., Feustel, M., Jung, M.: Liebigs Ann. Chem. *1981*, 1907
96. Stämpfli, U., Neuenschwander, M.: Helv. Chim. Acta *66*, 1427 (1983)
97. Buyle, R., Viehe, H. G., Tetrahedron *25*, 3447 (1969)
98. Speziale, A. J., Freeman, R. C.: Org. Synth. Coll. Vol. V, p. 387, J. Wiley, New York 1973
99. Himbert, G., Regitz, M.: Chem. Ber. *105*, 2963 (1972)
100. Speziale, A. J., Smith, L. R.: J. Org. Chem. *27*, 4361 (1962)
101. Speziale, A. J., Smith, L. R.: J. Am. Chem. Soc. *84*, 1868 (1962)
102. Partos, R. D., Speziale, A. J.: ibid. *87*, 5068 (1965)
103. Speziale, A. J., Taylor, R. J.: J. Org. Chem. *31*, 2450 (1966)
104. Himbert, G., Regitz, M.: Liebigs Ann. Chem. *1973*, 1505
105. Ficini, J., Barbara, C.: Bull. Soc. Chim. France *1964*, 871
106. Ficini, J., Barbara, C.: ibid. *1965*, 2787
107. Ficini, J., Barbara, C., Colodny, S., Dureault, A.: Tetrahedron Lett. *1968*, 943
108. ref. 2, p. 286
109. Zaugg, H. E., Swett, L. R., Stone, G. R.: J. Org. Chem. *23*, 1389 (1958)
110. Mahamoud, A., Galy, J. P., Vincent, E. J.: Synthesis *1981*, 917
111. ref. 15, p. 143
112. Hubert, A. J., Viehe, H. G.: J. Chem. Soc. C *1968*, 228
113. Hubert, A. J., Reimlinger, H.: ibid. *1968*, 606
114. Craig, J. C., Ekwuribe, N. N.: Tetrahdron Lett. *21*, 2587 (1980)
115. Verkrijsse, H. D., Bos, H. J. T., de Noten, L. J., Brandsma, L.: Recl. Trav. Chim. Pays-Bas *100*, 244 (1981)
116. Viehe, H. G., Hubert, A. J.: U.S. Patent 3,439,038 (1969) Union Carbide Corp.; C. A. *71* 13131 (1969)
117. Corey, E. J., Cane, D. E.: J. Org. Chem. *35*, 3405 (1970)
118. Klop, W., Klusener, P. A. A., Brandsma, L.: Recl. Trav. Chim. Pays-Bas *103*, 27 (1984)
119. Verboom, W., Everhardus, R. H., Bos, H. J. T., Brandsma, L.: ibid. *78*, 508 (1979)
120. Gallesloot, W. G., De Bie, M. J. A., Brandsma, L., Arens, J. F.: ibid. *89*, 575 (1970)
121. Tolchinskii, S. E., Dogadina, A. V., Maretina, I. A., Petrov, A. A.: Zh. Org. Khim. *15*, 1824 (1979)
122. Tolchinskii, S. E., Maretina, I. A., Petrov, A. A.: Zh. Org. Khim. *16*, 1149 (1980)
123. Verboom, W., Everhardus, R. H., Zwikker, J. W., Brandsma, L.: Recl. Trav. Chim. Pays-Bas *99*, 325 (1980)
124. Brandsma, L., van Rijn, P. E., Verkruijsse, H. D., von R. Schleyer, P.: Angew. Chem. *94*, 875 (1982); Angew. Chem. Internat. Ed. Engl. *21*, 862 (1982)
125. van Rijn, P. E., Klop, W., Verkruijsse, H. D., von R. Schleyer, P., Brandsma, L.: J. Chem. Soc. Chem. Commun. *1983*, 79
126. Pennanen, S. I.: Synthetic Commun. *10*, 373 (1980)
127. Ficini, J., Barbara, C., d'Angelo, J., Dureault, A.: Bull. Soc. Chim. France *1974*, 1535
128. Cadiot, P., Chodkiewitz, W.: in ref. 14, p. 597
129. ref. 2, p. 925
130. Ficini, J., Duréault, A., d'Angelo, J., Barbara, C.: Bull. Soc. Chim. France *1974*, 1528
131. Himbert, G., Henn, L., Hoge, R.: J. Organomet. Chem. *184*, 317 (1980)
132. Ficini, J., Falou, S., d'Angelo, J.: Tetrahedron Lett. *1977*, 1931
133. Ficini, J. Guingant, A., d'Angelo, J., Stork, G.: ibid. *24*, 907 (1983)
134. Ficini, J., Berlan, J., Schmidt, F., d'Angelo, J., Guingant, A.: ibid. *23*, 1821 (1982)
135. Genet, J. P., Kahn, Ph.: ibid. *21*, 1521 (1980)
136. Sato, Y., Kato, M., Aoyama, T., Sugiura, M., Shirai, H.: J. Org. Chem. *43*, 2466 (1978)
137. Himbert, G.: J. Chem. Soc. S *1978*, 104; (M) 1445
138. Himbert, G., Schwickenrath, W.: Liebigs Ann. Chem. *1981*, 1844
139. Feustel, M., Himbert, G.: ibid. *1982*, 196
140. Himbert, G., Frank, D., Regitz, M.: Chem. Ber. *109*, 370 (1976)
141. Himbert, G.: Angew. Chem. *91*, 432 (1979); Angew. Chem. Internat. Ed. Engl. *18*, 405 (1979)

142. Berger, H. O., Noth, H., Wrackmeyer, B.: J. Organomet. Chem. *145*, 17 (1978)
143. Himbert, G., Regitz, M.: Chem. Ber. *107*, 2513 (1974)
144. Schaumann, E., Lindstaedt, J., Forster, W. R.: ibid. *116*, 509 (1983)
145. Kuehne, M. E., Linde, H.: J. Org. Chem. *37*, 1846 (1972)
146. Ficini, J., Barbara, C., Duréault, A.: C. R. Acad. Sci. C *265*, 1496 (1967)
147. Himbert, G., Schwickerath, W.: Liebigs Ann. Chem. *1983*, 1185
148. Himbert, G., Schwickerath, W.: ibid. *1984*, 85
149. Himbert, G.: J. Chem. Res. S *1978*, 442; M 5240
150. Himbert, G.: Angew. Chem. *88*, 59 (1976); Angew. Chem. Internat. Ed. Engl. *15*, 51 (1976)
151. Himbert, G.: Liebigs Ann. Chem. *1979*, 829
152. Himbert, G., Henn, L.: Z. Naturforsch. *36b*, 218 (1981)
153. Himbert, G., Schwickerath, W.: Tetrahedron Lett. *1978*, 1951
154. Himbert, G.: J. Chem. Res. S 88; M 1201 (1979)
155. Henn, L., Himbert, G.: Chem. Ber. *114*, 1015 (1981)
156. Himbert, G.: Liebigs Ann. Chem. *1979*, 1828
157. Himbert, G., Henn, L.: Liebigs Ann. Chem. *1984*, 1358
158. Himbert, G., Umbach, H., Barz, M.: Z. Naturforsch. *39b*, 661 (1984)
159. ref. 2, p. 214
160. Cahiez, G., Bernard, D., Normant, J. F.: Synthesis *1976*, 245
161. Duhamel, L., Poirier, J. M.: J. Org. Chem. *44*, 3585 (1979)
162. unpublished results quoted in: ref. 2, p. 217
163. René, L., Janousek, Z., Viehe, H. G.: Synthesis *1982*, 645
164. Pielishowski, J., Popielarz, R.: Synthesis *1984*, 433
165. Ficini, J., Berlan, J., d'Angelo, J.: Tetrahedron Lett. *21*, 3055 (1980)
166. Wilcox, C., Breslow, R.: Tetrahedron Lett. *21*, 3241 (1980)
167. Barlow, M. G., Haszeldine, R. N., Simon, Ch.: J. Chem. Soc. Perkin I *1980*, 2254
168. Winter, H.-W., Wentrup, C.: Angew. Chem. *92*, 743 (1980)
169. Peterson, L. I.: U.S. 3.657.342; C. A. *77*, 48068v (1972) and C. A. *77*, 34173p (1972)
170. Boche, G., Bernheim, M., Niessner, M.: Angew. Chem. *95*, 48 (1983); Angew. Chem. Internat. Ed. Engl. *22*, 53 (1983)
171. ref. 2, p. 185
172. Bestmann, H. J., Kisielowski, L.: Chem. Ber. *116*, 1320 (1983)

Electrochemically Reduced Photoreversible Products of Pyrimidine and Purine Analogues

Barbara Czochralska[1], Monika Wrona[1], and David Shugar[1,2]

1 Department of Biophysics, Institute of Experimental Physics, University of Warsaw, 02-089 Warszawa, Poland

2 Institute of Biochemistry & Biophysics, Academy of Sciences, 02-532 Warszawa, Poland

Table of Contents

I Introduction

We present here a brief account of the specific dimerization, and other related, reactions undergone by a variety of purine and pyrimidine derivatives, and a number of related compounds, during the course of their electrochemical reduction at the surface of a mercury electrode. A characteristic feature of these reactions is the transfer of an electron to the compound, accompanied, or preceded, by its protonation. The resultant free radicals, generated by a one-electron reduction process, rapidly dimerize to products in which each of the monomeric components possesses an additional electron and an additional proton, relative to the parent monomer [1-7]. (See Scheme 1)

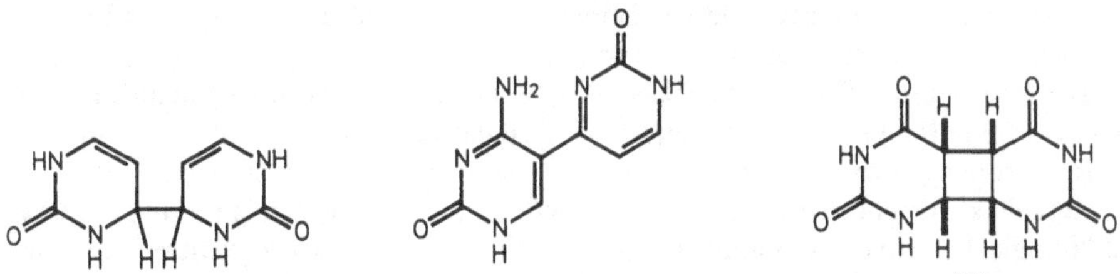

I II III

Scheme 1. Structures of pyrimidine dimers: (I) electrochemically generated hydrodimer of pyrimidone-2; (II) photochemically generated pyrimidine adducts; (III) photochemically generated cyclobutane dimers

Formally the electrochemically generated dimers are somewhat analogous to several purine and pyrimidine photoadducts obtained from DNA irradiated with ultraviolet [9,10] or ionizing [11] radiations, in that the monomers are linked by a single, C—C, bond. The electrochemically produced dimers differ, however, from the photoadducts, in that their monomeric constituents are in the reduced form, whereas those of the photoadducts are unaltered with respect to their redox properties. Furthermore, the electrochemically generated dimer products readily undergo photodissociation to quantitatively regenerate the parent monomers in relatively high quantum yields [2,6,7], a property previously considered specific for 2,4-diketopyrimidine cyclobutane photodimers [12]. It will be shown below that the photochemical transformation of the electrochemical dimer reduction products to the parent monomers is a photo-oxidation process accompanied by liberation of a proton.

The processes of electroreduction, and subsequent photochemical regeneration of the parent monomers, consequently form a closed cycle involving the transport of electrons and protons, as occurs in the reduction and oxidation of other systems, e.g. coenzymes such as NAD^+ and $NADP^+$ [13,14].

II Electroanalytical Methods in Biological Systems

II.1 General

Polarographic methods enable one to investigate biological compounds and systems under conditions approximating those in vivo, i.e. in aqueous electrolytes at defined pH values and temperatures. Most biological reactions proceed at interphases. Hence, bearing in mind the heterogeneity of electrochemical reactions at high potential gradients, the systems studied under polarographic conditions may be considered to approximate to those occurring at, or in the proximity of, cell membranes [15].

The foregoing analogy between in vivo and polarographic conditions has led to the postulate that studies on the steps involved in the electrochemical reactions of biologically active compounds may be extrapolated to some conclusions regarding reduction mechanisms in vivo. This derives support from photochemical studies where, in a number of instances, the mechanisms of photochemical transformations of biologically active compounds are strikingly similar to those of redox processes observed polarographically [16].

Furthermore, a series of recent investigations on the electrochemical and enzymatic oxidation of several purines [17,18] have demonstrated that both processes proceed, in a chemical sense, by virtually identical mechanisms. These studies provide a good illustration of how modern electrochemistry can provide a useful insight into the frequently complex mechanisms of enzymatic reactions.

Attempts at more rigorous correlations between processes occurring at an electrode and in vivo redox processes are somewhat speculative. Biological redox systems, e.g. those involving enzymes, are frequently more complex than electrode reactions. Nonetheless, current investigations on the mechanisms of electrode redox properties of biologically active compounds are of some help towards the clarification of in vivo processes, by delineation of the active groups involved in redox reactions, by studies on the mechanisms of electron transport, the influence of substituents, etc.

As summarized by Elving [15], there are several factors which support the use of an analogy between the electrochemical properties of biologically active substances (e.g. enzymes) and their biological and other properties. They are the following:

a) For both types of processes the electron transport reactions are heterogeneous in nature.
b) Stereochemical factors are of key importance in both processes.
c) Both processes may involve adsorption, often accompanied by the formation of adducts.
d) Both processes proceed in dilute solution, at comparable pH, ionic strength and temperature.

Generally, polarographic methods may also provide information regarding the kinetics and energetics of processes involving electron transport, interactions between compounds in solution, and the redox reactivities of specific functional groups. Occasionally polarographic procedures may also be utilized for the synthesis of biologically active compounds.

II.2 Electrochemistry of Nucleic Acids

Because of the complex behaviour to be expected for natural nucleic acids, it is only natural that considerable effort has been devoted to studies of the electrochemical properties of their monomeric units, and defined analogues of these, as well as of synthetic oligo- and polynucleotides. A variety of techniques has been applied for this purpose, and some of the details and findings are covered in several reviews [19–24]. Most investigations have dealt with electroreduction processes [15,20,24,25]. Only relatively recently has attention been directed to possible electrooxidation of nucleic acids and their constituents with the aid of the graphite electrode which, in comparison with the mercury electrode, possesses a much greater accessible range of positive potentials [26–29].

Most studies have hitherto concentrated on the electrochemical behaviour of DNA, including the mechanism of interfacial opening of the DNA helix [19,22,23,30] and the nature of the lesions induced in DNA by ionizing radiations [19,27,31]. It would now be useful to extend investigations to natural and synthetic RNA, profiting from the current availability of suitable models and the accompanying structural information about these forthcoming from X-ray diffraction, CD and NMR spectroscopy, etc.

Of the five common heterocyclic bases found in natural RNA and DNA, only adenine and cytosine are polarographically reducible in aqueous medium [19-23,32-38]. Palecek et al. [39-41] have reported that guanine, and its nucleosides and nucleotides, may also undergo electroreduction under well-defined conditions, viz. in acid and neutral media, where they exhibit an "anodic indentation" on oscillopolarographic plots of dE/dt vs f(E). The anodic signal probably originates from the oxidation of the product(s) of reduction of the guanine derivatives formed at, or beyond, background discharge potentials [39,40]. The results of Janik [41,42] on various guanine analogues suggest that it is the N(7)=C(8) bond of the imidazole ring which is reduced. This is the only reported example of electrochemical reduction of the imidazole ring; imidazole itself has been found to be polarographically inert [15].

By contrast, 7-methylguanosine readily undergoes electrochemical reduction, both in the free state and when incorporated into DNA [24,25]. Although the reduction product(s) of 7-methylguanosine have not been unequivocally identified, it has been proposed that the mechanism involves addition of two electrons and a proton, leading to saturation of the C(8)—N(7) bond [24]. This reaction is worthy of further study, bearing in mind that treatment of nucleic acids with some alkylating agents leads to alkylation of the N(7) of some guanosine residues. Furthermore, 7-methyl-GMP is a key component of the cap of various eukaryotic mRNA's. It is also one of the component residues of yeast $tRNA^{Phe}$ [43]. A procedure has been recently proposed for determining the extent of alkylation of DNA, following treatment with alkylating agents, and based on the electrooxidation of 7-methylguanine at the carbon electrode [44]. The actual mechanism of electrooxidation has not been clarified, but presumably the N(7)—C(8) bond of the imidazole ring is the electroactive centre [45].

The ability of adenine and cytosine to undergo reduction in aqueous medium is at least partially retained at the level of oligo- and polynucleotides. In both acid and neutral media, such residues in single-stranded RNA, DNA and synthetic polynucleotides, when subjected to d.c. polarography, undergo irreversible reduction in a protonated, adsorbed, state [27,46-52], with the transfer of four electrons to adenine residues, and two electrons to cytosine residues [27,37,53]. The coulometric data are, however, based on measurements on the monomeric units. But extrapolation of these to model oligonucleotides such as ApG, ApC, ApA, $(Ap)_3A$ and $(Ap)_5A$ appear to be consistent with uptake, in all instances, of 4 electrons by those adenine residues which are reduced [53], as will be shown below (Sect. IV). However, even at the level of small oligonucleotides, the process of electroreduction, and recording of the polarographic waves, become rather complex because of adsorption and formation of associates [54-57].

Whereas each adenine residue in a dinucleoside monophosphate undergoes reduction, the level and ease of reduction are apparently radically reduced with an increase in length of the chain. It is claimed that, for longer oligonucleotides, the maximal number of adenine residues reduced is three. In an analogous study on the mechanism of electroreduction of cytosine, its nucleosides and nucleotides, and dinucleoside phosphates such as CpC, CpU, UpC and CpG, Elving et al. [37] report that for most of these the electroreduction of cytosine residues proceeds via a 3e mechanism. Two exceptions were CMP and CpC, both reported to undergo 2e reduction. This rather anomalous finding is difficult to interpret and, if confirmed, is undoubtedly worthy of further study.

It should, however, be noted that in none of the foregoing investigations were attempts made to carry out macroelectrolysis with a view to identification of products. Even more surprising is the fact that there are no reported trials of the use of various commercially available nucleolytic enzymes to hydrolyze reduced oligonucleotide products, which might have rendered possible at least partial, if not complete, identification of the electrochemically modified monomeric units by simple paper or thin-layer chromatography in conjunction with UV spectroscopy.

One well-established observation is that, under conditions where single-stranded polynucleotides give rise to a d.c. polarographic reduction wave, both native DNA and other double-helical natural and synthetic polynucleotides are inactive [22,23,46,47,58,59,61]. This is readily interpretable in that, in such helical structures, the adenine and cytosine residues are located in the interior of the helix, and hydrogen bonded in complementary base pairs (see below). Z-DNA, in which cytosine residues are at the surface of the helix, is of obvious interest in this regard, and the B → Z transition in the synthetic poly(dG · dC) has been investigated with the aid of differential pulse polarography and UV spectroscopy [60].

Considerably less attention has been devoted to the polarographic behaviour of RNA. It has been shown that the relative electrochemical behaviours of ds-RNA and ss-RNA are similar to those of native ds-DNA and denatured ss-DNA [20,22,23], but this has not been adequately profited from to distinguish the single- from the double-stranded regions displayed by different types of RNA, such as tRNA, rRNA, etc. It is, in fact, somewhat surprising that so little attention has been devoted to the electrochemical behaviour of various tRNAs, bearing in mind present knowledge of their three-dimensional structures, and their content of modified and hypermodified bases, apart from adenine and cytosine, which are normally susceptible to electroreduction, e.g. thioracils. There is apparently only one reported attempt [62] to examine such an RNA, viz. $tRNA^{phenyl}$ which, however, does not contain bases, apart from adenine and cytosine (and, possibly, 7-methylguanine), susceptible to reduction. It was shown, with the aid of both a.c. and d.c. polarography, in the pH range 5 to 7, that this tRNA does undergo reduction and that the process is kinetically controlled and linked to the reduction of protonated residues of adenine and cytosine in the single-stranded regions of the molecule. CD spectroscopy also pointed to a change in conformation of the tRNA [62]. A rather unexpected observation was an increase in the d.c. current intensity for the same tRNA following removal of the —CCA terminus.

It has been observed that double-stranded helical DNA does exhibit detectable reduction with the use of differential pulse polarography (dpp) [27]. The resulting signal (I) is about 100-fold weaker, and occurs at a different potential, than the signal (II) exhibited by totally denatured DNA under the same conditions [27]. Similar results have been noted for ds-RNA and the single-stranded forms of poly(A) and poly(C) [27,48,63]. The foregoing behaviour has been ascribed to the existence in the helix of some non-paired bases. The number of such "labile" regions may be increased by various chemical, physical and enzymatic treatments, or heating to a temperature just below the level of the T_m (premelting temperature) [22,27,64], while simultaneously following the appearance of signal I by means of the dpp procedure. The high sensitivity of this procedure, exceeding that accessible by spectrophotometric methods, points to its possible utility in detection of "defects" in the structures of helical DNA and RNA, and to determination of the level of denatured regions in a given DNA. Sequaris et al. [65] have reported a procedure, based on voltammetry at the HMDE, for quantitative evaluation of the extent of labilization of DNA following γ-irradiation in solution.

DNA is strongly adsorbed at the surface of the mercury electrode within a strictly defined range of potentials, accompanied by partial, reversible, denaturation [27,61,66–69]. The extent of these conformational changes is dependent on the pH and ionic strength of the solution and, in particular, on the electrode potential and the time of interaction of the DNA with the charged electrode surface [27,61–69]. It is, consequently, pertinent to inquire why, if surface denaturation really occurs, native DNA appears to be inactive when examined by d.c. polarography, even with the use of a dropping electrode with a long life-time [70]. Furthermore, is the dpp method capable of revealing the actual modifications in conformation of the DNA in solution which do not result from interaction at the electrode surface? Two attempts have been made to clarify the foregoing.

Berg et al. [71] proposed that the adenine and cytosine residues in native DNA are reduced by a so-called "electron hopping" mechanism, the only condition for this being adsorption of protonated DNA at the electrode surface at the reduction potential of these bases. It was also assumed that the DNA is adsorbed in its A-form, exhibiting semi-conducting properties. There is consequently no surface denaturation of the DNA.

An alternative interpretation, advanced by Palecek [22] and Nurnberg [72], postulates that reduction of protonated DNA occurs only if adenine and cytosine residues are in direct contact with the electrode, hence when the DNA is constrained to a partially denatured form at the electrode surface, within a strictly defined range of potentials lower than that required for reduction of the free bases. Possible polarographic detection of these changes is, consequently, dependent on the nature of polarization of the electrode, hence on the type of electroanalytical technique employed. Methods which apply small voltage excursions during the lifetime of the drop, such as dpp and a.c. and d.c. polarography, disclose changes in conformation in the solution phase. By contrast, techniques which employ large voltage excursions during the drop lifetime, such as normal pulse polarography (NPP), linear sweep voltammetry (LSV) or other techniques in cooperation with stationary electrodes, are particularly advantageous for following the successive modifications in conformation resulting from interaction with the charged electrode surface [19,27,30]. This proposal is by far the simpler theoretically, and most experimental data are consistent with it. Reasonably good direct evidence for changes in conformation of DNA at the solid electrode/aqueous solution interface has been recently reported by Sequaris et al. [73], from spectroelectrochemical measurements with the aid of laser Raman spectroscopy. More adequate models for testing the foregoing might well include various tRNAs (for reasons cited above) and Z forms of DNA, now accessible as oligo and polynucleotides, and in some instances reversible to the B-form by modification of the ionic strength of the medium.

III Electroreduction of 2-Oxopyrimidines

III.1 Pyrimidone-2

The mechanism of electrochemical reduction of pyrimidone-2 is of interest from several points of view, not the least of which is the fact that it is an intermediate in the electroreduction of cytosine. The isolation and identication of the products of reduction of pyrimidone-2 and its N-methyl derivatives led, in fact, to the resolution of conflicting conclusions [33,74] regarding the mechanism of reduction of cytosine (see Sect. IV).

In an aqueous buffered medium, over the pH range 1–12, pyrimidone-2 exhibits a single one-electron wave. Preparative electrolysis, at a potential corresponding to the initial limiting current, led to formation of an insoluble product, isolated as a white amorphous powder, and shown by various physico-chemical criteria to correspond to a dimer consisting of two molecules of reduced pyrimidone-2. This was further confirmed by ^{1}H NMR spectroscopy, which also established the structure of the product as 6,6'(or 4,4')-bis-(3,6(4)-dihydropyrimidone-2), shown in Scheme 2, below. The structure of the dimer reduction product, and its solid state conformation, were subsequently further established by X-ray diffraction (see Sect. III.3.).

In aqueous 0.1 M $(CH_3)_4NBr$, pyrimidone-2 was found to exhibit two reduction waves of equal height, with $E_{1/2}$ values of —0.75 V and —1.55 V for waves I and II, respectively [1,2]. (Fig. 1) Preparative electrolysis under these conditions at the potential of wave I resulted in formation of the same dimer reduction product as in aqueous buffered medium. By contrast, electrolysis on wave II led to formation of two products, one of which was identical with that formed on wave I. The other, readily soluble in aqueous medium, was identified as 3,6-dihydropyrimidone-2, identical with that synthesized chemically and described earlier by Skaric [75].

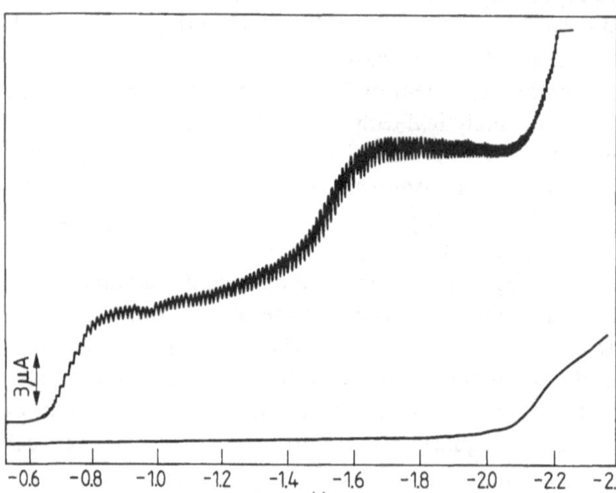

Fig. 1. Polarogram of pyrimidone-2 (2×10^{-3} M) in aqueous unbuffered 0.1 M $(CH_3)_4NBr$. Potential *vs* SCE, showing wave I ($E_{1/2}$ = —0.75 V) and wave II ($E_{1/2}$ = —1.53 V)

The dimer reduction product of wave I, in turn, exhibited a polarographic anodic wave with $E_{1/2} = -0.27$ V in neutral aqueous medium. Anodic electrolysis at the crest of this wave led to regeneration of the parent pyrimidone-2 [2)].

The pathway for electrochemical reduction of pyrimidone-2 is consequently as formulated in Scheme 2.

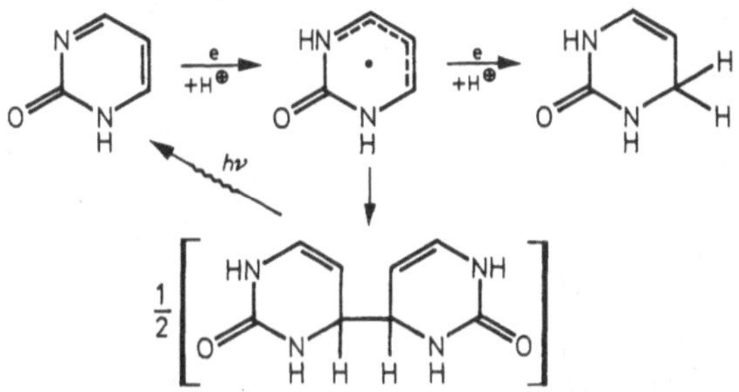

Scheme 2. Proposed redox pattern for pyrimidone-2

III.1.1 Photochemistry of the Pyrimidone-2 Dimer Reduction Product

The UV absorption spectra, in neutral aqueous medium, of pyrimidone-2 and its dimer reduction product, 6,6'-*bis*-(3,4-dihydropyrimidone-2), are exhibited in Fig. 2 Particularly interesting was the finding that irradiation of the dimer at 254 nm under these conditions led to the stepwise disappearance of its characteristic absorption spectrum, with the simultaneous appearance of the spectrum of the parent monomer. Additional evidence for the identity of the photoproduct with the parent pyrimidone-2 was furnished by chromatography and polarographic behaviour. The photochemical conversion reaction was shown to be quantitative (under these conditions pyrimidone-2 itself is quite radiation resistant), and to proceed with a quantum yield of ~0.1, both in the presence and absence of oxygen [2)]. This value was unchanged when irradiation was conducted in ^2H_2O; the absence of an isotope effect is clearly of relevance to the mechanism of the photodissociation reaction.

The foregoing is of obvious significance in relation to the photochemical behaviour of cyclobutane photodimers of natural pyrimidines such as uracil (III, Scheme 1), thymine, cytosine [12, 76)]. Photodissociation of such photodimers to the parent monomers, which proceed with quantum yields ranging from 0.5 to nearly unity, has, indeed, been employed as one of the criteria for identification of such dimers. The validity of this criterion is now, in the light of the behaviour of the dimer electroreduction product of pyrimidone-2, at best somewhat restricted, notwithstanding that the quantum yields for photodissociation of the latter are lower.

III.1.2 Tetrameric Photoproduct of Thymine and Pyrimidone-2

Of additional, and particular, interest is the finding that the dimer reduction product of pyrimidone-2 is an integral component of a tetrameric photoproduct resulting from the photodimerization of a photoadduct of thymine and pyrimidone-2 [77)], as shown in Scheme 3. This tetrameric photoproduct has, furthermore, been isolated from an irradiated frozen aqueous solution of cytosine and thymine [8)], and, more

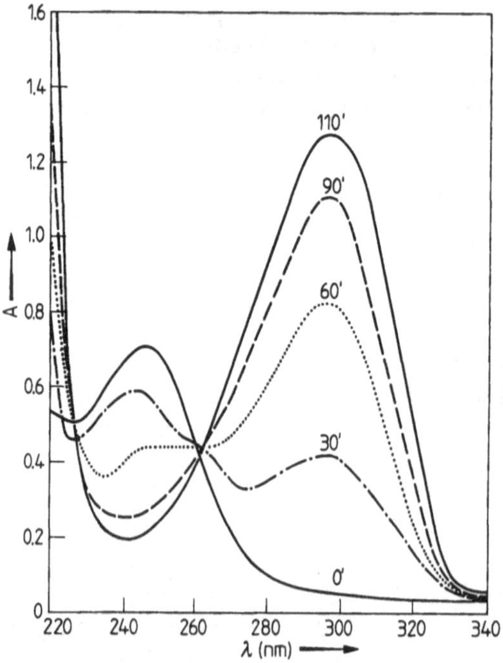

Fig. 2. Photochemical transformation, on irradiation at 254 nm in 0.01 M phosphate buffer (pH 7.2) of 1.7×10^{-4} M 6,6'-*bis*-(3,6-dihydropyrimidone-2), with quantitative regeneration of pyrimidone-2, as shown by isosbestic point at 262 nm (see text for further details). The curve marked "o" is that for the initial reduction product; figures beside the other curves represent the time of irradiation in min

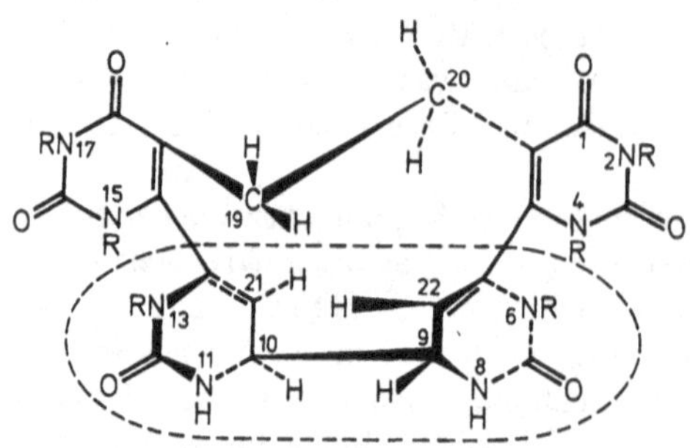

Scheme 3.

significantly, from UV-irradiated DNA [8], the latter fact underlining its biological importance. The possible susceptibility of the foregoing tetrameric photoproduct to UV irradiation has not been examined, and is clearly deserving of investigation. On the other hand, its solid state structure, determined by X-ray diffraction [78], rendered possible a direct comparison with the solid state structure of the dimer reduction product of another pyrimidone-2 analogue (see Sect. III.3, below).

III.2 N-Methyl Derivatives of Pyrimidone-2

Pyrimidone-2 is known to be predominatly, if not exclusively, in the keto form [79]. Monomethylation gives 1(3)-methylpyrimidone-2, which is clearly fixed in the 2-keto form. This also undergoes one-electron reduction, with formation of a single diffusion-controlled wave, the half-wave potential of which is pH-dependent, consistent with the fact that the free ring nitrogen may undergo protonation. The product of reduction, unlike that of pyrimidone-2, is quite soluble in water, and is readily extracted with chloroform to give an oil which is transformed into a pseudocrystalline form. The mass spectrum exhibited an intense peak at m/e = 222, corresponding to a dimer of the protonated form of reduced 1-methylpyrimidone-2, further supported by elementary analysis, which pointed to the presence of one additional proton per pyrimidine ring [3].

On irradiation in neutral aqueous medium, the foregoing reduction product was converted to the parent 1-methylpyrimidone-2. When this photochemical conversion

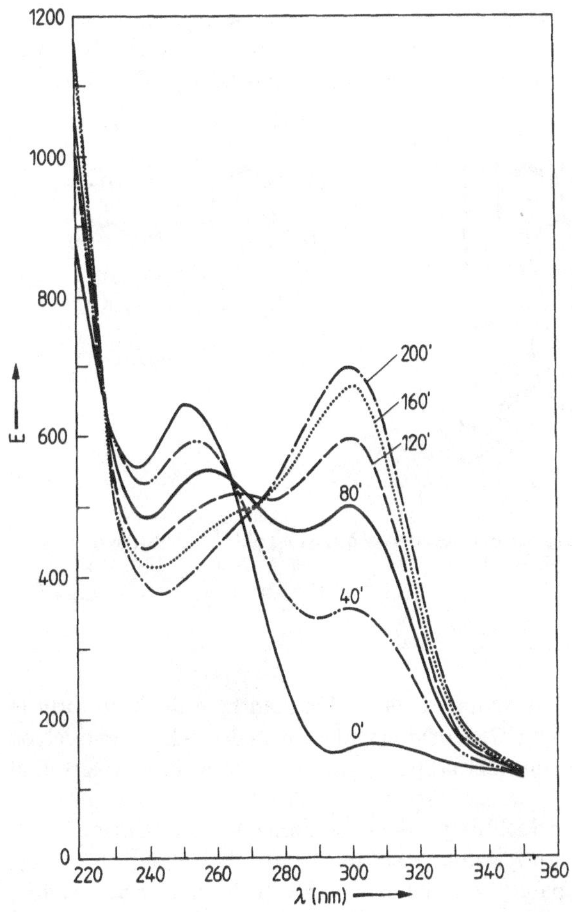

Fig. 3. Photochemical transformation, on irradiation at 254 nm in 0.01 M phosphate buffer pH 7.1, of 6,6'-*bis*-(1-methyl-3,6-dihydropyrimidone-2) with regeneration of 1-methylpyrimidone-2. The curve marked "o" is that for the initial reduction product; numbers beside other curves are times of irradiation in min.

reaction was followed spectrally, no well-defined isosbestic point was observed (see Fig. 3), as in the case of photodissociation of the dimer reduction product of pyrimidone-2 (see Fig. 2). This is readily interpreted as being due to the existence of two isomers of the reduction product of the 1-methyl derivative (Scheme 4), although attempts to separate these by chromatography were unsuccesful.

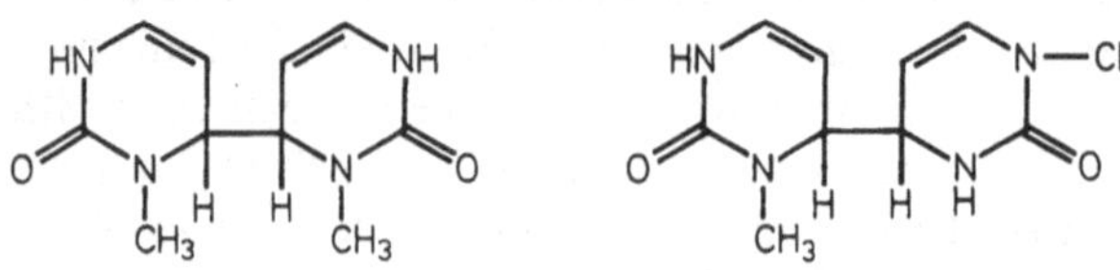

Scheme 4.

In contrast to the foregoing, 1,3-dimethylpyrimidone-2 undergoes one-electron reduction only in the cationic form (i.e. in the pH range 1–9). Macroelectrolysis at pH 4.5 resulted in formation of a single product, isolated as a pale yellow oil and identified as a dimer (Scheme 5) which was readily converted, on irradiation in neutral aqueous medium, to the parent 1,3-dimethylpyrimidone-2 [3].

On the other hand, electrolysis of the dimethyl derivative in the presence of $(CH_3)_4NBr$ as electrolyte, led to the appearance of two products, one of which was the photochemically dissociable dimer, the other the dihydroderivative of 1,3-dimethylpyrimidone-2 [3].

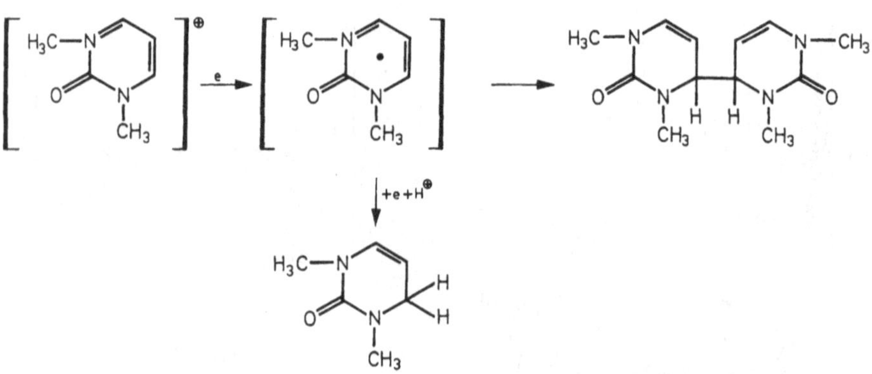

Scheme 5. Proposed reaction sequence for the electrochemical behaviour of 1,3-dimethyl pyrimidone-2

III.3 4,6-Dimethylpyrimidone-2

This compound, like the parent pyrimidone-2, is predominantly in the keto form in aqueous medium and, in the pH range 2–9, exhibits, like pyrimidone-2, a one-electron polarographic wave. In more alkaline media (pH 9–12), a second wave is observed at more negative potentials.

Macroelectrolysis on wave I yielded as product the dimer Id (see Scheme 6) [80], apparently identical with the photoproduct resulting from UV-irradiation of 4,6-dimethylpyrimidone-2 in isopropanol at 254 nm [81]. The dimer reduction product

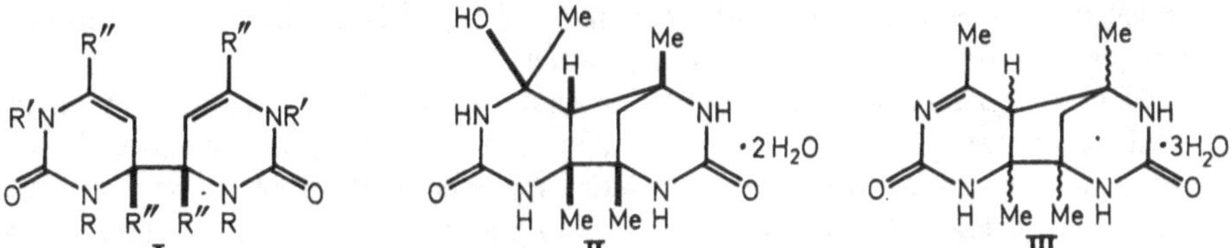

I II III

a R = R′ = R″ = H
b R = Me; R′ = R″ = H
c R = R′ = Me; R″ = H
d R = R′ = H ; R″ = Me

Scheme 6. Dimers I a–d resulting from electrochemical reduction of pyrimidone-2 and its methylated analogues; and the acid-catalyzed cyclization product of I d, shown by X-ray diffraction to be II, and not III as proposed by Pfoertner II [81]

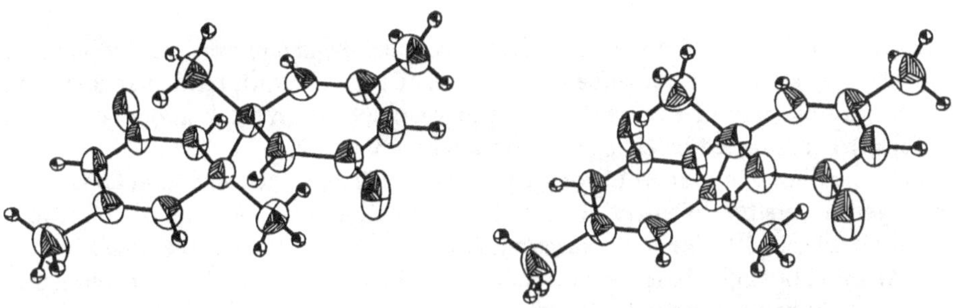

Fig. 4. Stereoscopic view of the dimer of 4,6-dimethylpyrimidone-2 (I d in Scheme 6).

was, by a fortunate coincidence, found to yield crystals suitable for X-ray diffraction; and Fig. 4 exhibits a stereochemical view of the solid state structure, showing it to have the form *meso*, with a center of symmetry in the dimer molecule [80]. Furthermore, the bond distances and bond angles (Scheme 7) are very similar to those for the corresponding dimer moiety of the tetrameric photoproduct (see Scheme 3) isolated from an irradiated solution of an adduct of thymine and pyrimidone-2 [78].

Irradiation of the dimer I d in neutral aqueous medium at 254 nm led to quantitative regeneration of the parent 4,6-diemthylpyrimidone-2 with a quantum yield of 0.03.

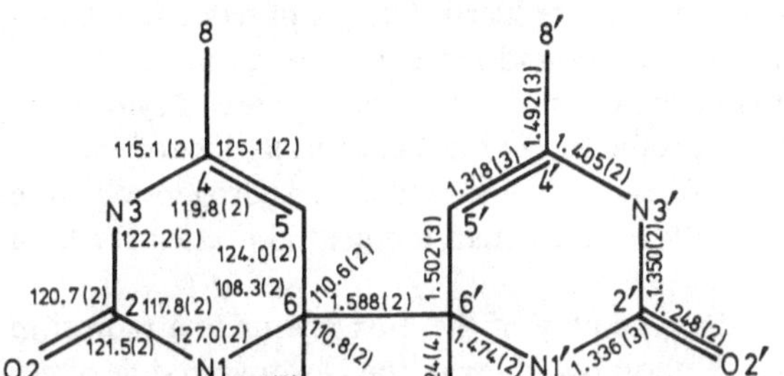

N1 - C6 - C6′ 108.2(2)
C5 - C6 - C7 110.3(2)

Scheme 7. Bond lengths (Å) and angles (deg) in dimer I d of Scheme 6.

This value is to be compared with that of 0.1 for photochemical regeneration of pyrimidone-2 from its dimer reduction product (Sect. III.1.1).

Particularly striking was the observation that storage of an aqueous solution of Id in the cold room for several days resulted in its spontaneous transformation, in the dark, to another product, which resembled that reported by Pfoertner [81] on treatment of Id with acid. The conversion is essentially quantitative in 3 hr at room temperature in 0.001 N HCl, and the resulting product again provided suitable crystals for X-ray diffraction. In contrast to the structure III assigned to this product by Pfoertner [81], the X-ray diffraction data showed the correct structure to be II, clearly the result of addition of a water molecule to Id, followed by cyclization (Scheme 6) [80].

III.4 Pyrimidone-2 in DMSO

Additional information about the electrochemical behaviour of pyrimidone-2 is furnished by studies in non-aqueous medium, $(CH_3)_2SO$, with particular attention to the possible role of free radicals and anionic species (radical anions, dianions, dissociated conjugated bases) as reaction intermediates [82].

In the absence of a source of hydrogen or hydroxide ions, the compound apparently undergoes reversible 1e reduction to yield a radical anion, which can dimerize to a dianionic species [82]. The latter reaction, however, is slower than the attack of the radical anion (a strong base) on non-reduced pyrimidone-2 to abstract a proton, and leading to the neutral free radical, which dimerizes much more rapidly than the corresponding radical anion, as follows:

$$RH + e \rightarrow \dot{R}H^-$$

$$\dot{R}H^- + RH \rightarrow \dot{R}H_2 + R^-$$

$$\downarrow$$

$$^1/_2\,(RH_2 - RH_2)$$

Evidence that the radical anion is indeed a stronger base than the dimer anion species is forthcoming from the observed higher reactivity between the former and a non-reduced molecule than between two molecules of the radical anion, plus the fact that only about 50% of the pyrimidone-2 is reduced. The site of reduction, based on the pathway previously established for pyrimidone-2 in aqueous medium [2,74], is clearly the N(3)=C(4) bond, with dimerization via a 6,6′ (or, because of symmetry, 4,4′) linkage. However, the reduction products were not chemically identified.

On addition of a strong acid, i.e. a freely available source of hydrogen ions, the protonated pyrimidone-2 is more readily reduced than the neutral species [82], as in the case of pyrimidine (Sect. VI).

The foregoing and, in particular, the reactivity of the 2-oxopyrimidine radical in DMSO, is relevant to the mechanism of photooxidation of the 2-oxopyrimidine dimer to the monomer on irradiation at 254 nm in DMSO [83]. The initial step in the latter process would be the formation of a protonated neutral free radical, $RH_2^{\cdot}$, which is rapidly oxidized (—0.8 V) to RH_2^+, followed by deprotonation.

IV Cytosine and some Derivatives

The polarographic behaviour of cytosine and several of its derivatives was initially examined by Smith & Elving [74] and Janik & Palecek [33], who proposed that 2-electron electroreduction proceeds initially at the 3,4 bond, with elimination of NH_3 to give pyrimidone-2. Smith & Elving [74] further proposed that reduction of pyrimidone-2 is a one-electron process, whereas Janik & Palecek [33] suggested a two-electron mechanism leading to formation of 3,4-dihydropyrimidone-2, which was unstable and subject to ring scission at the 3,4 bond. It was the subsequent isolation and identification of the reduction product(s) of pyrimidone-2, described above [3], which made possible a more precise analysis of the mechanism of electroreduction of cytosine and cytidine [1,84].

In aqueous buffered medium, over the pH range 2–9, cytosine exhibits a single weakly structured reduction wave [33,74]. The pH-dependence of $E_{1/2}$ indicates that this reduction proceeds via the protonated form [33,37]. The basic reduction pattern for cytosine involves: rapid protonation at N_3 to form the electroactive species, two-electron reduction at the 3,4 double bond, protonation of the latter (5×10^4 sec^{-1}), deamination (10 sec^{-1}) to regenerate the N(3)=C(4) bond and one-electron reduction at the latter site to form a free radical which dimerizes [37]. Electrolysis at pH 4.5 and 7.0 demonstrated quantitative liberation of NH_3 at the acid pH, but only 60% of the theoretically expected amount at neutral pH [1,84]. The foregoing is consistent with a three-electron reduction in acid medium [1], but not at neutral pH, where coulometric measurements at potential $E = -1.5$ V point to a four-electron wave (Table I).

Chromatography of the electrolysis products formed in acid medium demonstrated that the major one was the photodissociable dimer 6,6'-*bis*-(3,6-dihydropyrimidone-2) (70%), accompanied by 3,6-dihydropyrimidone-2 (15%) and about 15% of a product not fully identified, but the UV absorption spectrum of which was consistent with its being a 1:1 adduct of cytosine and reduced pyrimidone-2 [84].

Table I. Macrocoulometric determination of number of electrons transferred (n) and amount of NH_3 evolved during electrochemical reduction of pyrimidone-2, cytosine and cytidine

pH of electrolysis	Conc. (mM)	E_{red} (V)	n	NH_3 (%)
Pyrimidone-2				
4.5	1.25	−1.15	0.78	
4.5	5	−1.15	0.72	
7.03	5	−1.15	0.85	
Cytosine				
4.5	1	−1.50	3.62	69.7
4.5	5	−1.60	4.1	99.9
6.3	5	−1.67	3.91	51.0
7.03	5	−1.75	3.78	57.4
7.03	5	−1.8	3.70	63.0
Cytidine				
7.03	5	−1.65	3.87	82
7.03	5	−1.65	3.67	72

The electrolysis products in neutral medium included one additional absorbing substance, the chromatographic properties of which, together with its reaction with the Fink reagent to give a yellow product [85], pointed to its being 5,6-dihydrocytosine [86], further confirmed by its hydrolytic deamination to a product identical with 5,6-dihydrouracil.

From the foregoing it is clear that in acid medium there is a three-electron reduction involving elimination of ammonia, followed by one-electron reduction of the resulting pyrimidone-2.

In neutral medium, on the other hand, the neutral form undergoes preferential four-electron reduction of the bond systems C_3—C_4 and C_5—C_6, with the formation of the tetrahydro derivative, but the 2e-intermediate (5,6-dihydrocytosine) was also found. In neutral medium the reduction products undergo only partial (60 %) deamination, in agreement with data on the catalytic reduction of cytosine [86]. The scheme for electroreduction of cytosine would therefore be as follows (Scheme 8**a**, **b**) [1, 84].

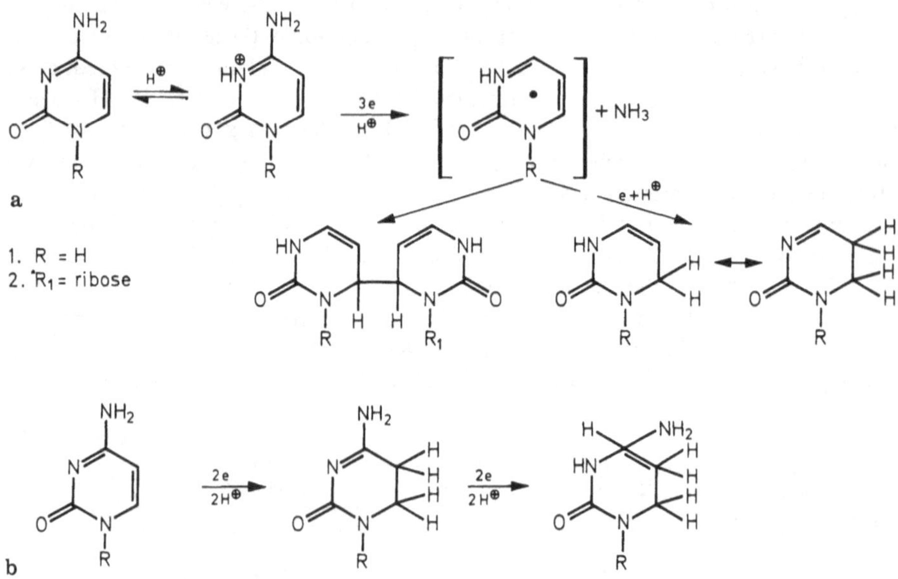

Scheme 8. Mechanism of electroreduction of cytosine: (**a**) in acid or neutral medium; (**b**) in neutral medium

IV.1 1-Methylcytosines

The polarographic behaviour of 1-methylcytosine and 1,N^4-dimethylcytosine at pH 3–7 was, as expected, similar to that for cytosine. Reduction led to formation, from each compound, of a dimer of 1-methylpyrimidone-2, accompanied by quantitative elimination of ammonia and dimethylamine, respectively. The dimer reduction product in each case underwent photochemical conversion to 1-methylpyrimidone-2 [84].

IV.2 Cytidine

In accordance with the foregoing, i.e. that substitution at N_1 does not affect the reduction pathway, the nucleoside and nucleotide react at the mercury electrode essentially like the bases [36,37] but adsorb more strongly than cytosine at a potential more positive than -1.6 V [37,48]. The $E_{1/2}$ for the reduction wave in the cytosine series becomes more positive in the order: base > nucleotide > nucleoside, and is linearly pH-dependent [37,53]. The mechanism for electrochemical reduction of cytosine, cytidine, CMP and CpC have been considered in terms of their structure, association in solution and adsorption [37]. It was concluded that the deamination step for CpC occurs very slowly or not at all [37].

Cytidine undergoes a one-step reduction in the pH range 2–7 [1,37] with formation of an irreversible kinetic-diffusion wave [37]. Coulometric determinations point to a 3-electron wave at pH 4.5 and a 4-electron wave at pH 7 [1,84].

In acid medium (pH 4.5) electrochemical reduction led to formation of two products. One of these was characterized as the 6,6′-dimer of the riboside of pyrimidone-2. It was photochemically converted to the riboside of pyrimidone-2. The second, on the basis of its chromatographic behaviour, UV spectrum, and reaction with the Fink reagent, was identified as the riboside of 3,6-dihydropyrimidone-2. The mechanism of electrochemical reduction of cytidine in acid medium is consequently analogous to that for 1-methylcytosine. The products of reduction of cytidine at pH 7 were shown chromatographically to contain 5,6-dihydrocytidine [1,84]. A comparison of the electrochemical and catalytic reduction products under analogous conditions at pH 7 demonstrated that both led to the same products, one of them 5,6-dihydrocytidine [84].

The foregoing findings lead to resolution of the earlier conflicting proposals regarding the mechanism of reduction of cytosine and cytidine [33,74]. Bearing in mind the pH-dependence of these reactions, leading to dimeric products in acid medium, and to di- and tetra- hydro derivatives in neutral medium, it is not surprising that Smith and Elving [74] concluded that the reduction process is 3e at pH 4.5, whereas Janik and Palecek [33] identified a 4e reduction at pH 7. Furthermore, the latter authors, who carried out the reduction of cytosine at potentials more negative than the wave crest, demonstrated that this led to products with only end absorption in the quartz ultraviolet, correctly interpreted as due to formation of the 3,4-dihydro derivative followed by opening of the ring at the 3,4 bond [33].

The foregoing proposed mechanism of reduction of cytosine, cytidine and their derivatives in acid and moderately alkaline media are further supported by the results of a recent study, by means of a.c. polarography, of the orientation of cytosine and cytidine at the mercury electrode. It was shown [87] that, whereas at acid pH (about 5) both of these are adsorbed via the positive charge on the ring N_3, so that the $N_3{=}C_4$ bond is adjacent to the electrode surface, the orientation in alkaline medium (pH 9) is modified so that the $C_5{=}C_6$ bond is located at the surface. This is consistent with the proposed mechanisms of reduction.

IV.3 Photoadducts of Cytosine and Pyrimidone-2

The redox properties of pyrimidone-2 and cytosine are of some relevance to the electrochemical behaviour of reported photoadducts of pyrimidone-2 with cytosine, viz.

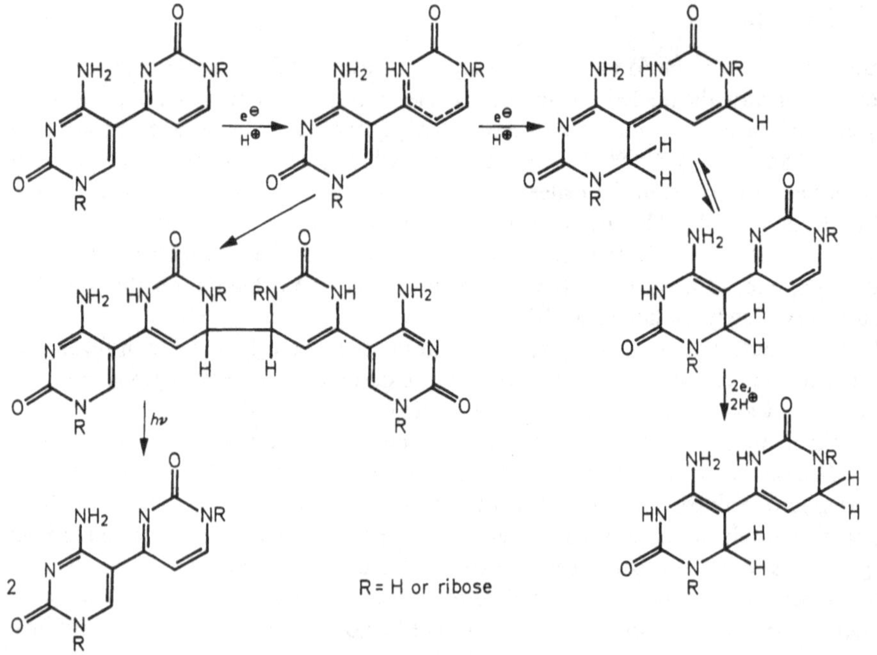

Scheme 9. Proposed reaction sequence for electrochemical reduction of Cyt(5-4)Pyo and Cyd(5-4)Pdo at the DME

Cyt(5-4)Pyo, and the riboside of pyrimidone-2 with cytidine [10,88,89], Cyd(5-4)Pdo (top left, Scheme 9). These photoadducts are of considerable biological interest, in that they have been isolated not only from UV-irradiated cytidine and poly(C) [10], but are also formed by irradiation at 335 nm of certain tRNA's [90,91].

Both the foregoing photoadducts exhibit three diffusion-controlled polarographic waves in aqueous medium in the pH range 1–12 [92]. The first two waves are due to successive one-electron additions, followed by the third, a two-electron reduction step. The pH-dependence of the initial one-electron reduction wave was found to be similar to that for pyrimidone-2. Following electrolysis, at the crest of wave I, of either of the photoadducts, the two one-electron waves disappeared and the reduction products exhibited UV absorption spectra with a band at 375 nm. A similar absorption band is exhibited following sodium borohydride reduction of Cyt(5-4)Pyo, leading to a product identified as 5-(4-pyrimidin-2-one)-3,6-dihydrocytosine. Mass spectroscopy revealed that the products of reduction on wave I were dimers consisting of two molecules of reduced photoadducts [92].

The proposed mechanistic pathway for electroreduction of the photoproducts is shown in Scheme 9.

The electrochemically reduced dimer products of each of the photoadducts readily photodissociated to the parent photoadducts, as described above (Sect. III.1) for pyrimidone-2 and its derivatives. From this it may be inferred that the two component rings of the dimer photoadducts are bonded *via* a 6,6′ linkage. By contrast, the dihydroderivative of Cyt(5-4)Pyo, formed by borohydride reduction, was not susceptible to photodissociation.

V Non-Fused Bipyrimidines

Discussion of the foregoing pyrimidine photoadducts would be incomplete without at least a brief reference to synthetic non-fused bipyrimidines with non-saturated rings. Various procedures initially developed for the synthesis of such compounds [10, 93–95], examples of which are shown in Scheme 10, were subsequently further stimulated by the observation that some of them appreciably amplify the in vitro cytotoxic activity of the antibiotic phleomycin [93]. This, in turn, led to the synthesis of additional analogues in which one or both components are 5- and 6-membered heterocyclic rings other than pyrimidine.

The mechanism of action of bipyrimidines as amplifiers of phleomycin cytotoxicity is unknown. However, initial investigations on the potentiating activities of non-fused heterobicyclic analogues were prompted by the proposal that the mode of action of phleomycin (and bleomycin) involves interaction, with DNA, of a side chain comprising a 2,4′-bithiazole moiety [93]. In searches for more effective amplifiers, attention might profitably be directed to preparation of analogues in which one of the rings has been reduced by chemical or electrochemical procedures. Furthermore, many of the hitherto synthesized bipyrimidines are obvious models for electrochemical studies. It may be anticipated that analogues in which one of the constituents is 2-oxo-, 2-amino- or 2-thio-pyrimidine may undergo electroreduction, with formation of "dimers" which, in effect, consist of four pyrimidine residues, hence similar to the electroreduction product of the photoadduct of 4,5′-bipirymidine, referred to in the previous section. For some bipyrimidine analogues, e.g. of the form 4.5′-bipyrimidine-2,2′-diamine or 4,5′-bipyrimidine-2,2′-oxo, electrochemical formation of more highly polymerized products may be envisaged.

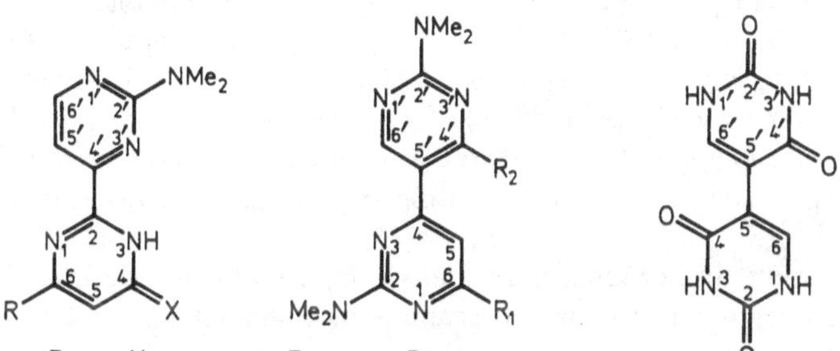

R	X
OH	O
Me	O
Me	S

2,4′

R_1	R_2
OMe	OMe
SMe	SMe
SMe	NHMe
SMe	N(Me)$_2$

4,5′

5,5′

Scheme 10. Bipirymidine structures

VI Pyrimidine

The polarographic reduction of pyrimidine itself [74, 96, 97] was most extensively described by Elving et al. [36, 74, 98, 99]. In aqueous medium it is reduced in the protonated

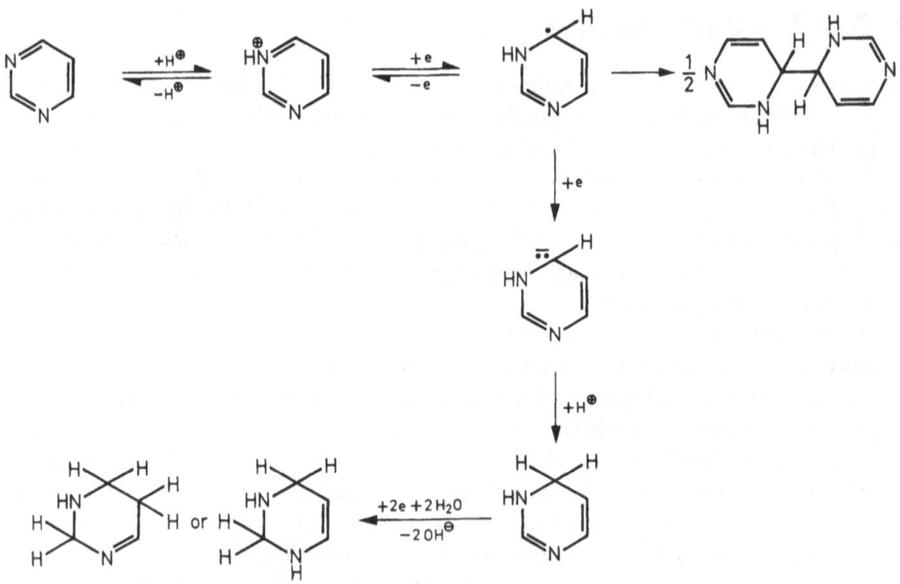

Scheme 11. Proposed reaction sequence for electrochemical reduction of pyrimidine

form [15, 74, 99], with a rather complex reduction pattern involving 5 waves over the pH range 0–12. The initial step is a 1e pH-dependent reduction (wave I) in the pH range 0–5, with formation of a free radical which may dimerize to 4,4′-pyrimidine, or undergoes further reduction *via* a le process (wave II) to 3,4-dihydropyrimidine. These two waves merge at pH 5 to a pH-dependent 2e wave III. At pH 7 a pH-independent wave IV appears, corresponding to the 2e reduction of 3,4-dihydropyrimidine to the tetrahydropyrimidine. Waves III and IV merge at about pH 9 to a pH-dependent 4e wave V (see Scheme 11). Cyclic voltammetry at the H.M.D.E. showed cathodic processes corresponding to each of the five waves [36]. The foregoing results were supported by Thevenot et al. [100, 101], who also noted that, at a pyrimidine concentration exceeding 1 mM, wave III was resolved into two waves, pointing to involvement of adsorption processes.

The postulated dimer reduction product is very unstable, even in the absence of oxygen. Attempts to identify it directly, and to examine its potential susceptibility to photooxidation, have been unsuccessful.

In aprotic medium, on the other hand, pyrimidine gives a reversible diffusion-controlled 1e wave at a very negative potential, with formation of a radical anion which is deactivated via two pathways: rapid formation of the anionic, probably 4,4′-, dimer, with a rate constant of 8×10^5 L mol^{-1} sec^{-1}, and proton abstraction from residual water in the medium at a much lower rate constant, 7 L mol^{-1} sec^{-1} [98]. This is rapidly followed by a further 1e reduction to produce, ultimately, 3,4-dihydropyrimidine [98]. In the presence of acid there is also a 1e reduction wave corresponding to formation of a free radical which, as in aqueous medium, dimerizes, most likely to 4,4′-*bis*-(3,4-dihydropyrimidine). Examination of the mechanism of reduction in acetonitrile in the presence of acids supported the conclusion that reduction of pyrimidine in aqueous medium is preceded by its protonation [98].

It must, nonetheless, be emphasized that the products of reduction of pyrimidine have not been unequivocally identified, largely due to their instability in the presence of air (oxygen). Furthermore, the UV absorption spectra of the reduction products of waves I and II (λ_{max} 284 nm, ε_{max} 1.5×10^3) are suggestive of rapid conversion (proton-catalyzed hydration?) of the products, since both the dimer and the dihydro derivative possess a reduced system of aromatic bonds relative to the parent pyrimidine, as a result of which the UV absorption maximum should be shifted to the violet, whereas it is, in fact, shifted 44 nm to the red (from 240 nm to 284 nm) for both products. Of possible relevance to this is the fact that the reduced rings of 4-aminopyrimidine [102] and nicotinamide [103] undergo acidic hydration to form products absorbing at 280 to 290 nm.

VII Aminopyrimidines

VII.1 2-Aminopyrimidines

Following initial studies by Cavalieri & Lowy [96], it was shown by Smith & Elving [74] that the polarographic behaviour of 2-aminopyrimidine in acid medium is similar to that for the parent pyrimidine, which exhibits three waves at the dropping mercury electrode [74]. The initial 1e step involves formation of a free radical which dimerizes; and, at the potential of the second 1e reduction step, 2-amino-3,4-dihydropyrimidine is formed. But, unlike pyrimidine, 2-aminopyrimidines do not undergo a second 2e reduction to tetrahydro derivatives. Wave III (pH 7–9) involves two electrons and two protons, and is due to the combined processes responsible for waves I and II at lower pH. Both Smith & Elving [74], and Sugino [104], found that reduction of 2-aminopyrimidine on a mercury electrode [74] and lead cathode [104] resulted in the formation of unstable products.

Since 2-iminopyrimidine is formally similar, as regards electronic structure, to 2-oxo- and 2-thio- pyrimidine, it might be expected to undergo 1e reduction with the formation of a photodissociable dimer. It was, in fact, found that the fixed imino form, 1-methyl-2-iminopyrimidine iodide [105], gives rise to two 1e waves in the pH range 3–12, the $E_{1/2}$ values and pH-dependence of which differed from that observed for 2-aminopyrimidine. Preparative electrolysis led to isolation and identification of the product of the first step as the dimer, based on NMR and mass spectroscopy, and polarographic behaviour. The ^1H NMR spectrum of the dimer of 1-methyl-1,2-dihydro-2-iminopyrimidine [105] was consistent with that for the dimeric reduction product of 2-oxopyrimidine [2]. However, the former exhibits two well-resolved signals for the methyl group, consistent with formation of two isomers, *cis* and *trans*. The spectrum of the reduction product of 2-oxopyrimidine excludes the existence of such isomeric products.

Irradiation at 254 nm of an aqueous neutral solution of the reduction product of wave I led to stepwise appearance of a new band in the UV with λ_{max} 305 nm, with a quantum yield of 0.08 (Fig. 5), with on overall yield of 80%. Chromatography and polarographic analysis demonstrated that the photoproduct was the parent 1-methyl-1,2-dihydro-2-iminopyrimidine [105].

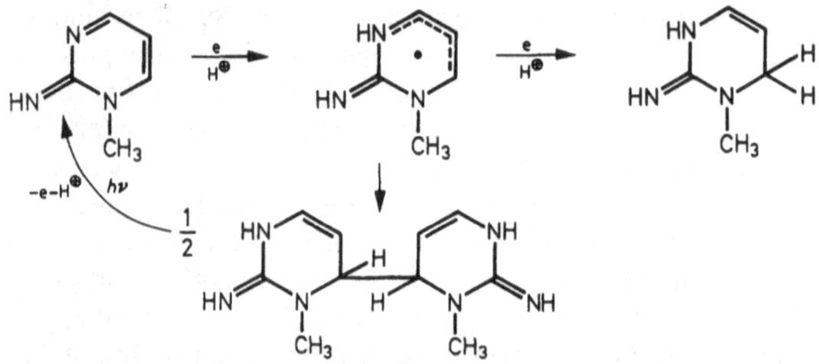

Scheme 12. Proposed pathway for electroreduction of 1-methyl-2-iminopyrimidine

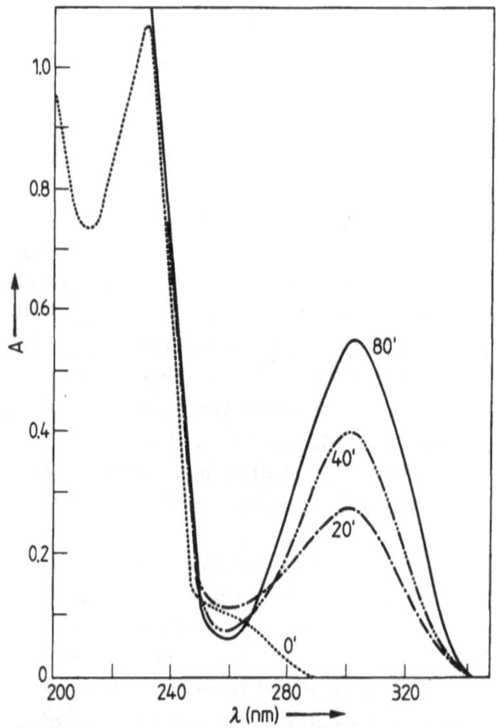

Fig. 5. Conversion of the dimer of 1-methyl-2-iminopyrimidine to the parent monomer on irradiation at 254 nm at neutral pH. The dimer was initially formed by electrochemical reduction at pH 5.0, and its UV absorption spectrum is that of curve *0*. Increasing times of irradiation (shown beside each curve) lead to generation of the monomer

Electrolysis at the peak of wave II led to the formation of two products readily separated by paper chromatography, both with λ_{max} 257 nm. Only one of these, identified as the reduction product of wave I, underwent photodissociation to the parent monomer (Scheme 12).

It is of additional interest that the product of reduction of wave I (1-electron) of 2-methylaminopyrimidine is a dimer (as shown by mass spectroscopy), which is photochemically converted in aqueous medium at 254 nm to 2-methylaminopyrimidine. Since the polarographic behaviour of 2-methylaminopyrimidine is similar to that of 2-aminopyrimidine, in agreement with the existence of both in the amino forms in aqueous medium [106, 107], it is not surprising that reduction of 2-aminopyrimidine on wave I also yields a photodissociable dimer, i.e. under the conditions of reduction employed earlier by Elving [74]. Hence the apparent lability of the product(s) of reduction of 2-aminopyrimidine noted by Smith and Elving [74] are at least partially explicable in terms of the light sensitivity of the reduction product.

VII.2 4-Aminopyrimidines

Since 4-aminopyrimidine is known to exist predominantly in the amino form, with a double bond between N(3) and C(4), it bears some structural resemblance to adenine, which may be considered a formal analogue of a 5,6-disubstituted 4-aminopyrimidine, and (at least to some extent) of cytosine, which is 4-amino-2-oxopyrimidine (Table II). This pointed to the possible utility of a detailed examination of the electrochemical behaviour of 4-aminopyrimidine, and some of its methylated analogues.

It was long ago noted that 4-aminopyrimidine exhibits two polarographic waves at pH 1,2 with apparent n-values of 2 and 1, and one wave between pH 2,3 and 6.8 [96].

Table II. Comparative reduction patterns in aqueous media for 4-aminopyrimidine and related compounds[a]

Compound	pH range	pH-dependence of $E_{1/2}$[b] (V)	n	$E_{1/2}$ at pH 4.2 (V)[c]
Pyrimidine	0.5–5	I −0.576–0.105 pH	1	−1.02
	3–5	II −1.142–0.011 pH	1	−1.19
	5–8	III −0.680–0.089 pH	2	(−1.05)
	7–8	IV −1.600–0.005 pH	2	
	9–13	V −0.805–0.079 pH	4	
4-Aminopyrimidine	0.4–2.0	I −1.070–0.025 pH	3	(−1.18)
	2.0–7.5	I −0.960–0.080 pH	3	−1.30
	0.4–2.0	II −1.207–0.007 pH	1	(−1.24)
2-Aminopyrimidine	2–2	−0.685–0.049 pH	1	
	4–7	I −0.425–0.121 pH	1	−0.93
	4–7	II −1.360–0.004 pH	1	−1.38
	7–9	III −0.680–0.090 pH	2	
4-Amino-2,6-dimethyl-pyrimidine	2–8	−1.130–0.073 pH	3 or 4	−1.44
4-Amino-2,5-dimethyl-pyrimidine	1–8	−1.06–0.076 pH	3	−1.38
2-Hydroxypyrimidine	2–9	−0.530–0.078 pH	1	−0.86
Cytosine (4-amino-2-hydroxypyrimidine)	4–6	−1.125–0.073 pH	3	−1.44
Purine	1–6	I −0.697–0.083 pH	2	−1.05
	1–6	II −0.902–0.080 pH	2	−1.24
Adenine	1–6	−0.975–0.090 pH	4	−1.35

[a] Data for 4-aminopyrimidine are from Ref. [102]; Sources of the data for the other compounds are taken from tables in Ref. [15].

[b] Roman numbers refer to the sequence of waves.

[c] Values in parentheses have been calculated by the adjacent equations even though the wave in question is not seen at pH 4.2.

The 2,5-dimethyl derivative was reported to undergo 3-electron reduction [108]. A more recent investigation, with the aid of various techniques, demonstrated that the lability of the primary electrolysis products, as well as of the secondary products, pointed to a much more complex reaction pathway than might have been anticipated [15, 102].

In acidic medium (pH 0–2), 4-aminopyrimidine undergoes a two-step reduction. The pH-dependent wave I corresponds to a 3e process; and wave II, corresponding to a 1e transfer, is accompanied by catalytic hydrogen discharge. Only wave I is observed between pH 2 and 7.5. Macroscale electrolysis at the potentials of waves I and II was followed electrochemically and by UV spectroscopy, with isolation and identification of wave I products [102].

Because of the important role of protonation in these reactions, it is perhaps pertinent to underline the difference in behaviour between 4-aminopyrimidines and adenine and cytosine. The latter have long been known to undergo protonation in acid medium at the ring nitrogen *ortho* to the amino group, with distribution of the charge between the ring nitrogen N(1) in the case of adenine, and N(3) in the case of cytosine, and the amino nitrogen (see Markowski et al. [109], and references therein). By contrast, 4-aminopyrimidines unsubstituted in the 1-position are reported not to protonate on the amino group (see Riand et al. [110], and references therein), but only on the ring nitrogens, and predominantly on the ring N(1), i.e. on the ring nitrogen *para* to the amino group. Two products have been isolated from the macroelectrolysis of 4-aminopyrimidine, one not absorbing the ultraviolet, whereas the other, ascribed to a hydrodimer formed following 1 e reduction, absorbs strongly in the 225 and 305 to 310 nm regions [102]. The initial 1e nucleophilic attack on the pyrimidine ring results

Table III. Transformation of electrolysis product of 4-aminopyrimidine[a] by ultraviolet irradiation

pH	Hg presence[d]	Irradiation[b] period min	Ultraviolet spectra of solutions[c]			
			Before irradiation		After irradiation	
			λ_{max}	ε_{max}	λ_{max}	ε_{max}
0.4	No	25	305	6500		—[e]
			225	2800		
0.4	Yes	25	305	6500	235	20000
			225	2800		
7.0	No	25	305	4800	300	3600[f]
			225	5200	225	4000[f]
7.0	Yes	15	305	4800	300	3600[f]
			225	5200	225	4000[f]

[a] Solutions were originally electrolyzed at pH 5.0

[b] Solutions were irradiated with a RPR 3000 A lamp with a photon intensity of 4×10^{17} quanta/ml/min.

[c] Values for the reduced solutions were calculated assuming a concentration of the reduction product equal to that of the original 4-aminopyrimidine before electrolysis. Values for dimers should be about twice as great.

[d] Solution of electrolysis product was irradiated in presence or absence of mercury drop

[e] — no absorption maximum observed

[f] Ratio of ε after irradiation to ε before irradiation equals 0.75–0.77.

in reduction of the $N_3=C_4$ bond and formation of the 4,4'-dihydrodimer [102]. At pH 6 and 7 the product seems to undergo ring fission, probably as a result of rearrangment [102,111] with hydrolysis and oxidation being other possible factors.

The proposed reaction pathway for the electrochemical reduction of 4-aminopyrimidine is rather complex and the reader is referred to the original publication [102].

Photochemical transformations were carried out to determine whether the 4-aminopyrimidine wave I reduction product is susceptible to photochemical oxidation, as in the case of other pyrimidine derivatives [2,3,5,96]. The resulting data are summarized in Table III. Irradiation at 300 nm or 254 nm of a freshly electrolyzed solution subsequently brought to pH 0.5, in the presence of mercury, did in fact lead to partial (about 50%) regeneration of 4-aminopyrimidine. Such partial regeneration, however, also occurred spontaneously in the presence, but not in the absence, of air (oxygen). The fact that this reaction occurred only in the presence of mercury is rather puzzling, but points to involvement of some adsorption process(es).

Purine nucleosides are known to form complexes with mercury (II) in aqueous solutions [112,113]. It is of some interest to the foregoing that mercuric chloride was shown to accelerate decomposition of monoprotonated adenine and purine derivatives [112,113].

VII.2.1 4-Amino-2,6-Dimethylpyrimidine

It was reported by Smith and Elving [74] that the single pH-dependent polarographic wave exhibited by 4-amino-2,6-dimethylpyrimidine in aqueous medium was due to a 4e process proceeding via an initial 2e — 2 H^+ reduction of the N(3)—C(4) bond to give the 3,4-dihydro derivative (Scheme 13). This, in turn, underwent reductive deamination to 2,6-dimethylpyrimidine, which was postulated to further undergo instantaneous reduction to 3,4-dihydro-2,6-dimethylpyrimidine [74].

It should, however, be noted that the crude electroreduction product of 4-amino-2,6-dimethylpyrimidine, when irradiated at 254 nm in neutral aqueous medium, was partially converted to a product similar to that reported by Wierzchowski and Shugar [114] as the major photochemical product of 4-amino-2,6-dimethylpyrimidine. It is therefore conceivable that one of the products of reduction is a dimer, which photochemically dissociates to the parent 4-amino-2,6-dimethylpyrimidine, and that this, in turn, undergoes further photochemical transformation (unpublished results).

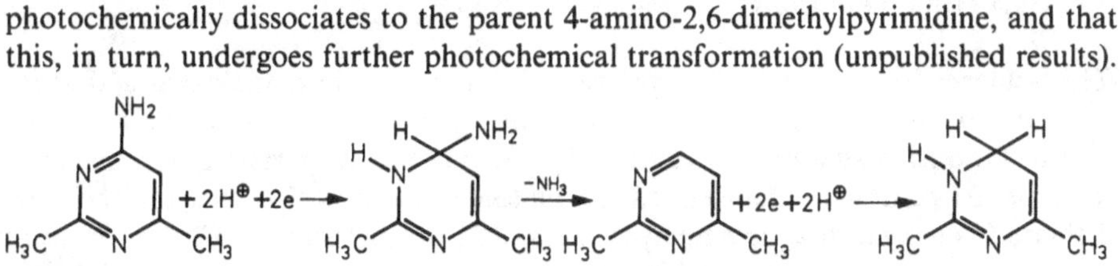

Scheme 13. Interpretation of the electrochemical and chemical behaviour observed for 4-amino-2,6-dimethyl pyrimidine (Ref. [74])

VIII Halogenopyrimidines

VIII.1 Chloropyrimidines

The electrochemical reduction of 2,4-dichloropyrimidine and 2,4,6-trichloropyrimidine in aqueous medium proceeds via cleavage of the C—Cl bonds at positions 4 and

4,6, respectively, with formation of 2-chloropyrimidine and chloride ion [115]. The reduction process is preceded by protonation of the ring.

The ease of reduction increases with the number of substituted chlorine atoms according to the decrease in energy of the lowest unoccupied C—Cl orbital. Addition of a chlorine at C(2) does not decrease the energy of the lowest unoccupied C—Cl orbitals, the energies of the anti-bonding C—Cl and C—C orbitals of 2-chloropyrimidines being of the same order of magnitude [116,117], so that the C—Cl bond of 2-chloropyrimidine is reduced at the same potential as the C=C bond of the pyrimidine ring. In acid medium (pH 1–3), 2-chloropyrimidine is reduced by a 3-electron process, leading to formation of a dimer reduction product [115]. At more elevated pH, 4–8, a 4-electron process results in formation of dihydropyrimidine, as for pyrimidine itself. The reduction pathways are shown in Scheme 14. The products are rapidly degraded on exposure to light and oxygen, and are coloured red like the reduction products of pyrimidine. Irradiation at 254 nm of the dimer reduction product does not lead to photodissociation to monomers, as in the case of the dimeric product of 2-oxo-pyrimidine. It is, nonetheless, of interest that irradiation of 2-chloropyrimidine in aqueous medium at 254 nm results in its conversion to pyrimidone-2, presumably by hydration of the 1,2 bond and elimination of HCl (unpublished obervations).

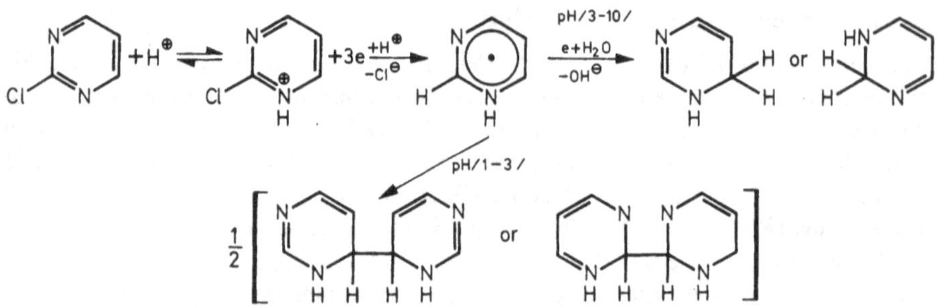

Scheme 14. Mechanism of electroreduction of 2-chloropyrimidine

VIII.2 5-Halogenouracils

The 5-chloro, 5-bromo and 5-iodo derivatives of uracil are base analogues of thymine and, in cellular systems, can replace the latter in DNA [118]. Furthermore, 5-iodo-2′-deoxyuridine is an antiherpes agent currently used for treatment of ocular herpes keratitis. By contrast, 5-fluorouracil can replace uracil in RNA, and, together with 5-fluoro-2′-deoxyuridine, is employed in tumour chemotherapy. All the foregoing are also known mutagens, and 5-bromouracil and its deoxyriboside are widely employed in studies on mutagenesis [119].

The electrochemical reduction of these compounds has been investigated with the aid of polarography, preparative electrolysis, coulometry and chemical analysis of products [120]. The chloro, bromo and iodo analogues, as well as their nucleosides and nucleotides, in buffered media, exhibit a single irreversible polarographic wave, involving a 2e reduction to uracil and the corresponding halogen ion. In the case of 5-fluorouracil there is a single non-reversible wave involving a 4e reduction, with formation of 5,6-dihydrouracil and the fluoride ion (Scheme 15). From the observed reduction potentials, $E_{1/2}$, it was concluded that the ease of reduction of the carbon-halogen bond decreases with the electronegativity of the halogen substituent, i.e. in the order I > Br > Cl > F, in accordance with expectations. The ease of reduction is somewhat enhanced when the N(1) is substituted, as in the case of nucleosides and nucleotides (Table IV).

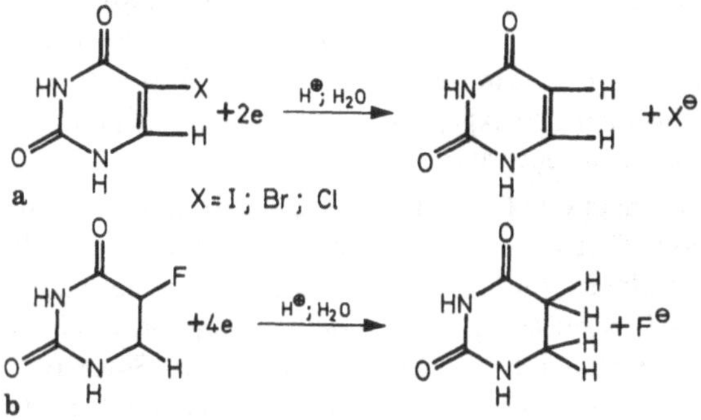

Scheme 15. Mechanism of electroreduction of 5-halogenouracils

Table IV. Polarographic data for the reduction of halogenated uracils and their glycosides

Compound	pH range	$-E_{1/2}$ (V)	pK_a^a	pK_a^b	Diffusion current constant (I_d)
5-I-uracil	1.0–8.4	1.02	8.25	8.40	4.3 ± 0.2
	8.4–13	1.02 + 0.1/pH			
5-I-uridine	1.0–8.9	0.92			
	8.9–13	0.92 + 0.1/pH	8.50	8.90	3.3 ± 0.2
Oligonucleotides of 5-I-uracil	7.4	0.96	—	—	
5-Br-uracil	6.1–8.6	1.62 + 0.02/pH			
	8.6–11	1.67 + 0.08/pH	8.05	8.60	4.7 ± 0.1
1-Me-5-Br-uracil	6.2–8.4	1.60 + 0.02/pH	8.30	8.40	3.8 ± 0.2
	8.4–11	1.65 + 0.07/pH			
1,3-DiMe-5-Br-uracil	6.1–13	1.60	—	—	2.9 ± 0.2
5-Br-deoxyuridine	6.2–8.3	1.48 + 0.04/pH	8.10	8.30	3.1 ± 0.2
	8.3–11	1.56 + 0.08/pH			
5,6-Dihydro-5-Br-uracil	3.5–11	—	—	—	—
5-Cl-uracil	6.0–7.9	1.75 + 0.01/pH	7.95	7.90	4.3 ± 0.1
	7.9–10	1.77 + 0.05/pH			
1,3-DiMe-5-Cl-uracil	7.0–10	1.80	—	—	4.5 ± 0.3
5-Cl-uridine	6.0–8.6	1.65 + 0.02/pH	8.50	8.60	3.3 ± 0.4
	8.6–11	1.70 + 0.07/pH			
Oligonucleotides of 5-Cl-uracil	7.4	1.75	—	—	—
5-F-uracil	7.4	1.845	7.8	—	8.9 ± 0.8

[a] By spectrophotometric titration; data from Berens and Shugar [125],
[b] By polarography; data from Wrona and Czochralska [120]

The facility with which the bromo, chloro and iodo analogues undergo reduction, with maintenance of the 5,6-double bond and replacement of the halogen by hydrogen, has been exploited for the electrochemical synthesis of 5-(^{3}H)-uracil by conducting the reaction in tritiated water [121]. This procedure should be equally effective for the synthesis of labelled uracil nucleosides and nucleotides.

The observation that oligonucleotides containing 5-chlorouracil residues readily undergo reduction [120] points to the utility of a more extensive investigation of the behaviour of halogenated uracil oligo- and polynucleotides. Furthermore, it is rather surprising that no efforts have been made to examine the electrochemical reduction of 5-halogenocytosines, although this might be expected to be more complex because of possible accompanying reduction of the cytosine ring.

Relevant to the mechanism of electroreduction of 5-halogenouracils are reported results on the photochemical transformation of these compounds in aqueous medium [122-124]. The latter reactions have also been studied at the level of polynucleotides. In particular, replacement of thymine in DNA by 5-iodouracil or 5-bromouracil (both of which are thymine analogues) enhances the sensitivity to UV irradiation [126,127]. Since both of these analogues exhibit absorption at wavelengths to the red of 310 nm, where the natural bases do not absorb, the use of such irradiation wavelengths results in photochemical reactions involving these halogeno uracils. One of these reactions leads to the formation of 5,5′-diuracil [122-124], i.e. a dimer with a C—C linkage between two pyrimidine rings. Efforts to clarify the mechanism of this reaction have been based largely on a study of the photochemical behaviour of aqueous solutions of 5-bromouracil and 5-bromouridine. The resulting products are diuracil, uracil and, to a lesser extent, barbituric acid, isoorotic acid and oxalic acid. It has been proposed that the reaction pathway involves homolytic cleavage of the C—Br bond, with formation of uracil and Br radicals. The uracil radical may dimerize to 5,5′-diuracil, or extract a hydrogen atom from a neighbouring molecule to form uracil [123]. It is not clear, however, what is the fate of the Br radicals, or the mechanism of formation of uracil from uracil radicals. In particular it is not clear which molecules are the hydrogen donors, water molecules being excluded as donors on the basis of other observations.

At least in the case of 5-bromouracil, both the photochemical and electrochemical reactions lead to cleavage of the carbon-halogen bond with formation of similar products, the photochemical reaction involving homolytic cleavage, and the electrochemical reaction rather heterolysis cleavage. It is, furthermore, of special interest that dehalogenation of incorporated 5-bromouracil in DNA may occur in the absence of irradiation, with the consequent formation of uracil residues. The biological consequences of this reaction in mutagenesis and repair have been extensively described [128], and references therein).

As in the case of electrochemical reduction, the photochemical transformation of 5-fluorouracil derivatives differs from that of the other 5-halogeno uracils. The primary photoproduct of 5-fluorouracil, its glycosides and poly(5-FU) is the photohydrate. However, at shorter wavelengths of irradiation, e.g. 254 nm where the photohydrate exhibits absorption, there is elimination of HF from the 5,6 bond and formation of barbituric acid [129-131]. There is also some evidence for acetone photosensitized formation of cyclobutane dimers of 5-fluorouracil [132], as well as dimer formation in irradiated poly(5-FU) [133].

IX Thiopyrimidine Analogues

IX.1 Thiouracils

Interest in thio derivatives of ketopyrimidines stems in part from the biological activities of some of these, as well as their appearance in various tRNAs.

2-Thiouracil, the neutral form of which is 2-thione-4-keto [134, 135], does not undergo electroreduction in aqueous medium. However it is strongly adsorbed at the surface of the mercury electrode [136], and gives rise to an anodic wave due to formation of a mercury salt [137]. By contrast, the fixed tautomeric forms, 2-thiol and 4-enol, with additional conjugated double bond systems in the ring, are readily reducible [136] e.g. the 4-enol, 4-ethoxy-2-thiouracil, undergoes a 4e, $4\,H^+$ process, involving 2e reduction of the N(3)—C(4) bond, elimination of the ethoxy substituent to form 2-thiopyrimidine, and 2e, $2\,H^+$ reduction of the latter to the 3,4-dihydro derivative. The 2-thiol species, e.g. 2-methylthiouracil, which may exist in two prototropic forms, with a hydrogen on N(1) or N(3), (see Scheme 16) undergoes a two-step reduction

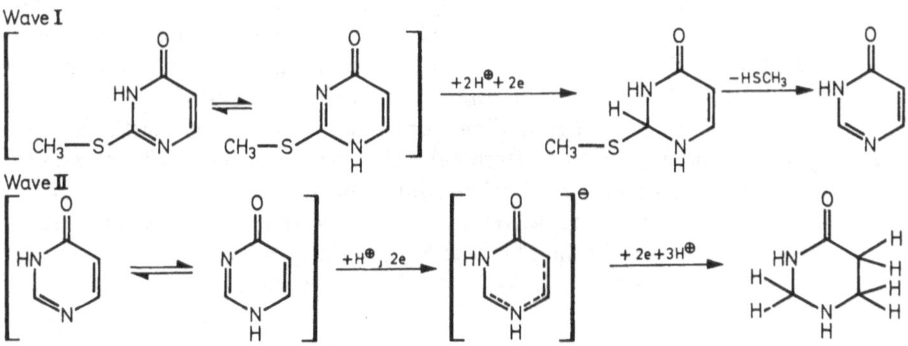

Scheme 16. Proposed reaction sequence for the electrochemical behaviour of 2-methylthiouracil

identical for both forms, with elimination of CH_3SH and formation of 4-oxopyrimidine; the latter, in turn, is further reduced to 1,2,5,6-tetrahydro-4-ketopyrimidine [136].

4-Thiouracil: The neutral form of this compound in aqueous medium is 2-keto-4-thione [138]. In acid medium it, and its N-substituted derivatives, including 4-thiouridine, give rise to a typical catalytic hydrogen evolution wave [139], which disappears at pH values above 6. Simultaneously a second catalytic wave appears at pH > 4.2. Both waves possess pronounced surface characteristics, probably due to adsorption of the molecules with differing orientations at the electrode surface [139]. The second catalytic wave overlaps that accompanying reduction of 4-thiouracil, which is a 4e, $4\,H^+$ process leading to formation of 5,6-dihydropyrimidone-2, via a transition state intermediate, 5,6-dihydro-4-thiouracil [139], which is converted to the final product by a 2e, $2\,H^+$ process (Scheme 17). The electrochemical properties of 4-thiouracil, and its N-substituted derivative differ from those of other reducible natural bases, such as adenine and cytosine, so that the former may serve as a "probe" for analytical studies of tRNA structure in solution [139].

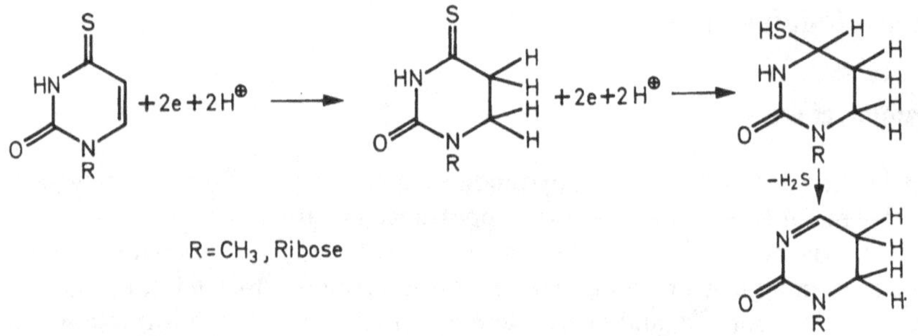

Scheme 17. Proposed pathways for the electrochemical reduction of 4-thiouracil

The polarographic reduction of the fixed thiol form of 4-thiouracil, i.e. 4-methylthiouracil, its 1-methyl analogue and 4-methylthiouridine, has been studied by means of d.c. and a.c. polarography, cyclic voltammetry at the HMDE, and preparative electrolysis [5]. The mechanism, similar for all three compounds, is a two-step reaction in the pH range 1–6; at higher pH values these merge into a single step. The first step, a 2e reduction, leads to elimination of $HSCH_3$. The second, a 1e reduction, results in formation of a free radical, followed rapidly either by dimerization to 4,4′(6,6′)-bis-(3,6-dihydropyrimidine-2), or further 1e reduction to 3,4(6)-dihydropyrimidone-2. The products are therefore identical with those resulting from reduation of pyrimidone-2 and cytosine. Furthermore, in the case of 4-methylthiouridine, the ribose moiety is unaffected, so that the dimer formed is that of the riboside (Scheme 18).

As might have been anticipated from the behaviour of 2-thiouracil (see above), 2,4-dithiouracil exhibits polarographic behaviour similar to that for 4-thiouracil [140]. In the pH range 3–11 it gives three waves, the first two of which are due to the catalytic activity of dithiouracil, and the third to the catalytic activity of the reduction product(s) formed at the potential of wave II. As for 4-thiouracil, wave II overlaps the electro-

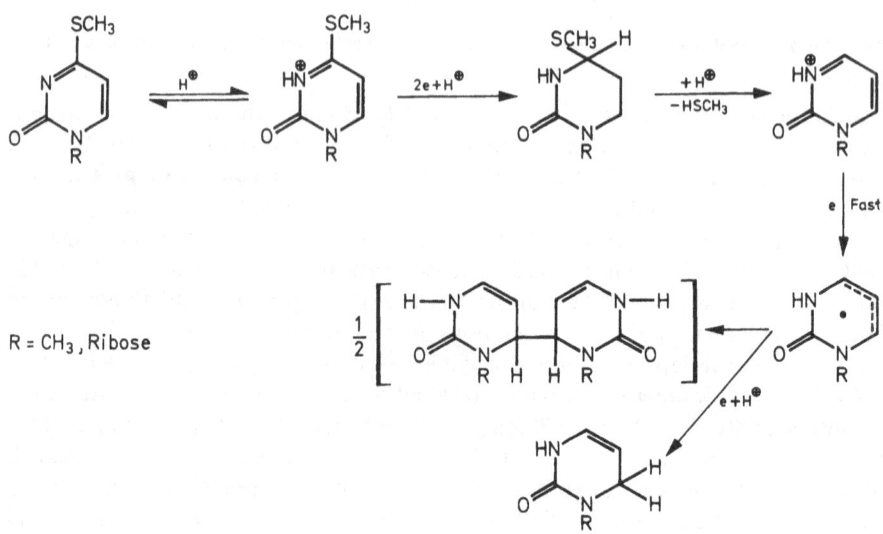

Scheme 18. Proposed pathways for the electrochemical reduction of 4-thiomethyluracil derivatives

reduction process of the dithiouracil, which leads to elimination of S^4 and formation of two products. One of these was shown to be the photoreversible dimer (4,4'(6,6')-*bis*-(3,4(6)-dihydro-2-thiopyrimidone-4). The second, major, product exhibits the same UV spectrum as the dimer, but is not photoreversible; its structure is under investigation, but it is most likely the 3,4-dihydroderivative of 2-thiouracil [140].

IX.2 Thiopyrimidines

2-Thiopyrimidine, which exists in aqueous medium predominantly in the thione form [141] together with its 4,6-dimethyl- and 1,4,6-trimethyl-derivatives, the latter of which is fixed in the thione form, are readily reducible [6,7,142] by a mechanism similar to that for pyrimidone-2, and presented in Scheme 19.

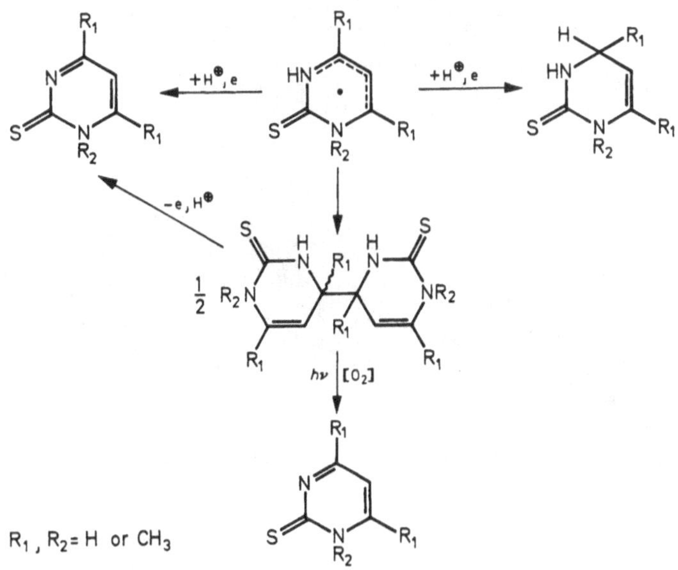

Scheme 19. Proposed reaction sequence for the electrochemical behaviour of 2-thiopyrimidine derivatives

The initial 1e, 1 H^+ reduction step leads to a free radical, which rapidly dimerizes to products isolated and identified as 4,4'(6,6')-*bis*(3,4(6)-dihydropyrimidone-2), or to the corresponding 4,6-dimethyl-, and 1,4,6-trimethyl-derivatives [6,7], These dimers are readily photodissociated to quantitatively regenerate the parent monomers in high quantum yields, 0.25 to 0.35, at 254 nm. The initial free radical may also undergo further reduction, observed only for non-methylated 2-thiopyrimidine, via a 1e process, to 3,4(6)-dihydro-2-thiopyrimidine [7].

The foregoing shows that modification of $C_2{=}O$ to $C_2{=}S$ in the pyrimidine does not affect the reduction pathway. However, for 4,6-dimethyl-2-methylthio-pyrimidine, which is in the fixed thiol form, the electroactive centre is shifted from C_2 to C_4 [7,136], and electroreduction leads to elimination of the sulfur substituent (Scheme 20).

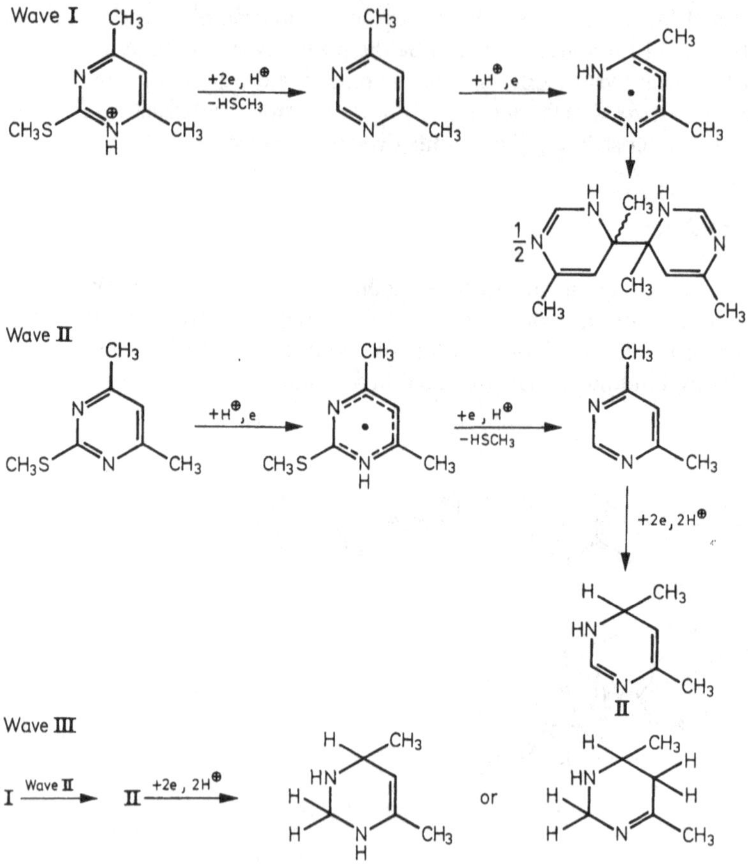

Scheme 20. Proposed reaction sequence for electrochemical reduction of 4,6-dimethyl-2-thiomethyl-pyrimidine at the DME

Table V. Comparison of $U_{1/2}^{Red}$ (or $E_{1/2}$) potentials and LUMO values for pyrimidine derivatives [142]

Molecule	pK_1[a]	$-U_{1/2}^{Red}$ (pH 7)[b] (V)	LUMO energy (eV)[c]
Pyrimidine[d]	1.31	1.303	0.115
2-Oxopyrimidine[d]	1.85	1.080	0.069
2-Thiopyrimidine (I)	1.35	1.018	0.037
4,6-Dimethyl-2-thiopyrimidine (II)	2.80	1.210	0.050
1,4,6-Trimethyl-2-thiopyrimidine (III)	3.15	1.262	0.055
4,6-Dimethyl-2-thiomethyl-pyrimidine (IV)	2.50	1.580	0.090

[a] pK_1 values taken from Ref. [141, 172]
[b] $U_{1/2}^{Red}$ at 0.5 mM concen.
[c] Calculated by the CNDO/2 method.
[d] $U_{1/2}^{Red}$ values taken from Ref. [15]

Hence, in general, electroreduction leads to removal of a thiol substituent at either C_2 or C_4, and a thione substituent only at C_4. A photodissociable dimer is formed only from those analogues, during the initial reduction step, with a thione or carbonyl at C_2, and a double bond at $N_3{=}C_4$ which permits the protonation of the ring N_3.

The experimental observations on the mechanism(s) of electroreduction of 2-thiopyrimidines have been interpreted on the basis of their electronic structures as calculated with the aid of the CNDO/2 and Hückel procedures [142]. The energies of LUMO (lowest unoccupied molecular orbitals), calculated for pyrimidine and its 2-oxo- and 2-thioderivatives, were compared with the reduction half-wave potentials (Table V). These show that the presence of a carbonyl or thione substituent at C_2 enhances the electron acceptor properties of the molecule, which are correlated with formation of a dimer susceptible to photooxidation.

IX.3 Mechanism of Photodissociation of Dihydrodimers of 2-Thiopyrimidine and 2-Ketopyrimidine

The photodissociation of the 2-thiopyrimidine dihydrodimers, D_{1-3} (Scheme 21) is really a photooxidation process since, formally, the observed products, the parent 2-thiopyrimidines, are regenerated by removal of two electrons and two protons. The same applies to photodissociation of the dimer of 2-oxopyrimidine (D_4).

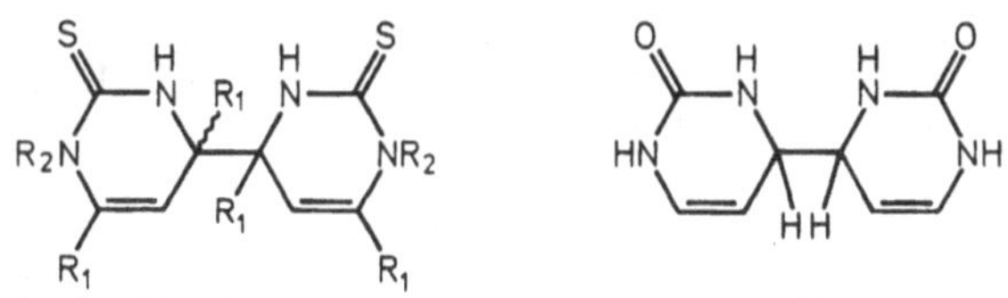

D_1: $R_1 = R_2 = H$
D_2: $R_1 = CH_3$; $R_2 = H$
D_3: $R_1 = R_2 = CH_3$

Scheme 21.

The photochemical conversion of the thiopyrimidine dimers to the parent thiopyrimidines is most readily followed by the regeneration of the spectra of the latter (Fig. 6), as for the ketopyrimidine dihydrodimers (Fig. 2). The reaction may also be followed electrochemically by measurements of the reduction wave of the regenerated parent pyrimidine. The behaviour of the thiopyrimidine dihydrodimers differs essentially from that of the ketopyrimidine dihydrodimers [140] in that, under aerobic conditions, the rates of conversion of the former are considerably more rapid and exhibit an oxygen effect (Fig. **7a**, **b**). The corresponding quantum yields under aerobic conditions are given in Table VI. Under anaerobic conditions the rates of photooxidation of D_{1-3} decrease considerably (Fig. **7a**), whereas that of D_4 is barely affected and is somewhat more complex (Fig. **7b**).

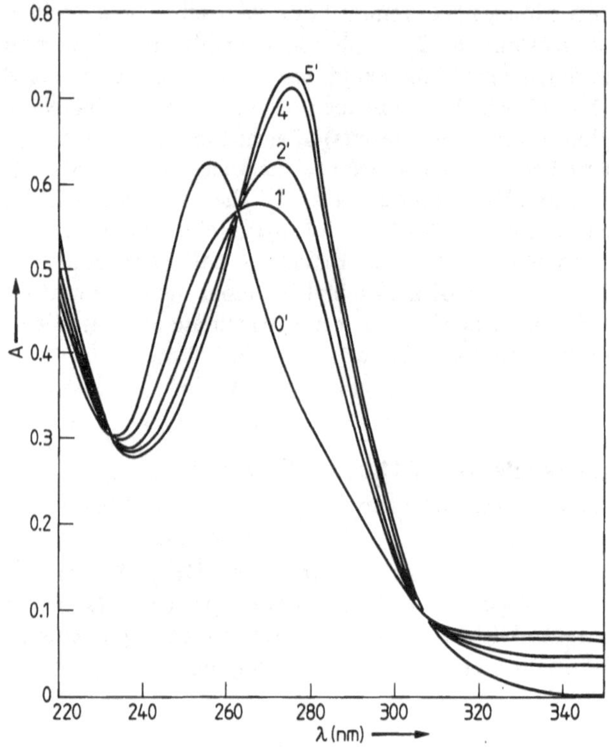

Fig. 6. Irradiation at 254 nm of dimer of 2-thiopyrimidine at pH 7.0, leading to its conversion to the parent monomer. Curve 0 is the absorption spectrum of the dimer, and curve 5 of the monomer. Intermediate curves correspond to increasing times of irradiation in min

Table VI. Quantum yields (∅) for photochemical oxidation under aerobic conditions at pH 7, and oxidation peak potentials (E_p^{ox}) for electrochemical oxidation at pH 7.3, of the dihydrodimers D_1–D_4 shown in Scheme 21.

Dimer	∅	E_p^{ox} (V)
D_1	0.24	0.270
D_2	0.35	0.235
D_3	0.23	0.230
D_4	0.02	0.450

Two mechanisms may be envisaged for the photooxidation of D_{1-4}: (a) via an intermediate state, a diradical, and (b) through formation of monomeric free radicals, as follows:

Mechanism (a)

$$\mathrm{HR{-}RH} \xrightarrow{h\nu} \mathrm{HR{-}RH^*}$$

$$\mathrm{HR{-}RH^*} \xrightarrow{O_2} {}^{-}\!\cdot\mathrm{R{-}R}\cdot^{-} + \mathrm{H_2O_2}$$

$${}^{-}\!\cdot\mathrm{R{-}R}\cdot^{-} \rightarrow 2\,\mathrm{R}$$

Mechanism (b)

$$\mathrm{HR{-}RH} \xrightarrow{h\nu} \mathrm{HR{-}RH^*}$$

$$\mathrm{HR{-}RH^*} \rightarrow 2\mathrm{RH}\cdot$$

$$\cdot\mathrm{RH}\; \begin{cases} \xrightarrow[O_2]{-e,\,-H^+} \mathrm{R} + 1/2\,\mathrm{H_2O_2} \\ \xrightarrow{\cdot \mathrm{RH}} \mathrm{HR{-}RH} \end{cases}$$

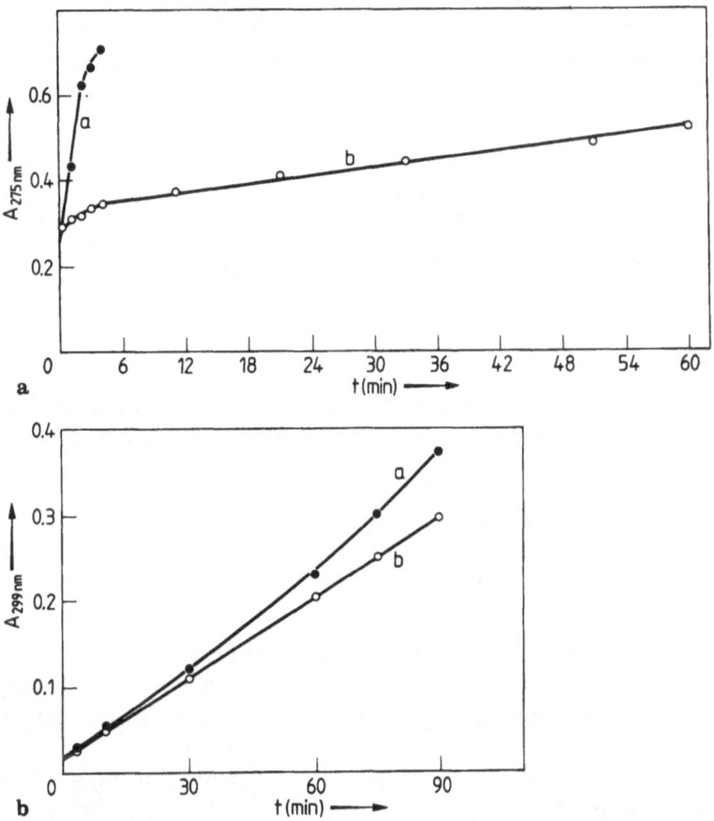

Fig. 7. **(a)** Course of appearance of 275 nm absorption band of 2-thiopyrimidine with time of irradiation of dimer at 254 nm at pH 6: a in presence of O_2, b in absence of O_2
(b) Course of appearance of 299 nm absorption band of pyrimidone-2 with time of irradiation of dimer at pH 7: a in presence of O_2, b in absence of O_2

Both mechanisms require an acceptor of protons and electrons which, as shown by preliminary data, may be molecular oxygen. Attempts to distinguish between the two mechanisms were based on the use of other oxidizing agents and kinetic studies [143]. In the case of mechanism (b), and in the presence of an excess of oxygen, the bimolecular reaction of dimerization of free radicals competes with the oxidation reaction. Hence, with an increase in the initial concentration of dimer, one should observe inhibition of the photooxidation of the latter as a result of an increase in the rate of dimerization of the free radicals, leading to regeneration of the dimer.

Electrochemical Oxidation of Dimers

The dimers D_{1-4} readily undergo electrochemical, irreversible, oxidation under anaerobic conditions on the pyrolytic graphite electrode, with oxidation peak potentials, determined by linear sweep voltammetry on 1 mM solutions, as shown in Table VI. Note the much lower values of E_p^{ox} for dimers D_{1-3}, relative to D_4, testifying to the greater susceptibility of the former to oxidation.

Preparative electrolysis at potentials corresponding to the oxidation peaks led to quantitative regeneration of monomers. In the case of D_1 the reaction proceeds beyond the stage of the parent monomer, since both exhibit almost identical oxidation potentials. The following pathway may therefore be pictured for electrochemical oxidation of the four dimers:

$$HR{-}RH \xrightarrow{-2e,\,-2H^+} \cdot R{-}R\cdot \rightarrow 2\,R$$

The absence of reversibility is merely a reflection of the very rapid dissociation of the intermediate dimer diradicals.

Stability of Dimers D_{1-4}

In aqueous medium, in the presence of oxygen, these dimers undergo rapid autooxidation to regenerate the parent monomers, probably by the following pathway (but see below):

$$HR{-}RH + O_2 \rightarrow {}^{-}\!\cdot R{-}R\cdot^{-} + H_2O_2$$
$$\downarrow$$
$$2\,R$$

In addition to H_2O_2, there may also be formation of the radical intermediates $O_2^{\dot{-}}$ and $O_2H^{\dot{}}$, all of which may accelerate the reaction because of their strong oxidizing properties. Dimer D_4 is more stable under these conditions and is also resistant to H_2O_2.

Dimers D_{1-3} are also rapidly oxidized by H_2O_2 to the parent monomers, which are further oxidized, as follows:

$$HR{-}RH \xrightarrow{H_2O_2} 2\,R \xrightarrow{H_2O_2} ?$$

The final products of this reaction were not identified.

Enzymatic Oxidation

The hydrodimers D_{1-3} are also oxidized enzymatically with the horseradish peroxidase-H_2O_2 system, the reactions being considerably more rapid in this case. Under these conditions, the methylated dimers D_2 and D_3 are quantitatively transformed to the parent monomers. In the case of dimer D_1, however, formation of the monomer is not observed, since the latter appears to be instantaneously oxidized. This is most likely due to the fact that both 2-thiopyrimidine and its dihydrodimer possess identical electrochemical oxidation potentials, whereas the methylated 2-thiopyrimidines exhibit higher oxidation potentials than their dihydrodimers [142]. The mechanism of the peroxidase-catalyzed oxidation of the foregoing dimers may be represented as follows:

$$\text{Peroxidase}\,(+3) + H_2O_2 \rightarrow \text{compound I}\,(+5)$$
$$\text{Compound I} + HR{-}RH \rightarrow {}^{\dot{-}}R{-}RH + \text{compound II}\,(+4)$$
$$\text{Compound II} + {}^{\dot{-}}R{-}RH \rightarrow {}^{\dot{-}}R{-}R^{\dot{-}} + \text{Peroxidase}\,(+3) + 2\,H_2O$$
$${}^{\dot{-}}R{-}R^{\dot{-}} \rightarrow 2\,R$$

The dihydrodimer D_4, by contrast, is inactive in the peroxidase system. Summing up, the oxidation of the dimers D_{1-3} by O_2, H_2O_2 and the peroxidase system proceeds via formation of an intermediate diradical, i.e. mechanism (a).

Mechanism of Photochemical Oxidation

With all four dimers the initial rates of photooxidation decreased with increasing concentration, and the reaction proceeds according to mechanism (b), with homolytic cleavage of the C—C bond between two pyrimidine rings and formation of free radicals. The latter may then react to regenerate the dihydrodimers or undergo oxidation to the parent monomer. With an increase in concentration of the dimers, the bimolecular reaction leading to regeneration of dimers will predominate, leading to apparent inhibition of the photooxidation reaction.

For the dimers D_{1-3} the electron and proton acceptor in the oxidation reaction is oxygen, in the absence of which photooxidation is inhibited. It should be noted that the maximum concentration of dimer employed in these reactions was 0.5×10^{-4} M, as compared to the oxygen concentration in an aerated solution of 3×10^{-4} M, so that an excess of oxygen prevailed in all instances.

X Uracil Derivatives in Non-Aqueous Medium

Both uracil and thymine, which are inactive electrochemically in aqueous medium, are polarographically reducible in non-aqueous media such as DMSO and DMF [144-146]; but these processes have not been sufficiently well examined to elucidate the nature of the reactions involved, especially in the case of thymine. Cummings and Elving [145] demonstrated that uracil in DMSO undergoes a 1e reduction at a potential $E_{1/2} = -2.3$ V vs. SCE. The resulting anion radical undergoes two reactions, viz. dimerization to the dianion of the dimer, and a competing reaction involving attachment of a proton extracted from a non-reacting uracil molecule, with formation of a uracil anion and a neutral free radical. The latter, in turn, either rapidly dimerizes, or undergoes a further 1e reduction to a dihydro derivative of uracil. Analogous studies showed that uridine in DMSO undergoes non-reversible electroreduction at a less negative potential than uracil [147], interpreted as due to the electron-withdrawing, effect of the ribose moiety, and postulated to proceed via four 1e steps leading to formation of several different radicals, dimers and hydroderivatives of uridine. Each of the steps was presumed to be similar to the 1e reduction of uracil in DMSO [145,147].

However, none of the postulated intermediates or products were unequivocally identified, but were characterized on the basis of electrochemical parameters, such as the number of electrons participating in the reduction process, and its concentration-dependence. From a comparison of the results with data derived from pulse radiolysis studies on uracil in aqueous medium [148], and from calculated parameters for the electron charge distribution in uracil, the authors [145,147] proposed that the electro-active site in uracil is C(2) or C(4), for thymine C(4), and for uridine both C(2) and C(4). For uridine, however, it was not excluded that the third and fourth 1e steps involve reduction, not of the carbonyl group at C(4), but of the 5,6 double bond. No prediction was advanced as to the site(s) involved in dimerization, since this would have required calculations of the electron charge distribution in the molecular radicals. The suggestion regarding C(2) or C(4) as the probable sites of reduction is at variance with the results for the photochemical reduction of uracil (and uridine) in aqueous medium, which proceeds at the 5,6 bond; and both chemical and biological reduction reactions, which also involve the 5,6 bond, the biological process being a 2e reduction via NAD(P)H.

Electrochemical activity of poly(U) in non-aqueous medium has been recently reported by Brabec and Glazers [149], but without elucidation of products or mechanism.

It is of interest, in relation to the foregoing, that exposure to ionizing radiations of uracil, thymine and their nucleosides in aqueous medium was reported to lead to formation of dimeric products linked by a single C—C bond, presumed to be 5-5′ [150,151]. More recent experiments by Nishimoto et al. [152] involved the γ-irradiation of thymidine in neutral aqueous medium in the presence of sodium formate. This led

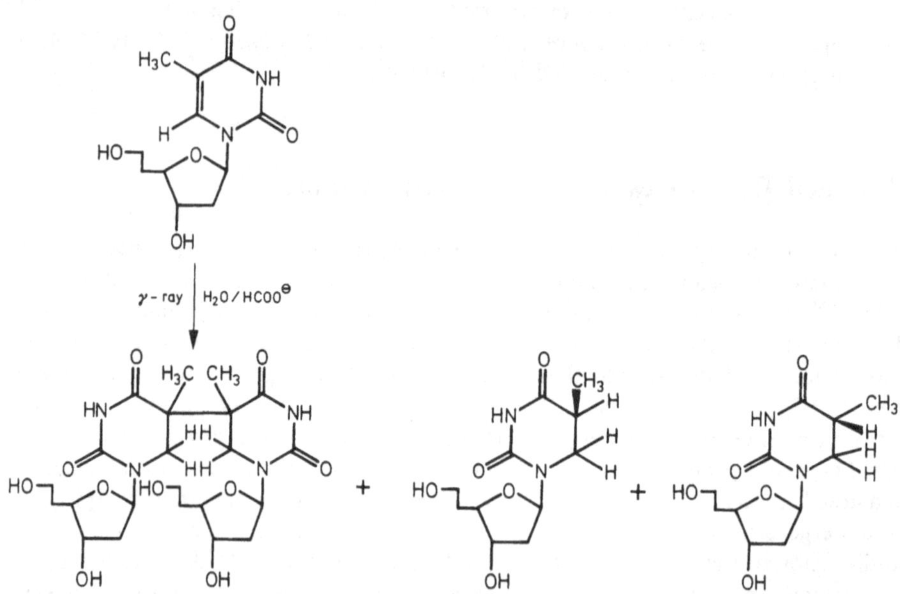

Scheme 22.

to the formation in 55% yield of an approximately equimolar mixture of the two isomeric forms of 5,6-dihydrothymidine, and a one-electron reduction product, via a hydrated electron, identified by NMR spectroscopy as 5,5'-*bis*-5,6-dihydrothymidine (Scheme 22). Formally the latter product differs from those generated electrochemically only in that the latter are linked via a 6,6' bond, and clearly underlines the potential biological significance of both the 5,5'- and 6,6'-linked dimers.

XI Purine Analogues

XI.1 Purine

Imidazole itself is polarographically inert [15,99,153] in aqueous medium. It is consequently not unexpected that electrochemical reduction of purine and some of its analogues proceeds via hydrogenation of the pyrimidine ring. The purine ring may be considered as a 5,6-disubstituted pyrimidine, and the presence of the imidazole ring "substituent" results in an increased electron density in the pyrimidine moiety relative to that of pyrimidine itself [154]; hence the initial step in the electrochemical reduction of purine in aqueous medium proceeds less readily than for the parent pyrimidine.

Purine exhibits two 2e diffusion-controlled irreversible reduction waves in aqueous buffered medium (over the pH range 2–12), corresponding to sequential reduction of the N(1)=C(6) and N(3)=C(2) bonds (Scheme 23). The potential required for initial addition of an electron to purine is so much higher than for pyrimidine [153] that the free radical species formed is instantaneously reduced, the result being a two-electron wave. Elving et al. [15,36,99,153] proposed that the initial reduction step (wave I) of purine involves a very rapid protonation of N(1) and successive one-electron transfer to the N(1)=C(6) bond, with formation of 1,6-dihydropurine (Scheme 23). Subsequent reduction of the N(3)=C(2) bond (corresponding to wave II) apparently proceeds by a similar mechanism (Scheme 23), with presumed formation of 1,3,4,6-tetrahydropurine, but this has not been experimentally established.

It is of additional interest that the product of reduction of wave I, 1,6-dihydropurine, undergoes slow reoxidation on exposure to air (see below), to regenerate purine [15], via a postulated intermediate, viz. 1-hydroxy-6-hydropurine. We shall revert to the properties of 1,6-dihydropurine, below (Sect. XI.2).

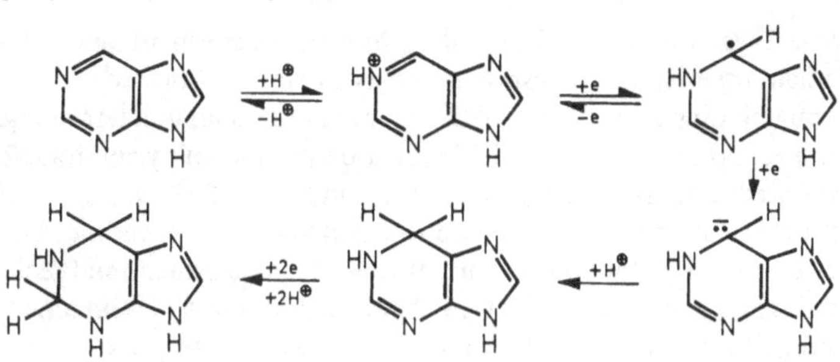

Scheme 23. Interpretation of the electrochemical and chemical behaviour observed for purine (Ref. [153])

6-Substituted purines, such as 6-alkylaminopurines and 6-methylpurine, were studied using dc and ac polarography to evaluate the effects of substitution on the electrochemical behaviour of these compounds [155]. When the 1,6 and 2,3 double bonds are present, a 4e polarographic wave, due to reduction of these bonds, is generally observed [155].

XI.2 2-Oxopurine

In view of the earlier demonstration that pyrimidone-2 undergoes one-electron reduction, with formation of a dimer identified as 6,6′-*bis*-3,6-dihydropyrimidine-2, which is suceptible to quantitative photodissociation to the parent pyrimidone-2, and bearing in mind that 2-oxopurine may be considered a formal analogue of a 5,6-disubstituted pyrimidone-2, it appeared of interest to examine whether an analogous reaction sequence occurs with 2-oxopurine.

The electrochemical reduction of 2-oxopurine in aqueous buffered medium (over the pH range 1–12) was found to proceed via two successive one-electron transfers. [4]. The initial one-electron transfer is accompanied by transfer of a proton, with formation of a free radical which rapidly dimerizes. At a more negative potential (wave II), the reduction leads to formation of 1,6-dihydro-2-oxopurine (Scheme 24).

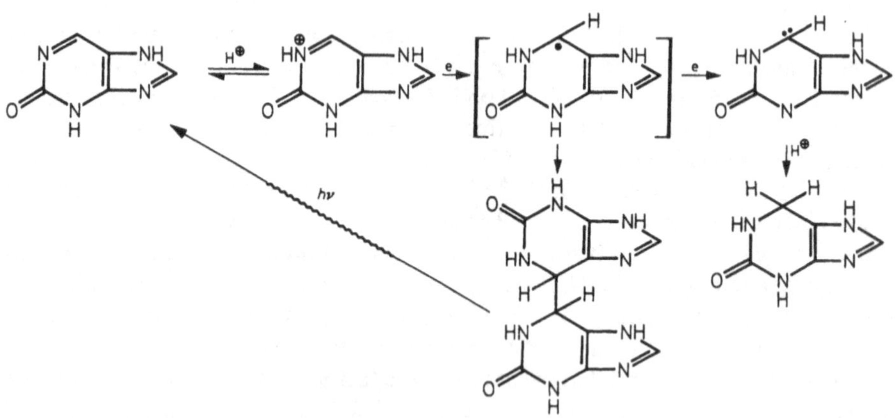

Scheme 24. Proposed reaction pathway for the reduction of 2-oxopurine in aqueous medium

The product of wave I was isolated and identified by various spectroscopic techniques as follows: elementary analysis corresponded to a dihydrated dimer of reduced 2-oxopurine; such strong binding of water has been observed in crystals of a tetrameric photoproduct of thymine, and the electrochemical reduction product of pyrimidone-2. Its UV spectrum exhibited a single short-wavelength band (λ_{max} 235 nm, ε_{max} 4.5 $\times 10^3$), consistent with elimination of one of the double bonds in the pyrimidine ring; and, in fact, the spectrum of the reduction product is that of a substituted imidazole. The IR spectrum exhibited an intense band at 1652 cm^{-1}, pointing to maintenance of the 2-keto structure. The mass spectrum exhibited several intense peaks at m/e > 200, plus an intense peak corresponding to the parent 2-oxopurine, indicating that the reduction product is a dimer. The ^1H NMR spectrum of the product in CD_3COOD exhibited two singlets at 4.63 ppm and 8.09 ppm, due to the ring H(6) and H(8), respectively.

The foregoing suggests that the structure of the reduction product is a dimer, as shown in Scheme 22, and supported by comparison with the dimer reduction product of pyrimidone-2 (Scheme 2), since the reduced pyrimidine rings in both dimers should be similar. The foregoing interpretation is confirmed by the results of earlier studies on 1:1 photoadducts of alcohols to purine [156,157]; the products were identified as substituted 1,6-dihydropurines, and the site of addition of the alcohols was readily established as C(6) on the basis of the ^1H NMR spectra prior to, and following, selective deuteration at C(6) or C(8) [156,157].

The mechanism of electroreduction of 2-oxopurine (Scheme 24) is therefore similar to that for pyrimidone-2, consistent with the fact that the imidazole ring does not undergo electroreduction.

In neutral aqueous medium the dimer reduction product, but not the product of reduction on wave II, underwent photodissociation to the parent 2-oxopurine, with a quantum yield at 254 nm of 0.03, as compared to 0.1 for photodissociation of pyrimidine-2.

The dimer reduction product of 2-oxopurine also quantitatively regenerated the parent monomer in an alkali catalyzed reaction at room temperature [4]. It is of interest that similar alkaline lability is exhibited by the *trans-anti* cyclobutane photodimer of thymine [158]. The alkaline lability of the purine dihydrodimer may be interpreted as follows: the ionized form of the dimer (in alkaline medium) is formally analogous to adducts of two radicals generated by ionizing radiation by the addition of an electron and a proton to pyrimidines [159] or purines [160] ionized in alkaline medium. It was shown by Rao & Hayon [159] that ketone radicals are in the "enol" form and are strong reducers, i.e. they may undergo rapid oxidation. In the case of the ionized 2-oxopurine hydrodimer, this may lead to transfer of two electrons to some acceptor, with concomitant conversion of the dimer to monomers, as follows:

$$HR - RH + 2\,OH^- \rightarrow R^- - R^- + 2\,H_2O$$
$$R^- - R^- \rightleftharpoons 2\,R^{\overline{\cdot}}$$
$$2\,R^{\overline{\cdot}} + 2\,A \rightarrow 2\,A\cdot + 2\,R^-$$

The properties of the reduction product of 2-oxopurine may be relevant to the observation that the reduction product of wave I of purine undergoes slow oxidation to the parent purine in the presence of air (oxygen). The regeneration process was suggested to involve formation of 1-hydroxy-6-hydropurine [15,153].

In contrast to the stability of 1,6-dihydro-2-oxopurine at neutral pH, 1,6-dihydropurine is slowly reoxidized in the presence of air, the reaction being very slow at neutral pH. When a solution of 1,6-dihydropurine was prepared by electrolysis of purine at pH 5.3, as described by Smith & Elving [153], and brought to pH 7 and irradiated at 254 nm, the UV absorption maxima of the dihydropurine at 240 nm and 292 nm were gradually replaced by the single band at 263 nm, characteristic of the parent purine itself. This photochemical conversion was quantitative and proceeded with a quantum yield of 0.03. Under these conditions a control non-irradiated solution exhibited barely perceptible spontaneous reoxidation to the parent purine (unpublished results).

A comparison of the structure of 1,6-dihydropurine with those of the 1:1 photoadducts of alcohols to purine reported by Linschitz & Connoly [156,157], which are formally identical, suggests that the latter should be equally susceptible to photodissociation to the parent purine and alcohol. The same probably applies to the photoadducts of alcohols to 6-methylpurine.

Of considerable interest in relation to the results of Linschitz & Connolly was the subsequent finding of Wolfenden et al. [161] that the photochemical adduct of methanol to purine ribonucleoside is a potent inhibitor of adenosine deaminase, in large part because of its structural resemblance to the presumed transition state intermediate involved in the deaminase reaction.

XI.3 Adenine

Of some relevance to this review is adenine, which is a component of nucleic acids, and of key nucleotide coenzymes like NAD^+. This base, and its nucleosides and nucleotides [35], exhibit a single reduction wave, with about the same total current as the parent purine, of the magnitude expected for a 4e process [15,153]. On controlled-potential electrolysis, however, it undergoes a 6e reduction to give the same product as does purine in its overall 4e reduction [15,35,36,153], as shown in Scheme 25. The reduction of adenine is accompanied by catalytic hydrogen evolution, so that it is not possible to determine directly the number of electrons involved.

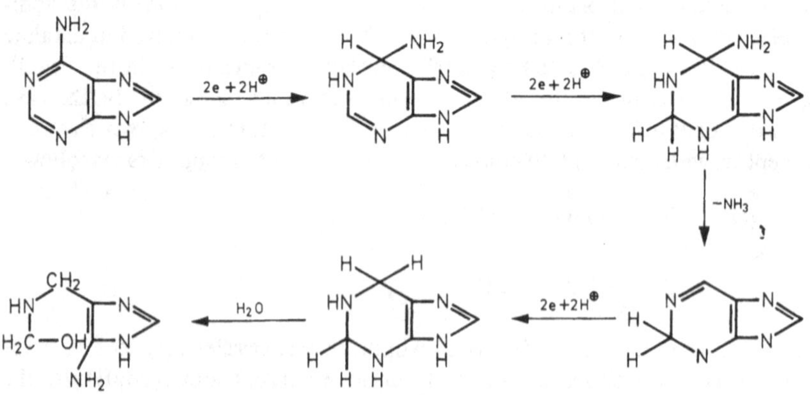

Scheme 25. Interpretation of the electrochemical and chemical behaviour observed for adenine (Ref. [153])

The mechanism of reduction is postulated to involve a 2e reduction of the 1,6 double bond, and immediate 2e reduction of the product, to give 1,2,3,6-tetrahydroadenine. This, in turn, undergoes reductive deamination to form 2,3-dihydropurine, which is further reduced to 1,2,3,6-tetrahydropurine, followed by cleavage of the 2,3 bond with formation of the same diazotizable amine as in purine reduction [153,161,162]. One reported observation in apparent conflict with this scheme is the detection of purine as an intermediate [163]. On cyclic voltammetry at the H.M.D.F., adenine exhibits a single pH-dependent cathodic peak, with no anodic peak on the return scan [36].

An additional characteristic, common to all purine analogues, and ascribed to the presence of the imidazole ring is the strong adsorption of adenine and its nucleosides and nucleotides at the electrode surface [35–37, 53, 153], with a standard free energy for adsorption of 29–38 kJ/M [37, 53].

Bearing in mind that one of the products of electroreduction of 4-aminopyrimidine, which may be regarded as a formal analogue of adenine, undergoes photodissociation [102], it might be expected that similar behaviour would be exhibited by the product of the potential one-electron reduction step of the neutral molecule of adenine. It is therefore of interest that the product of the one-electron reduction of adenine in DMF (non-protic solvent) exhibits a UV spectrum with a maximum at 300 nm [164], like the product of reduction of 4-aminopyrimidine [104].

Adenine N_1-oxide gives a pattern of three polarographic waves over the pH range of 1,4 to 5,6 [15]. Wave I is produced by the two-electron reduction of the N-oxide to adenine [15, 165a], whose own reduction at more negative potential gives rise to wave II. Wave III arises only from a catalytic hydrogen process.

Electrochemical reduction of 8-bromo-AMP proceeds at a potential more positive than that for the parent AMP, and leads to reduction of the C—Br bond with formation of AMP. This reaction also proceeds at the level of 8-bromo-NAD^+ [165b].

XI.4 Thiopurines

The reduction pathway for 2-thiopurine at the DME, and the pyrolytic graphite electrode (PGE), embraces three steps [166]. The first is a one-electron reduction to a free radical which, below pH 5, undergoes dimerization and/or a second one-electron reduction to the 1,6-dihydro derivative. Above pH 5 the second wave is not observed and the free radical dimerizes to 6,6'-*bis*-(1,6-dihydro-2-thiopurine) which, in turn, was postulated to undergo further reduction to 1,6-dihydropurine, accounting for the third wave [166].

The consequent similarities in polarographic reduction of 2-thiopurine and 2-oxopurine are analogous to the similarities in photochemical reduction of pyrimidone-2 and pyrimidone-2-thione, and of photodissociation of their dimer reduction products. The dimer reduction product of 2-thiopurine, obtained by electrolysis at pH 5, undergoes photodissociation at 254 nm to the parent monomer with a quantum yield comparable to that for the dimer reduction product of 4,6-dimethylpyrimidine-2-thione (Scheme 26).

The electrochemical reduction mechanism of 6-thiopurine [167] is pH-dependent and rather complex, with formation of 4 waves in the pH range 0–9 (Scheme 27). Coulometry at the potential of wave I revealed rapid release of H_2S, but catalytic production of hydrogen ions prevented determination of the electron number. The postulated proposed mechanism of reduction was linked to protonation and, as for 4-thiouracil [139], involves the thione form, the final product being the tetrahydroderivative.

A number of 6-thio alkyl and carboxyalkyl purines were reported to be polarographically reducible via one 4e wave or two 2e waves, except for analogues with a 2-oxo substituent, in which case there is no 2,3 double bond and only one 2e or two 1e waves are observed [168]. No mechanism for reduction was formulated but, from the known behaviour of 6-thiopurine and 4-methylthiouracil [167, 5], it is most likely

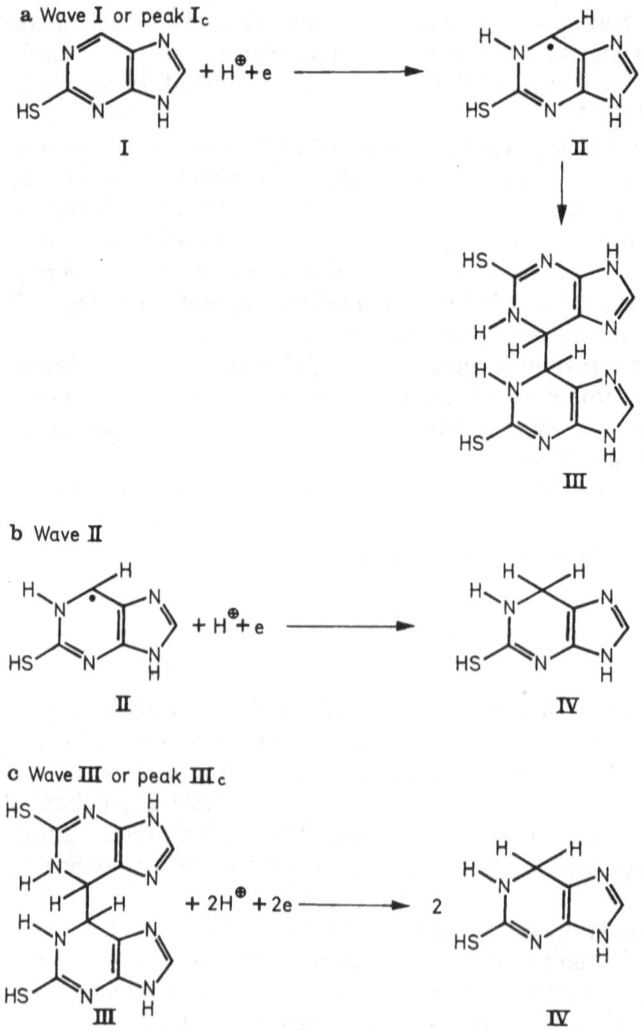

Scheme 26. Interpretation of the electrochemical and chemical behaviour observed for 2-thiopurine (Ref. [166]) at the DME

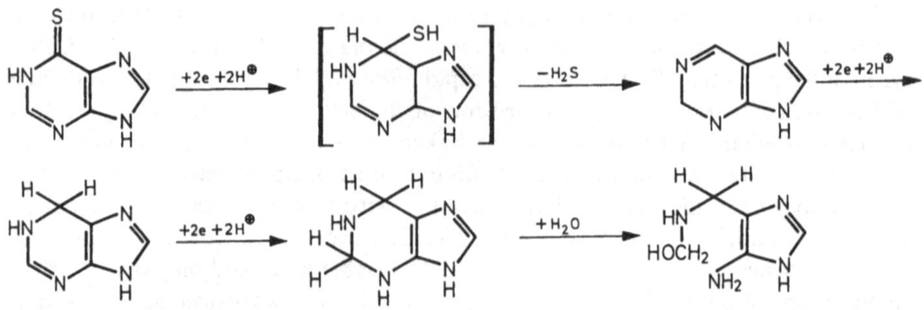

Scheme 27. Interpretation of the electrochemical and chemical behaviour observed for 6-thiopurine

that reduction proceeds in the pyrimidine ring with elimination of sulphur, e.g. methyl sulfide. In the case of the 2-oxo derivatives, one might expect formation of photodissociable dimer products, as for 4-methylthiouracil [5].

XI.5 Purines in Non-Aqueous Media

In non-aqueous media (DMF, DMSO, acetonitrile), purine and some of its 6-substituted derivatives (adenine, 6-methylpurine, 6-methylaminopurine, 6-methoxypurine, 6-dimethylaminopurine) undergo an initial 1e reduction to form the corresponding anionic free radical, followed by dimerization with rate constants of 10^3–10^5 $l\,mol^{-1}$ [164]. The free radical nature of the 1e reduction product was independently established by ESR spectroscopy [169]. This behaviour is to be contrasted with the 2e and 4e reductions undergone by these compounds in aqueous media [15,36,99,155].

On addition of weak proton donors, such as water and benzoic acid [164], the 1e reduction product is further reduced at the potential of its formation, due to production of more readily reduced protonated species, so that reduction attains the level of a 4e process at a mole ratio of acid to purine of 4 (the total faradaic for these purines in aqueous media is 4). In the presence of a strong acid (perchloric), purine and 6-methylpurine exhibit two 2e waves and other 6-substituted purines a simple 4e wave. The effect of substitution at the 6-position on ease of reducibility is the same in neutral [164] and protonated purines [155].

The following scheme has been proposed for reduction of purine and 6-substituted purines in non-aqueous media [164]:

$$R + e \rightarrow R^{\cdot}$$
$$2\,R^{\bar{\cdot}} \rightarrow R_2^{2-}$$
$$R_2^{2-} + 2\,H^+ \rightarrow R_2H_2$$

and is similar to that proposed for the electrochemical reduction of pyrimidine and other azabenzenes [98,170] in nonaquous media, but is in marked contrast to the behaviour of purines in aqueous media [4,15,36,99], where free radical formation is not observed because of prior protonation and initial multiple electron (2e or 4e) reduction.

However, the structures of the resulting dimers were not directly identified, but were postulated to involve a 6,6′ linkage for purine, and a 2,2′ (or 8,8′) linkage for the adenine dimer. This proposal was based on theoretically calculated values of the free valencies of the carbon atoms, which were assessed as being highest for C(6) in purine, and C(2) and C(8) in adenine [171]. It is doubtful whether such reasoning is applicable in this instance since the carbon atom(s) with the highest free valency in the free radical forms may be quite different from those in the ground state.

XII Acknowledgments

The preparation of this review, and the results of original investigations reported, were supported by the Ministry of Science and Higher Education (MR.I.5) and the Polish Cancer Research Program (PR-6).

XIII. References

1. Czochralska, B., Shugar, D.: Experientia (Suppl.) *18*, 251 (1971)
2. Czochralska, B., Shugar, D.: Biochim. Biophys. Acta *281* 1 (1972)
3. Czochralska, B., Wrona, M., Shugar, D.: Bioelectrochem. Bioenerget. *1*, 40 (1974)
4. Czochralska, B., Fritsche, H., Shugar, D.: Z. Naturforsch. *32c*, 488 (1977)
5. Wrona, M., Czochralska, B.: J. Electroanal. Chem. Interfac. Electrochem. *48*, 433 (1973)
6. Wrona, M., Giziewicz, J., Shugar, D.: Nucleic Acids Res. *2*, 2209 (1975)
7. Wrona, M.: Bioelectrochem. Bioenergetics *6*, 243 (1979)
8. Varghese, A. J., Wang, S. Y.: Science *156*, 955 (1967)
9. Khattak, M. N., Wang, S. Y.: ibid. *163*, 1341 (1969)
10. Rhodes, D. F., Wang, S. Y.: Biochemistry *10*, 4603 (1971)
11. Shragge, P. C., Varghese, A. J., Hunt, J., Greenstock, C. L.: Radiat. Res. *60*, 25 (1974)
12. Mc Laren, A. D., Shugar, D.: Photochemistry of Proteins and Nucleic Acids, Pergamon Press, Oxford 1964
13. Lehninger, A. L.: Biochemistry, Worth Publishers Inc., New York 1972
14. Kosower, E. M.: in Free Radicals in Biology (W. A. Pryer, ed.) Vol. II Chap. I, Academic Press, New York 1976
15. Elving, P. J., Reilly, J. E. O., Schmackel, C. O.: in Methods of Biochemical Analysis, (D. Glick, ed.), *21*, 1973
16. Bresnahan, W. T., Moiroux, J., Samec, Z., Elving, P. J.: Bioelectrochem. Bioenerg. *7*, 125 (1980)
17. Wrona, M., Dryhurst, G.: Biochim. Biophys. Acta *570* 371 (1979)
18. Wrona, M. Z., Goyal, R. N., Dryhurst, G.: Bioelectrochem. Bioenerg. *7*, 433 (1980)
19. Nürnberg, H. W.: in Bioelectrochemistry I (G. Milazzo, M. Blank, ed.), Plenum Press, London — New York 1983, pp. 183
20. Palecek, E.: in Methods in Enzymology *21*, 3 (1971)
21. Elving, P. J.: in Topics in Bioelectrochem. and Bioenergetics (G. Milazzo, ed.), J. Wiley, London (1976) *1*, 179
22. Palecek, E.: Prog. Nucleic Acids Res. Mol. Biol. *18*, 151 (1976)
23. Dryhurst, G.: Electrochemistry of Biological Molecules, Academic Press, New York 1977, p. 71
24. Sequaris, J. M., Reynaud, J. A.: J. Electroanal. Chem. *63*, 207 (1975)
25. Sequaris, J. A., Reynaud, J. A., Malfoy, B.: ibid. *77*, 67 (1977)
26. Brajter-Toth, A., Goyal, R. N., Wrona, M. Z., Lacava, T., Nguyen, N. T., Dryhurst, G.: Bioelectrochem. Bioenerg. *8*, 413 (1981)
27. Palecek, E.: Topics in Bioelectrochemistry and Bioenergetics, (G. Milazzo, ed.), J. Wiley, London *5*, 65 (1983)
28. Brabec, V., Dryhurst, G.: J. Electroanal. Chem. Interfac. Electrochem. *89*, 161 (1978)
29. Brabec, V., Dryhurst, G.: Studia Biophys. *67*, 23 (1978)
30. a) Valenta, P., Nürnberg, H. W.: Biophys. Struct. Mechanism *1*, 17 (1974)
    b) Palecek, E.: Coll. Czech. Chem. Commun. *39*, 3449 (1974)
31. a) Lukasova, E., Palecek, E.: Radiat. Res. *47*, 51 (1971)
    b) Vorlickova, M., Palecek, E.: Biochim. Biophys. Acta *517*, 308 (1978)
    c) Sequaris, J. M., Valenta, P., Nürnberg, H. W.: Int. J. Radiat. Biol. *42*, 407 (1982)
32. Brabec, V.: Bioelectrochem. Bioenerg. *8*, 437 (1981)
33. Janik, B., Palecek, E.: Arch. Biochem. Biophys. *105*, 225 (1964)
34. Janik, B., Elving, P. J.: Chem. Rev. *68*, 295 (1968)
35. Janik, B., Elving, P. J.: J. Am. Chem. Soc. *92*, 235 (1970)
36. Dryhurst, G., Elving, P. J.: Talanta *16*, 855 (1969)
37. Webb, J. W., Janik, B., Elving, P. J.: J. Am. Chem. Soc. *95*, 991 (1973)
38. Cummings, T. E., Jensen, M. A., Elving, P. J.: Bioelectrochem. Bioenergetics *4*, 425 (1977)
39. Palecek, E.: Nature (London) *188*, 656 (1960)
40. Palecek, E.: Biochim. Biophys. Acta *51*, 1 (1961)
41. Janik, B., Palecek, E.: Chem. Zvesti *16*, 406 (1961)
42. Janik, B.: Z. Naturforsch. *21b*, 1117 (1966)
43. Hall, R. S.: The Modified Nucleosides in Nucleic Acids, Columbia Univ. Press, New York 1971

44. Sequaris, J.-M., Valenta, P., Nürnberg, H. W.: J. Electroanal. Chem. *122*, 263 (1981)
45. Dryhurst, G., Pace, G. F.: ibid *115*, 1014 (1968)
46. Palecek, E.: J. Mol. Biol. *20*, 263 (1966)
47. Palecek, E., Vetterl, V.: Biopolymers *6*, 917 (1968)
48. Palecek, E.: J. Electroanal. Chem. *22*, 347 (1969)
49. Brabec, V., Palecek, E.: Biophysik *6*, 290 (1970)
50. Valenta, P., Gragmann, P.: J. Electroanal. Chem., Interfac. Electrochem. *49*, 41 (1974)
51. Valenta, P., Nürnberg, H. W.: ibid. *49*, 55 (1974)
52. Filipski, J., Chmielewski, J., Chorąży, M.: Biochim. Biophys. Acta *232*, 45 (1971)
53. Webb, J. W., Janik, B., Elving, P. J.: J. Am. Chem. Soc. *95*, 8495 (1973)
54. Temerk, Y. M., Valenta, P., Nürnberg, H. W.: J. Electroanal. Chem. *100*, 77 (1979)
55. Temerk, Y. M., Valenta, P., Nürnberg, H. W.: Bioelectrochem. Bioenerg. *7*, 705 (1980)
56. Temerk, Y. M., Valenta, P., Nürnberg, H. W.: J. Electroanal. Chem. *100*, 289 (1980)
57. Temerk, Y. M., Valenta, P., Nürnberg, H. W.: ibid. *131*, 265 (1982)
58. Palecek, E.: in Prog. Nucleic Acids Res. Mol. Biol. *9*, 31 (1968)
59. Palecek, E., Jelen, F.: Bioelectrochem. Bioenergetics *4*, 369 (1977)
60. Sequaris, J.-M., Kaba, M. L., Valenta, P.: Bioelectrochem. Bioenerg. (in press)
61. Valenta, P., Nürnberg, H. W., Klahre, P.: ibid. *2*, 204 (1975)
62. Reynaud, J. A.: ibid. *3*, 561 (1976)
63. Palecek, E., Doskocil, J.: Anal. Biochem. *60*, 518 (1974)
64. Lukasova, E., Jelen, F., Palecek, E.: Gen. Physiol. Biophys. *1*, 53 (1982)
65. Sequaris, J.-M., Valenta, P., Nürnberg, H. W.: Int. J. Radiat. Biol. *42*, 407 (1982)
66. Palecek, E., Vetterl, V.: Bioelectrochem. Bioenergetics *4*, 361 (1977)
67. Brabec, V., Palecek, E.: J. Electroanal. Chem. Interfac. Electrochem. *88*, 373 (1978)
68. Sequaris, J.-M., Malfoy, B., Valenta, P., Nürnberg, H. W.: Bioelectrochem. Bioenerg. *3*, 461 (1976)
69. Malfoy, B., Sequaris, J.-M., Valenta, P., Nürnberg, H. W.: J. Electroanal. Chem. Interfac. Electrochem. *75*, 455 (1977)
70. Brabec, V., Palecek, E.: Biophys. Chem. *4*, 79 (1976)
71. Berg, H., Flemming, J., Horn, G.: Bioelectrochem. Bioenergetics *2*, 287 (1975)
72. Valenta, P., Nürnberg, H. W., Klahre, P.: ibid. 1, 487 (1974)
73. a) Sequaris, J. M., Koglin, E., Valenta, P., Nürnberg, H. W.: Ber. Bunsenges. Phys. Chem. *85*, 512 (1981)
    b) Sequaris, J.-M., Valenta, P., Nürnberg, H. W., Malfoy, B.: Bioelectrochem. Bioenerg. *5*, 483 (1978)
74. Smith, D. L., Elving, P. J.: J. Am. Chem. Soc. *84*, 2471 (1962)
75. Skaric, V., Gaspert, B., Skaric, D.: Croatica Chem. Acta *36*, 87 (1964)
76. Wang, S. Y. (ed.): Photochemistry and Photobiology of Nucleic Acids, Vol. I, Academic Press, New York 1976
77. Wang, S. Y., Rhodes, D. F.: J. Am. Chem. Soc. *93*, 2554 (1971)
78. Filippen, J. L., Giraldi, R. D. Karle, I. L., Phoades, D. F., Wang, S. Y.: ibid. *93*, 2556 (1971)
79. Brown, D. J., Foster, W. V.: J. Chem. Soc. 4911 (1965)
80. Czochralska, B., Shugar, D., Arora, S. K., Bates, R. B., Cutler, R. S.: J. Am. Chem. Soc. *99*, 2583 (1977)
81. Pfoertner, K. H.: Helvetica Chim. Acta *58*, 865 (1975)
82. Wasa, T., Elving, P. J.: J. Electroanal. Chem. *91*, 249 (1978)
83. Kosmala, J., Czochralska, B.: unpublished results
84. Czochralska, B.: Habilitation thesis, Warsaw 1975
85. Fink, R. M., Cline, R. E., Mc Ganghey, C., Fink, K.: Anal. Chem. *28*, 4 (1956)
86. Janion, C., Shugar, D.: Acta Biochim. Polon. *14*, 293 (1971)
87. Vetterl, V., Pokorny, J.: Bioelectrochem. Bioenerg. *7*, 517 (1980)
88. Bergstrom, D. E., Leonard, N. J.: Biochemistry *11*, 1 (1972)
89. Bergstrom, D. E., Leonard, N. J.: J. Am. Chem. Soc. *94*, 6178 (1972)
90. Yaniv, M., Favre, A., Barelli, B. G.: Nature (London) *223* 1331 (1972)
91. Favre, A., Michelson, A. M., Yaniv, M.: J. Mol. Biol., *58*, 367 (1971)
92. Czochralska, B., Hahn, B. S., Wang, S. Y.: Bioelectrochem. Bioenerg. *3*, 41 (1976)
93. Brown, D. J., Cowden, W. B., Grigg, G. W., Kovalak, D.: Aust. J. Chem. *33*, 2291 (1980)

94. Brown, D. J., Cowden, W. B., Strękowski, L.: ibid. *34*, 1353 (1981)
95. Kowalewski, A., Strękowski, L., Szajda, M., Walenciak, K., Brown, D. J.: ibid. *34*, 2629 (1981)
96. Cavalieri, L. F., Lowy, B. A.: Arch. Biochem. Biophys. *35*, 158 (1952)
97. Hamer, D., Waldron, D. M., Woodhouse, D. L.: ibid. *47*, 272 (1953)
98. O'Reilly, J. E., Elving, P. J.: J. Am. Chem. Soc. *93*, 1871 (1971)
99. Elving, P. J., Pace, S. J., O'Reilly, J. E.: ibid. *95*, 647 (1973)
100. Thevenot, D., Hammouya, G., Buvet, R.: Comp. Rend., Ser. C, *268*, 1488 (1969)
101. Thevenot, D., Hammouya, G., Buvet, R.: J. Chim. Physicochim. Biol. *66*, 1903 (1969)
102. Czochralska, B., Elving, P. J.: Electrochim. Acta *26*, 1755 (1981)
103. Miller, M., Czochralska, B., Shugar, D.: Bioelectrochem. Bioenerg. *9*, 287 (1982)
104. Sugino, K., Shivai, K., Sekina, T., Aoo, K.: J. Electrochem. Soc. *104*, 667 (1957)
105. Czochralska, B., Sierakowski, H.: VIII Scandinavian Meet., Sundbjerg Denmark (1979) 45
106. Brown, D. J., Short, L. N.: J. Chem. Soc. 1953, 331
107. Boarland, M. P. V., Omie, J. F. W.: # 1952, 3716
108. Asahi, Y.: Yakugaku Zasshi *80*, 1222 (1960)
109. Markowski, V., Sullivan, G. R., Roberts, J. D.: J. Am. Chem. Soc. *99*, 714 (1977)
110. Riand, J., Chonon, M. T., Lumbrosa-Bader, N.: ibid. *99*, 6338 (1977)
111. Kotchetkova, J. K., Budowsky, E. J.: Organiczeskaia Chimia Nukleinovych Kislot, Chimia, Moscow 1970
112. Simpson, R. B.: J. Am. Chem. Soc. *86*, 2069 (1964)
113. Lonnberg, H., Kappi, R., Heikkinen, E.: Acta Chim. Scand. *B35*, 589 (1981)
114. Wierzchowski, K. L., Shugar, D.: Photochem. Photobiol. *2*, 377 (1963)
115. Czochralska, B.: Roczniki Chemii *44*, 2207 (1970)
116. Elving, P. J., Pullmann, B.: Adv. Chem. Phys. *3*, 1 (1961)
117. Fukui, N., Marokuma, K., Kato, M., Yonezawa, T.: Bull. Chem. Soc. Japan *36*, 217 (1963)
118. Lozeron, H. A., Gordon, M. P.: Biochemistry *3*, 507 (1964)
119. Freese, E.: J. Mol. Biol. *1*, 87 (1959)
120. Wrona, M., Czochralska, B.: Acta Biochim. Polon. *17*, 351 (1970)
121. Bratu, C.: J. Labelled Compounds *7*, 161 (1970)
122. Ishihara, H., Wang, S. Y.: Biochemistry *5*, 2307 (1966)
123. Ishihara, H., Wang, S. Y.: Nature *210*, 1222 (1966)
124. Sasson, S., Wang, S. Y.: Photochem. Photobiol. *26*, 357 (1977)
125. Berens, K., Shugar, D.: Acta Biochim. Polon. *10*, 25 (1963)
126. Greer, S. J.: J. Gen. Microbiol. *22*, 618 (1960)
127. Djordjevic, B., Szybalski, W.: J. Exptl. Med. *112* (1960)
128. Szyszko, J., Pietrzykowska, I., Twardowski, T., Shugar, D.: Mutation Res. *108*, 13 (1983)
129. Lozeron, H. A., Gordon, M. P., Gabriel, T., Tantz, W., Duschinsky, R.: Biochemistry *3*, 1844 (1964)
130. Fikus, M., Wierzchowski, K. L., Shugar, D.: Biochem. Biophys. Res. Commun. *16*, 478 (1964)
131. Fikus, M., Wierzchowski, K. L., Shugar, D.: Photochem. Photobiol. *4*, 521 (1965)
132. Schul, S., Lee, S. M.: ibid. *29*, 1035 (1979)
133. Wang, S. Y.: in Photochem. Photobiol. of Nucleic Acids (Wang, S. Y., ed.), Vol. 1, pp. 296, New York 1976
134. Shugar, D., Fox, J. J.: Bull. Soc. Chim. Belg. *61*, 293 (1952)
135. Psoda, A., Shugar, D.: Acta Biochim. Polonica *26*, 55 (1979)
136. Wrona, M.: Bioelectrochem. Bioenerget. *10*, 169 (1983)
137. Horn, G., Zuman, P.: Coll. Czech. Chem. Commun. *25*, 3401 (1960)
138. Psoda, A., Kazimierczuk, Z., Shugar, D.: J. Am. Chem. Soc. *96*, 6832 (1974)
139. Wrona, M., Czochralska, B., Shugar, D.: J. Electroanal. Chem. *68*, 355 (1976)
140. Wrona, M.: Ph. D. Thesis, Univ. of Warsaw 1977
141. Stanownik, B., Tisler, M.: Arzneim.-Forsch. *14*, 1004 (1964)
142. Wrona, M., Geller, M.: Bioelectrochem. Bioenerg. *6*, 263 (1979)
143. Evans, R. F.: Rev. Pure and Appl. Chem. *15*, 23 (1965)
144. Komenda, J.: Symp. on Electrochemical Analysis of Nucleic Acids, Lisewsky Dour, Czechoslovakia 1975, Abstracts p. 9
145. Cummings, T. E., Elving, P. J.: J. Electroanal. Chem. *94*, 123 (1978)
146. Cummings, T. E., Elving, P. J.: ibid. *102*, 237 (1979)

147. Bresnahan, W. T., Cummings, T. E., Elving, P. J.: Electrochim. Acta *26*, 691 (1981)
148. Hayon, E.: J. Chem. Phys. *51*, 4881 (1969)
149. Brabec, V., Glazers, V.: Gen. Physiol. Biophys. *2*, 193 (1983)
150. Shragge, P. C., Varghese, A. J., Hunt, J. W., Greenstoc, C. L.:
a) Chem. Comm. 1974, 736
b) Radiat. Res. *60*, 250 (1974)
151. a) Cadet, J., Teoule, R.: Inst. J. Appl. Radiat. Isot. *22*, 273 (1971)
b) Arai, I., Hanna, R., Daves, G. D. Jr.: J. Am. Chem. Soc. *103*, 7684 (1981)
152. Nishimoto, S., Hiroshi, I., Nakanichi, K., Kagiya, T.: ibid. *105*, 6740 (1983)
153. Smith, D. L., Elving, P. J.: ibid. *84*, 1412 (1962)
154. Fischer-Hjalmars, J., Nag-Chandlurvi, J.: Acta Chem. Scand. *23*, 2963 (1969)
155. Janik, B., Elving, P. J.: J. Electrochem. Soc. *116*, 1087 (1969)
156. Linschitz, H., Conolly, J. S.: J. Am. Chem. Soc. *90*, 2979 (1968)
157. Conolly, J. S., Linschitz, H.: Photochem. Photobiol. *7*, 791 (1962)
158. Herbert, M. A., La Blanc, J. C., Weinblaum, D., Johns, H. E.: ibid. *9*, 33 (1969)
159. Rao, P. S., Hayon, E.: J. Am. Chem. Soc. *96*, 1287 (1974)
160. Van der Vorst, A., Lion, I.: Biochim. Biophys. Acta *238*, 417 (1971)
161. Wolfenden, R., Wentworth, D. F., Mitchel, G. N.: Biochemistry *16*, 5071 (1977)
162. Elving, P. J., Webb, J. W.: The Purines — Theory and Experiment, Jerusalem Symp. on Quantum Chemistry and Biochem., Jerusalem 1972
163. Kwee, S., Lund, H.: Acta Chem. Scand. *26*, 1195 (1972)
164. Santhanam, K. S. V., Elving, P. J.: J. Am. Chem. Soc. *96*, 1653 (1974)
165. a) Warner, C. R., Elving, P. J.: Coll. Czech. Chem. Commun. *30*, 4210 (1965)
b) Czochralska, B., Bojarska, E., Nürnberg, H. W., Valenta, P.: Bioelectrochem. Bioenergetics (in press)
166. Dryhurst, G.: J. Electroanal. Chem. *28*, 33 (1970)
167. Dryhurst, G.: J. Electrochem. Soc. *116*, 1098 (1969)
168. Vachek, J., Kakal, B., Cerny, A., Semousky, M.: Pharmazie *23*, 444 (1968)
169. Sevilla, M. D.: J. Phys. Chem. *74*, 805 (1970)
170. O'Reilly, J. E., Elving, P. J.: J. Am. Chem. Soc. *94*, 7941 (1970)
171. Pullmann, B., Pullmann, A.: Proc. Nat. Acad. Sci. US. *45* 135 (1959)
172. Brown, D. J.: The Pyrimidines, Interscience, New York 1962

High-Pressure Synthesis of Cryptands and Complexing Behaviour of Chiral Cryptands

Janusz Jurczak and Marek Pietraszkiewicz

Institute of Organic Chemistry, Institute of Physical Chemistry,
Polish Academy of Sciences, 01-224 Warszawa, Poland

Table of Contents

1 Introduction and Nomenclature

1.1 Introductory Remarks, Scope and Limitations

There has been growing interest in the synthetic macrocyclic molecular receptors since Pedersen's pioneering work on crown ether synthesis [1]. Separate reports, dealing with crown ether chemistry, have evolved gradually into new, highly consistent knowledge, often termed as 'supramolecular chemistry'. Many books [2-6] and review articles [7-18] have been devoted to this rapidly expanding area which has also been the subject of international meetings for a number of years. The first issue of a specialist journal [19] appeared in 1983.

The main featur of macrocyclic receptors is that they possess an appropriate array of heteroatoms capable of binding small molecules and cations. It was soon recognized that they form stable complexes with a wide range of species — inorganic and organic cations and anions, uncharged molecules, transition metal complexes, and C—H acids. Introduction of nitrogen into macrocyclic structures as bridge-head atoms created a new branch of host molecules called "cryptands"-compounds with unique complexing properties. One may note that the incorporation of nitrogen pivoting atoms into receptor molecules offer an enormous number of possible structures. Thus, the molecular cavity can be modified in a highly predictable manner, providing molecular architectures for size, shape and chiral recognition. Obviously, they do not act only as selective complexers but also function as metallo-catalysts, enzyme mimics, metallo-enzyme analogues, etc.

In order to avoid an overloading, this review is limited to bicyclic cryptands possessing nitrogen atoms in bridge-head fashion. A more detailed discussion, concerning the complexing behaviour of chiral cryptands will be given in the end of this article.

1.2. Nomenclature and Abbreviations[1]

Although the IUPAC nomenclature is recommended in the majority of journals, it can be seen clearly that the use of jargon in respect to crown ethers and cryptands enjoys a great popularity. Not surprisingly, since their exact and complicated IUPAC names are difficult to mention frequently in the text. Common abbreviations can be found almost in all review articles [2], however, for the convenience sake, we draw attention to some of them to which is referred here. Chart 1 depicts simple examples of N,N'-dimethyl diazacoronands, cryptands and more elaborated cryptands incorporating carbohydrate units. The abbreviations below each formula are easy to follow.

---

[1] For more detailed information on the nomenclature of organic neutral ligands and host-guest systems see Ref. 19.

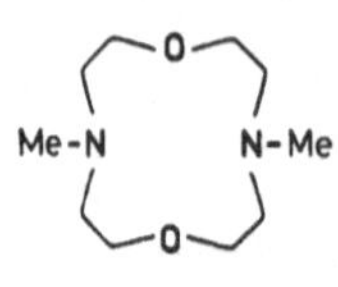

$[12]$-N_2O_2

1

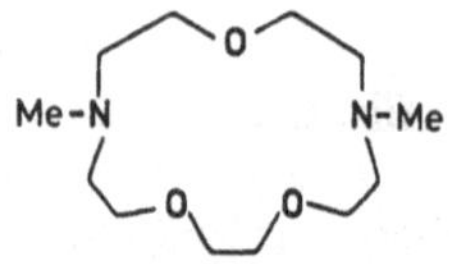

$[15]$-N_2O_3

2

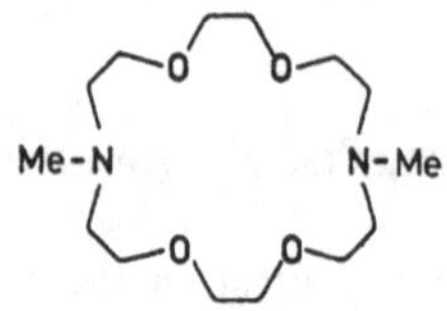

$[18]$-N_2O_4

3

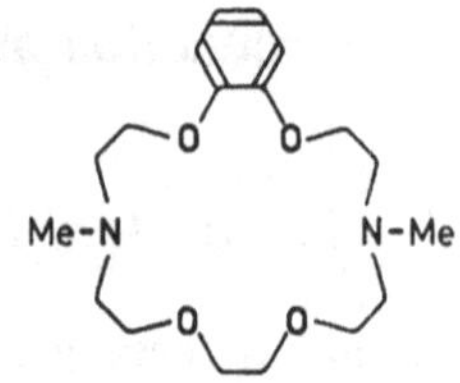

B$[18]$-N_2O_4

4

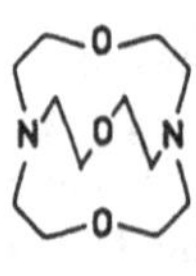

$[1.1.1]$

5

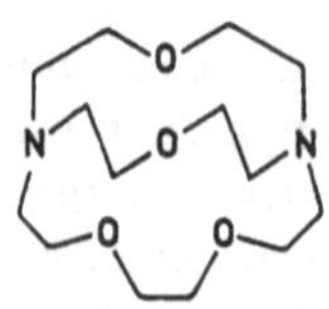

$[2.1.1]$

6

$[2.2.2]$

7

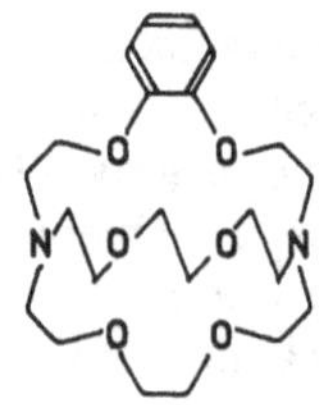

$[2_B.2.2]$

8

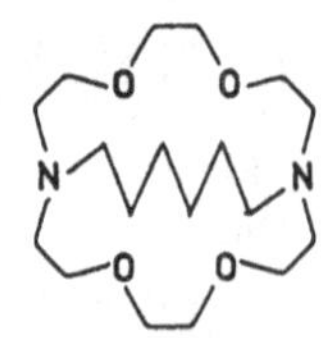

$[2.2.C_6]$

9

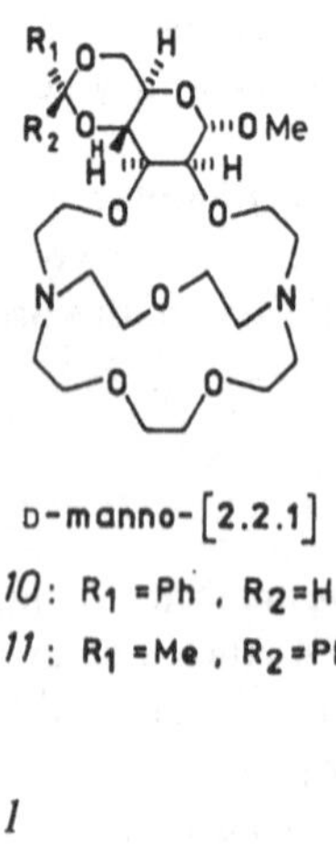

D-manno-[2.2.1]

10: R_1 = Ph, R_2 = H

11: R_1 = Me, R_2 = Ph

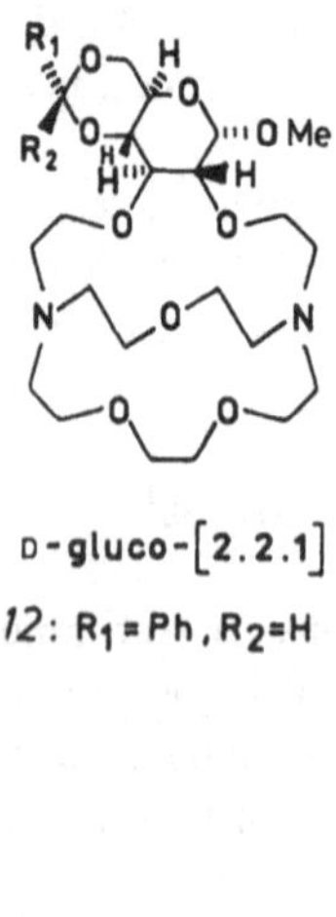

D-gluco-[2.2.1]

12: R_1 = Ph, R_2 = H

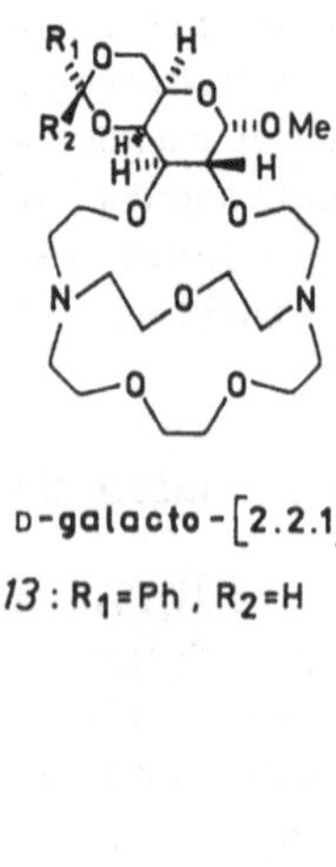

D-galacto-[2.2.1]

13: R_1 = Ph, R_2 = H

Chart 1

The letter B preceding the macrocycle descriptor (see *4*) means that the macrocycle [18]-N_2O_4 is fused with a benzo ring. Similarly, B in [2_B.2.2] refers to a benzo fused cryptand [2.2.2]. In the case of cryptand [2.2.C_6] (*9*), C_6 depicts the number of methylene units between the two nitrogen bridge heads [20]. Concerning the more elaborated cryptands e.g. *10* or *11* which bear carbohydrate units, we will refer in the text to D-manno-[2.2.1]-cryptands etc.

2 Cryptand Syntheses: Survey of Methods

2.1 High-Dilution Technique

The most significant contribution into progress of cryptand chemistry is associated with Lehn's research group in Strasbourg. The first report on the synthesis of a [2.2.2]cryptand appeared in 1969 [21]. This synthesis is presented in Scheme 1.

Scheme 1

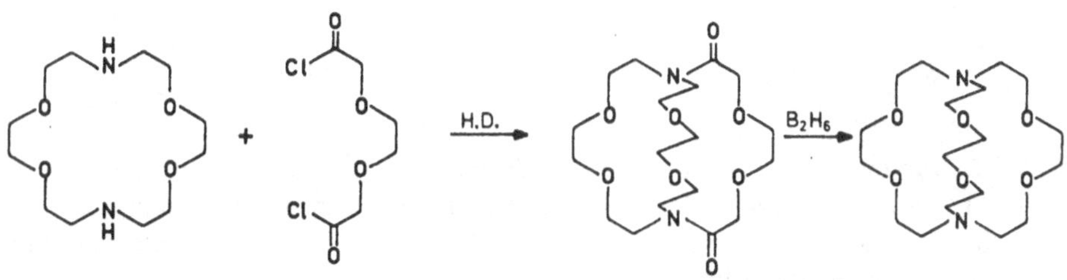

14 *15* *16* *7*

Macrocyclic diamine *14* and triglycolic acid dichloride *15* were condensed under high-dilution conditions to form *16* in 45% yield. Subsequent reduction of *16* using diborane yielded the [2.2.2]cryptand as the bis-borane adduct. Acidic work-up with 6 N hydrochloric acid afforded *7* in 90% yield.

The principle of the high-dilution (h.d.) method consists in the condensation of macrocyclic diamines with highly reactive acid dichlorides or, alternatively, 2,4-dinitrophenyl diesters in an inert solvent like toluene [22]. The yields of the resulting diamides are usually good, although [1.1.1]cryptand was obtained in this manner only in 10% yield, accompanied by a dimer (30% yield) [23].

The high-dilution method (for technical details see for example Ref. 21) has been used not only to form simple cryptands, but also in the syntheses of more complex systems — cylindrical macrotricyclic molecules [24], lateral macrobicycles [25], spherical macrotetracyclic systems [26], and speleands [27]. Nevertheless, the important drawback of the method is the last step, e.g. reduction with diborane. That means only compounds which do not interfere with this reagent can be formed in this manner. Also, even a small contamination of the solvents and reactants with water has a significant influence on the yield.

2.2 Template Synthesis

Template effects play an important role in most of the cyclizations leading to "all-oxygen" macrocycles. Usually, alkali or alkaline earth metal cations serve as a template ion. Transition metal cations are useful in the syntheses of macrocyclic polyamines. The template effect is less pronounced in the case when formation of mixed nitrogen/oxygen macrocycles are formed in comparison with their "all-oxygen" analogues.

In 1980, Kulstad and Malmsten [28] reported an alkali metal ion promoted synthesis of [2.2.2]cryptand from 1-iodo-8-chloro-3,6-dioxaoctane (*17*) and 1,8-diamino-3,6-dioxaoctane (*18*) (Scheme 2).

Scheme 2

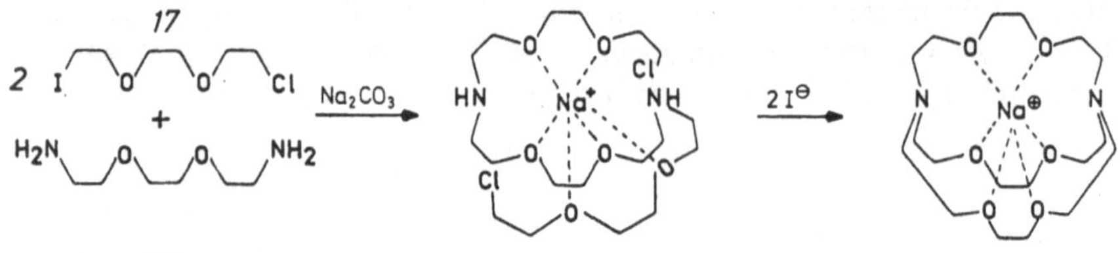

18 *19* *20*

[2.2.2]Cryptand was obtained as the corresponding sodium cryptate *20* in 27% yield. The cation-free cryptand was isolated by passing an acidic solution of the complex through cation- and anion-exchangers. However, the overall yield was not reported. When sodium carbonate was replaced by potassium carbonate, no detectable amount of cryptate $\{K^+ \subset [2.2.2]\}$ was observed; this confirms the involvent of a template effect, although this method is rather limited to simple mononucleating cryptands.

A template effect due to NH intramolecular hydrogen bonding in the synthesis of [1.1.1]cryptand has been reported by Montanari et al. [29]. (Scheme 3).

Scheme 3

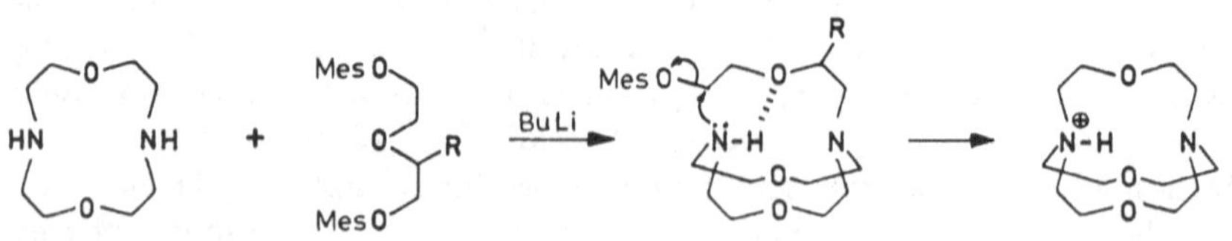

21 *22* *23* *24*

Diaza[12]coronand-4 (*21*) was condensed with diethylene glycol bismesylate *22* in the presence of butyllithium. Precipitation, occuring during the reaction course, afforded the proton cryptate *24* $\{H^+ \subset [1.1.1]\}$ in 40% yield. It should be noted that [1.1.1] was obtained only in 10% yield via the high-dilution method [23]. Lithium promoted cyclization was excluded (as an alternative mechanism) by an additional experiment in which KH served as a base instead of BuLi. Identical yield was achieved, indicating that intramolecular hydrogen bonding was responsible of the cyclization.

In conclusion, a specific template effect plays only a limited role in the syntheses of cryptands. In part, inconveniences are due to a unique strong complexation of metal ions by cryptands, thus difficult removal of them from the cavity affects significantly the yield. On the other hand, the scope of the method is limited to mononucleating elipsoidal cryptands, only.

2.3 Quaternization-Demethylation Procedure

An ingenous method for the preparation of the lateral macrobicycles has been developed by Newkome and co-workers [30]. The method consists on bis-quaternization of the macrocycles *25* with bis(2-iodoethoxy)ethane (*26*) in boiling acetonitrile to afford the bis-quaternary salts *27* in ca. 40% yield. Those were not isolated, but directly demethylated with L-selectride® to form the desired macrobicycles *28* in a 40% overall yield (Scheme 4).

Scheme 4

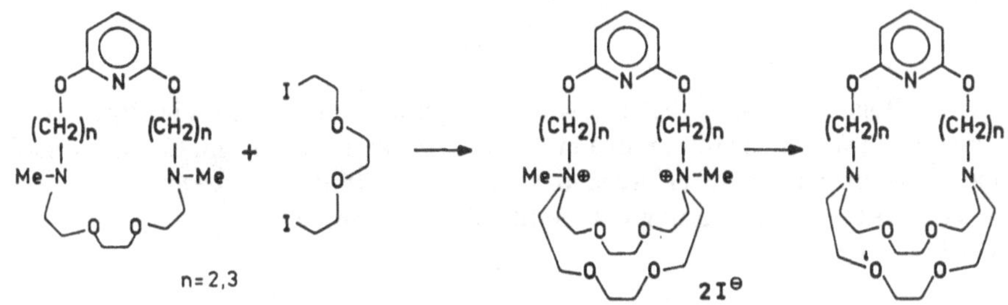

25 *26* *27* *28*

Probably, L-selectride® might have been replaced by milder reagents, for instance nucleophiles like PPh_3, thiourea, etc. [31], therefore extending the range of diverse structures containing easily reducible functional groups.

2.4 Miscellaneous

There are methods in which neither high-dilution conditions nor template effects have been involved. They will be reported below.

Nucleophilic substitution in 2,6-dihalopyridines by selected alkoxides can provide a cryptand skeleton. For instance, condensation of 2,6-dichloropyridine and triethanolamine in the presence of sodium hydride afforded the cryptand *29* in 3% yield [32].

Fo. *29*

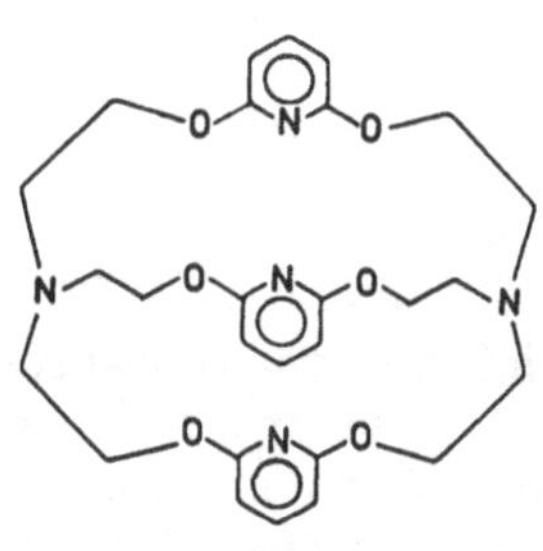

29

Condensation of diaza[18]coronand-6 (*30*) with triethylene glycol bismesylate (*31*) led to the formation of [2.2.2]cryptands *32*. Macrobicycle *32a* (25% yield) was isolated as $\{K^+ \subset [2.2.2]\}I^-$ and *32b* as the free cryptand in 26% yield [33] (Scheme 5).

Scheme 5

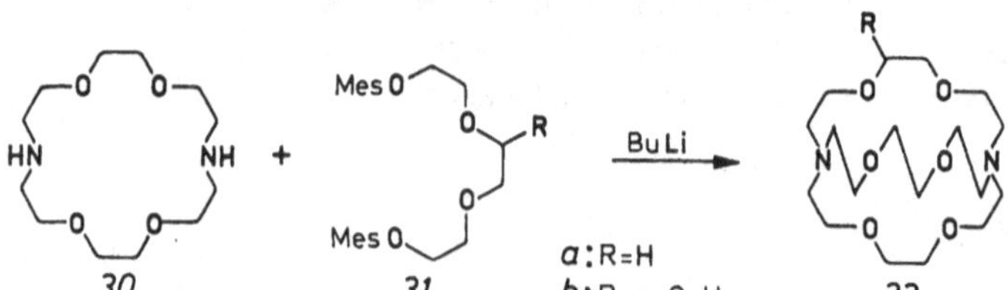

Recently, Bogatsky et al. [34] have reported the synthesis of cryptands incorporating urea moieties in which one of the urea nitrogens serve as a pivot atom. Cyclization of dianion *34* with diamine *18* afforded the macrocycle *35* in 21% yield, which was converted into the corresponding oxo-cryptand *36* (Scheme 6).

Scheme 6

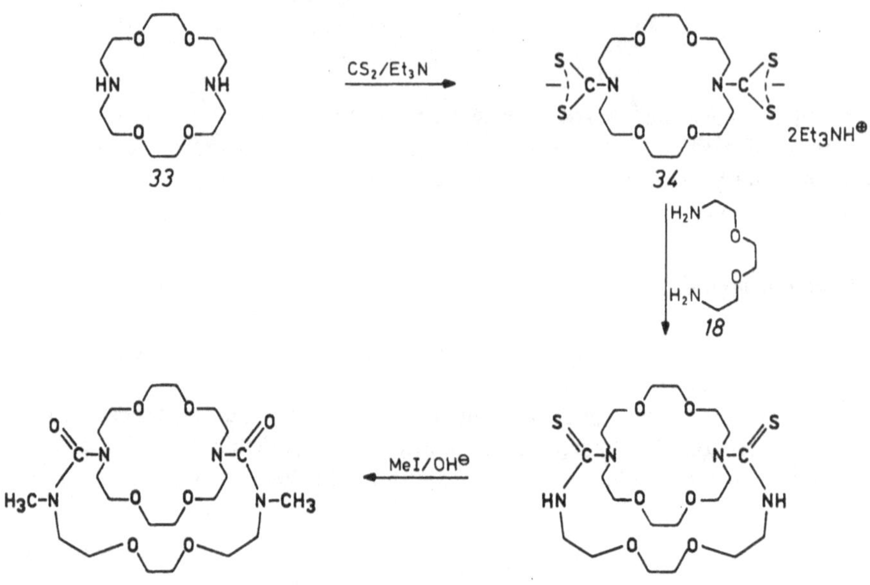

2.5 Concluding Remarks

All the methods presented in this Chapter have their merits and drawbacks. The most important factor in the synthesis of cryptands is the yield. As it was shown, the high-dilution technique is superior to all remaining methods in this respect (except for the [1.1.1] cryptand case). Furthermore, the high-dilution method has been applied successfully in the synthesis of more complicated macropolycyclic structures. Thus, it remains the general method for the synthesis of a variety of cryptands.

3 High-Pressure Approach for the Synthesis of Cryptands

3.1 Influence of Pressure on the Menshutkin Reaction. Mechanical Outfit

To begin with, a short background introduction into high-pressure chemistry is given making this Chapter easily understandable.

Pressure times volume (P · V) has the dimension of energy. Thus, the application of pressure constitutes as nonthermal means of doing work on a system. All reactions are characterized by the volume of activation ($\Delta V^{\neq}$) which is defined as the difference in volume between the reactants and the transition state; this quantity is expressed in $cm^3 \cdot mol^{-1}$. The volume of reaction (ΔV_{rxn}) is given by the difference in volume between products and reactants. Pressure will influence reaction rates according to the sign and magnitude of $\Delta V^{\neq}$. As a consequence, three types of reactions in a liquid phase can be distinguished according to the changes in $\Delta V^{\neq}$:
1) Reactions with $\Delta V^{\neq} > 0$; they are inhibited with increasing pressure.
2) Reactions where $\Delta V^{\neq} \approx 0$; their pressure effect is weak or negligible.
3) Reactions with $\Delta V^{\neq} < 0$; they are accelerated with increasing pressure.
Pressure also will influence reaction equilibria according to the sign and magnitude of ΔV_{rxn}; e.g. if $\Delta V_{rxn} < 0$, an increase of pressure will shift the equilibrium toward the products [35].

The formation of charged products from neutral substrates usually results in increased solvation. Thus, when the charge develops along the reaction coordinate, $\Delta V^{\neq}$ is often observed to be highly negative. For example, the formation of a quaternary ammonium salt from a tertiary amine and an alkylating agent (Menshutkin reaction) will have a $\Delta V^{\neq}$ in the -20 to -50 cm^3/mol range. Such reactions should be dramatically accelerated by the application of a few kbar of pressure. This can be illustrated by the quaternization of substituted pyridines with various alkyl iodides [36] (Scheme 7). Activation volumes for some Menshutkin reactions are compared in Table 1.

Scheme 7

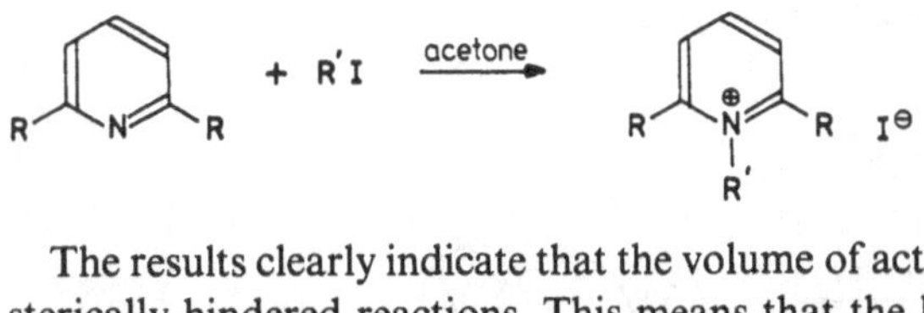

The results clearly indicate that the volume of activation is more negative for higher sterically hindered reactions. This means that the high-pressure approach should be especially effective in cases where highly hindered amines are quaternized. Indeed, the high-pressure reaction of methyl iodide with α-isolupanine (*37*) which is characterized by the "cisoid" arrangement of the N-atoms, leads in 99% yield to the respective methiodide *38* (Scheme 8). It is noteworthy to mention that the quaterniza-

Table 1. Activation volumes for some Menshutkin reactions at 25 °C in acetone

Iodide R′	Amine R	$\Delta V^{\neq}$ ($cm^3\ mol^{-1}$)
CH_3	H	−21.9
	CH_3	−24.4
	C_2H_5	−27.3
	$CH(CH_3)_2$	−30.2
C_2H_5	H	−23.3
	CH_3	−23.9
	C_2H_5	−28.9
	$CH(CH_3)_2$	−35.0
$CH(CH_3)_2$	H	−26.5
	CH_3	−28.2

tion reaction of compound *37* when carried out at room temperature and under atmospheric pressure over a period of 168 h, leads to the methiodide *38* only in a yield of ≈5%, accompanied by a few unidentified side products [37].

Scheme 8

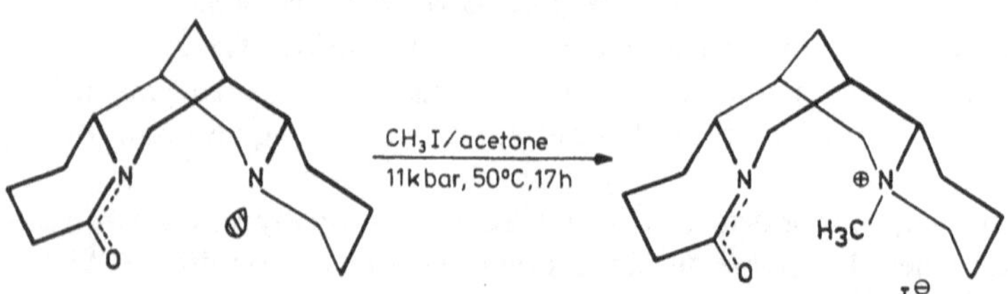

37 38

The results presented above inspired further investigations toward a double-quaternization of N,N′-dimethyl diazacoronands which might lead to the formation of the corresponding cryptand frameworks. This idea was supported by earlier successful findings concerning the double-quaternization of the alkaloid sparteine (*39*) [38]. Molecular models showed that the methylene group fit perfectly between the two nitrogen atoms of the sparteine molecule. Indeed, the double-quaternization of *39* with methylene iodide under high pressure furnished the corresponding bis-quaternary salt *40* in quantitative yield [38] (Scheme 9).

Scheme 9

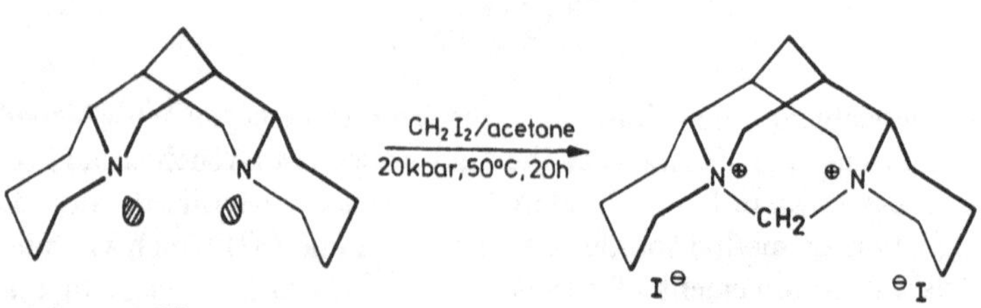

39 40

All high-pressure reactions were performed in a piston-cylinder apparatus, whose initial working volume is 10 ml. Details of this apparatus [39] are shown in Fig. 1 and explained below.

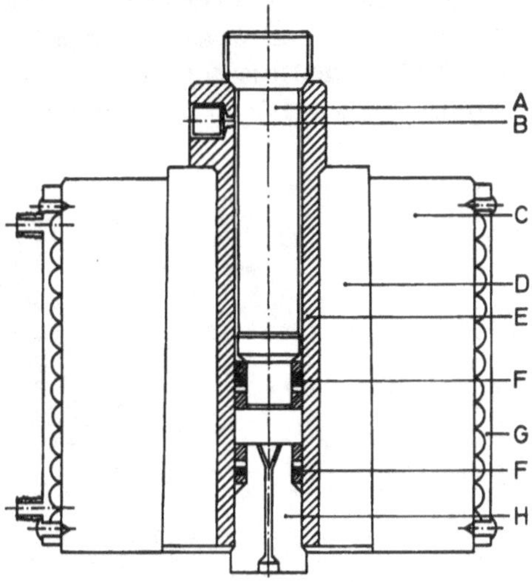

Fig. 1. High-pressure apparatus

The high-pressure vessel consists of two external steel rings (C, D) in which an internal brass vessel (E) is placed. The internal vessel is pressed into the supporting ring by a hydraulic piston in a well controlled way depending on the internal pressure. The cylindrical high-pressure volume is closed from below with a steel stopper (H). All electrical connections (manganine manometer, thermocouple, etc.) are led through a conical electrode placed in the stopper, as indicated in Fig. 1. The internal vessel is closed from above by a mobile piston (A), supported by another independent hydraulic force. Sealing of the mobile piston and the stopper are attained using resin O-rings and brass sealing rings (F). For reactions performed at temperatures higher than 20 °C, an external heating jacket (G) supplied by flowing water serves for thermostatic conditions. It could be maintained within ± 1 °C. The reaction can be done under any gaseous atmosphere. If so, the gas is introduced into the working volume through the capillary inlet (B), being initially compressed to 1–1.5 kbar in a separate multiplier. The pressure inside the working volume is measured by a calibrated manganine coil with accuracy of ± 0.1 kbar. A typical high-pressure procedure is as follows: the high-pressure apparatus, closed on the bottom with a stopper, is filled with the reaction mixture and a mobile piston is inserted. Then the whole assembly is placed between the pistons of a hydraulic press and the pressure is elevated up to the desired value (for example 10 kbar). The reaction mixture is kept under these conditions for an appropriate time. The pressure is released and the reaction mixture is worked up.

3.2 Double-Quaternization as a Method for the Formation of Cryptand Frameworks

Although, the first successful high-pressure double-quaternizations were performed on chiral diazacoronands, for the clarity sake, we give preferential treatment to the experimental results passing from simple cryptands to more complicated structures.

B[18]-N_2O_4 (*4*) turned out to be a very useful compound for testing the double-quaternization reactions, since it possesses four aromatic protons which are well separated in the ^{1}H NMR. Thus, the structures of the products can be easily established by means of ^{1}H NMR spectroscopy. Representative example of this double-quaternization method is illustrated in Scheme 10. For preparative performance, e.g. a 1:1 mixture of *4* and bis(2-iodoethyl)ether (*41*) (ca. 0.08 M solution in acetone) was exposed to the pressure of 10 kbar [40] for 20 h. The bisquaternary salt *42* precipitated during the reaction course quantitatively.

Scheme 10

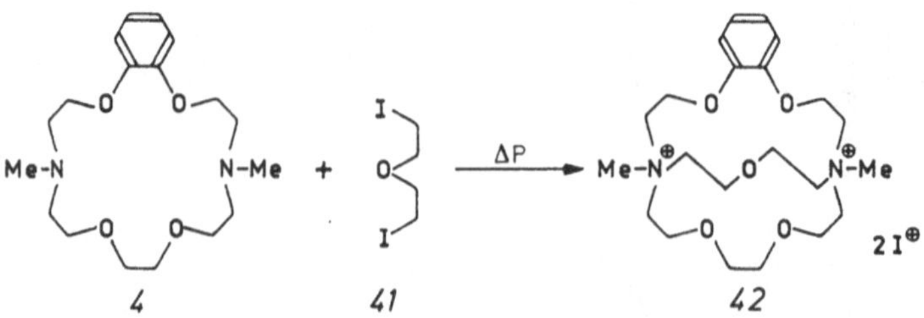

An additional experiment was done using two equivalents of bis(2-iodoethyl) ether under exactly the same conditions. The product ^{1}H NMR spectrum was found identical with those in the former case. The conclusion is, that under those conditions the transition state leading to the cryptand skeleton is highly favoured over the formation of linear polymers or the 1:2 adduct.

3.3 Triphenylphosphine-Assisted Demethylation of Quaternary Salts

Tse-Lok Ho [31] has explored the effectivness of the following four nucleophiles in dequaternization reactions: sodium azide, sodium thiosulphate, thiourea and triphenylphosphine. Among which, triphenylphosphine turned out to be the most effective demethylating agent. A strong solvent dependence on dequaternization has been demonstrated, indicating dimethylformamide as a superior solvent in all cases. In this manner, (2_B.2.1]cryptand (*44*) was obtained from bisquaternary salt *43* in 87% yield (Scheme 11).

Scheme 11

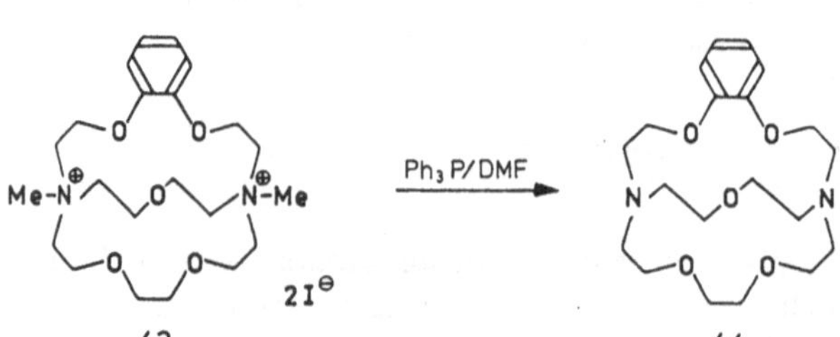

3.4 Reactions of Dimethyl Diazacoronands with Diiodoethers as Bridging Components

In order to demonstrate the utility of the high-pressure method, four N,N'-dimethyl diazacoronands were selected as the model compounds: [12]-N_2O_2 (*1*), [15]-N_2O_3 (*2*), [18]-N_2O_4 (*3*), and B[18]-N_2O_4 (*4*). Bis(2-iodoethyl) ether (*41*) and 1,2-bis(2-iodoethoxy)ethane (*26*) serve as bridging components. Although, the chosen combinations of bridging components and diazacoronands in principle provide eight possibilities for reaction, some of them led to identical compounds; for instance, [12]-N_2O_2 + *26* → [2.1.1] or [15]-N_2O_3 + *41* → [2.1.1]. A representative example showing the formation of [2.2.2]cryptand (*7*) is depicted in Scheme 12.

Scheme 12

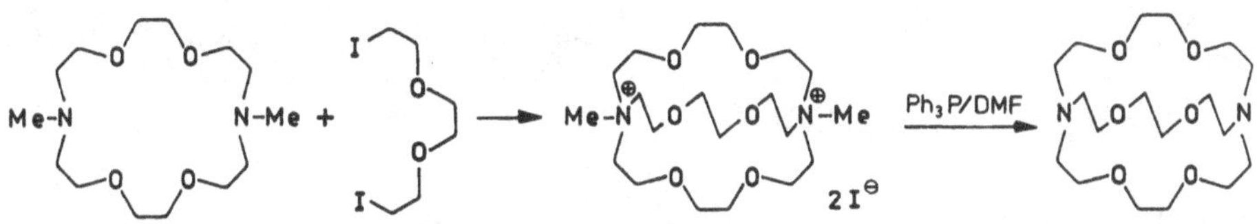

3 26 45 7

All quaternization reactions were performed under the same conditions, e.g., in ca. 0.08 M solution in acetone, at ambient temperature, for 20 h, and a pressure of 10 kbar was maintained during the reaction course. Without exception colourless crystalline solids precipitate in the reaction cell, and after a simple work-up (washing with hexane and drying in vaccuo) the isolated bisquaternary salts were analytically pure. Demethylation was carried out by heating the respective salt with triphenylphosphine in boiling DMF, followed by column chromatography on alumina. Yields obtained for the quaternization and demethylation reactions are listed in Table 2.

Table 2. Synthesis of simple cryptands

Diazacoronand	Bridging Component	Resulting Cryptand	Yield of Quaternization (%)	Yield of Demethylation (%)
[12]-N_2O_2	*41*	[1.1.1]	95	59
[12]-N_2O_2	*26*	[2.1.1]	81	51
[15]-N_2O_3	*41*	[2.1.1]	91	72
[15]-N_2O_3	*26*	[2.2.1]	82	62
[18]-N_2O_4	*41*	[2.2.1]	91	80
[18]-N_2O_4	*26*	[2.2.2]	94	65
B[18]-N_2O_4	*41*	[2_B.2.1]	100	87
B[18]-N_2O_4	*26*	[2_B.2.2]	100	71

3.5 Reactions Between Dimethyl Diazacoronands and α,ω-Diiodoalkanes

Very successful results in the synthesis of simple cryptands stimulated more systematic studies in this field. First of all, it was interesting to test the influence of the length of the bridging component on the yield obtained for the double-quaternization reaction [41].

The same diazacoronands were selected as above, and several α,ω-diiodoalkanes (from C_3 to C_{10}, except of C_7 and C_9) were used as suppliers of the bridging components. A typical example is provided in Scheme 13.

Scheme 13

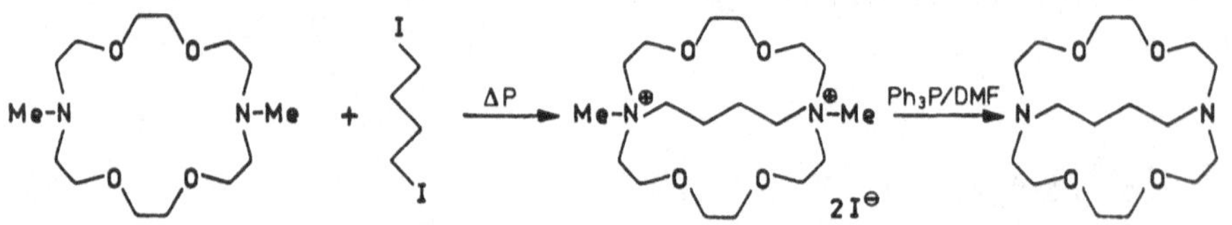

3 *46* *47* *48*

[18]-N_2O_4 (*3*) was reacted with 1,4-diiodobutane (*46*) under high pressure to give the bisquaternary salt *47* as a precipitate which was demethylated in the usual manner to afford [2.2.C_4]cryptand (*48*). All reactions were carried out under the same conditions, e.g. ca. 0.08 M solution in acetone, at ambient temperature, for 20 h, and under 10 kbar pressure. Twentytwo different bisquaternary salts have been obtained in this way. The yields are presented in Table 3.

Table 3. Yields of bis-quaternary salts obtained from N,N′-dimethyl diazacoronands and α,ω-diiodoalkanes

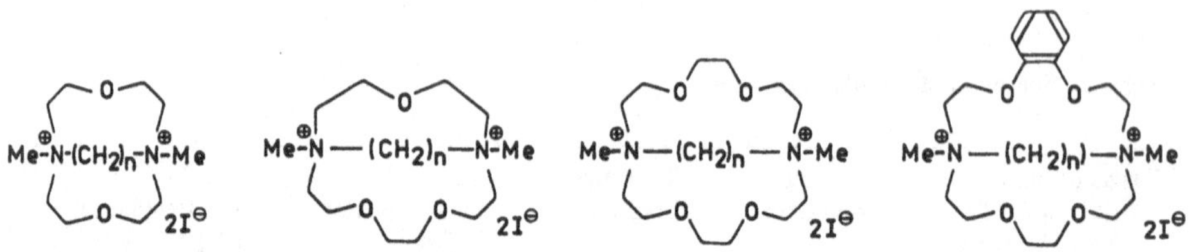

n	Yield (%)	n	Yield (%)	n	Yield (%)	n	Yield (%)
3	91	3	90	3	100	3	87
4	86	4	96	4	97	4	89
5	72	5	89	5	76	5	91
6	70	6	93	6	63	6	92
8	59	8	58	8	58	8	54
				10	56	10	55

As the result of this study we see: the yield of the double-quaternization product decreases with increasing length of the α,ω-diiodoalkane.

3.6 Synthesis of Chiral Cryptands Incorporating Carbohydrate Units

All chiral crown ethers incorporating one carbohydrate subunit possess two diastereotopic faces of the macrocyclic ring. They are able to form diastereoisomeric complexes with primary alkylammonium cations. Since nonbonding interactions are responsible of the chiral recognition of optically active species, it would be desirable to form monofacial ligands in which the inclusion of a chiral molecule (or chiral ion) could proceed from the sterically hindered side only. This special molecular architecture may be followed by the fusion of the cryptand framework and the chiral unit.

Initially, it was assumed that the diazacoronand with secondary nitrogen functions incorporating D-mannopyranosidic unit (*49*) could serve as a precursor for the synthesis of the chiral [2.2.1]cryptand (*10*) according high-dilution method [43]. Indeed, high-dilution reaction of *49* with diglycolic acid dichloride provided the bisamide *50* in 65% yield (Scheme 14).

Scheme 14

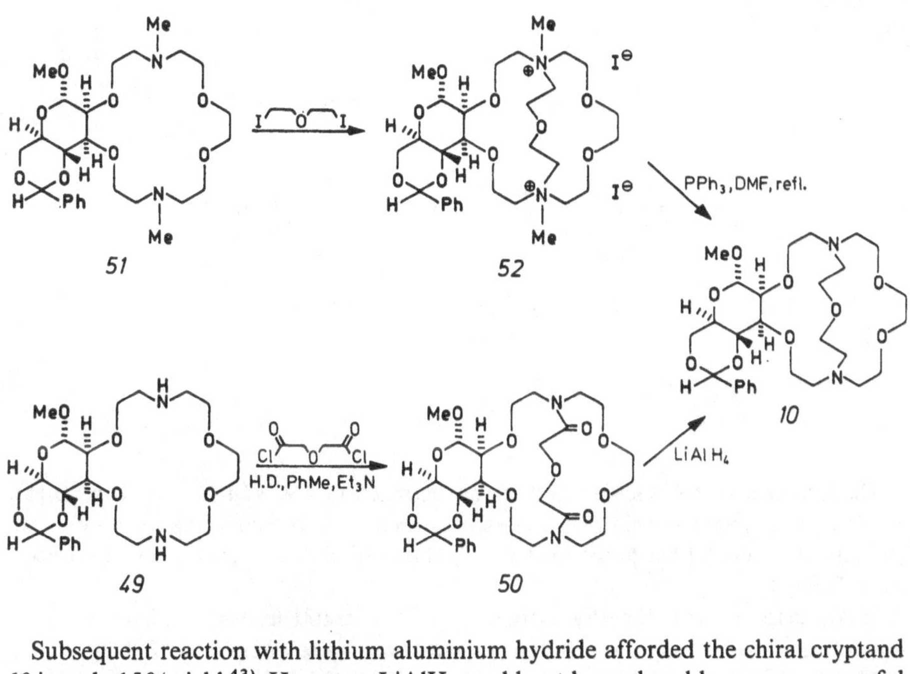

Subsequent reaction with lithium aluminium hydride afforded the chiral cryptand *10* in only 15% yield [43]. However, $LiAlH_4$ could not be replaced by a more powerful agent, like diborane which will cleave the acetal group.

In the light of this drawback with regard to the reduction step, double-quaternization reactions under high-pressure became interest. In this respect three chiral N,N'-dimethyl diazacoronands incorporating D-manno-, D-gluco-, and D-galactopyranosidic subunits [44,45] served as precursors in the double-quaternization reactions [42,

[43]. Three of them possessed 4,6-O-benzylidene protection of a sugar moiety and the fourth, of D-manno configuration, was blocked by the 4,6-O-[(S)-phenylethylidene) protection group [46].

1:1 Mixtures of the four mentioned chiral diazacoronands with bis(2-iodoethyl) ether in ca. 0.08 M solution in acetone were exposed to the pressure of 8 kbar for 20 h at ambient temperature. In all cases the quaternary salts were obtained in quantitative yield as colourless precipitates. Demethylation was followed in the usual manner to give the chiral cryptands *10*, *11*, *12*, and *13* in very good yields: 87, 72, 82, and 79, respectively.

3.7 The Role of Solvent and Leaving Group in the High-Pressure Double-Quaternization Method

In order to gain more knowledge about the nature of double-quaternization reactions, several additional experiments were thought [47]. It was assumed that the solvent polarity is an important factor which will influence the rate of a double-quaternization reaction. In the second instance, the question was asked, whether bis (2-halogenoethyl) ethers, compared with bis(2-tosyloxyethyl) ether as a bridging component, will differ in the reactivity (cf. Scheme 15).

Scheme 15

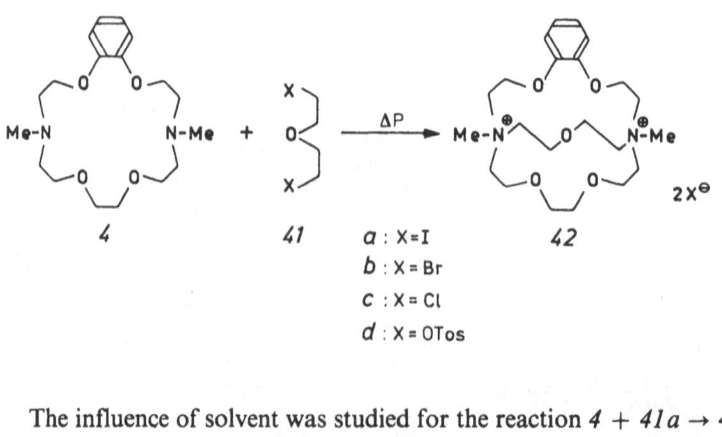

4 *41* *42*

a : X = I
b : X = Br
c : X = Cl
d : X = OTos

The influence of solvent was studied for the reaction *4* + *41a* → *42a* (Scheme 15). Six different aprotic solvents were chosen as a reaction medium. All reactions were performed under 10 kbar pressure at 30 °C. The yield of bisquaternary salt are collected in Table 4.

In all cases, except dimethylformamide and acetonitrile, white microcrystalline salts precipitated during the reaction. All reactions were completed after 20 h. The yields differed substantially after 4 h, indicating that the high-pressure double quaternization was highly accelerated in polar solvents like dimethylformamide or acetonitrile. From practical point of view, it would be desirable to perform a double-quaternization reaction in dimethylformamide within a short time and to demethylate the resulting compound directly, since dimethylformamide serves as good reaction medium for the following demethylation reaction.

Table 4. The influence of solvent on double quaternization for *4* + *41a*; 10 kbar, 30 °C

Solvent	Reaction time (h)	
	20	4
	Yield (%)	
DMF	100	100
MeCN	100	95
Me_2CO	100	65
AcOEt	100	53
Et_2O	100	30
PhMe	100	44

The influence of the leaving group (X) was demonstrated in a set of experiments. Compound B[18]-N_2O_4 (*4*) was reacted with four different bridging components *41a–d* in acetone and acetonitrile under standard conditions: 10 kbar, 30 °C, 20 h. The yield of the appropriate bisquaternary salts are listed in Table 5.

Table 5. The influence of leaving group on double quaternization; 10 kbar, 30 °C, 20 h

Reaction	Solvent	
	Me_2CO	MeCN
	Yield (%)	
4 + *41a*	100	100
4 + *41b*	85	100
4 + *41c*	8	52
4 + *41d*	70	100

It is clear that the results are in agreement with the well known order of reactivity giving by I > Br > Tos ≫ Cl. Evidently, the best bridging component is bis(2-iodoethyl) ether (*41a*). Not only it will give the best yield for *42*, but also it was found that bisquaternary diiodide (*42a*) is easier to demethylate than the corresponding dibromide (*42b*), dichloride (*42c*) or bistosylate (*42d*). Furthermore, the bisquaternary diiodides are non-hygroscopic solids, which allow all analytical treatments without problems.

The final conclusion is drawn as follows: cryptands can be prepared in very good yield by the high-pressure double-quaternization method using highly polar solvents like dimethylformamide. The corresponding bisquaternary salts as intermediates need not to be isolated before demethylation to give the cryptands.

4 Complexing Behaviour of the Chiral Cryptands

4.1 Complexing Properties of Chiral Cryptands Incorporating the Binaphthyl Unit

A short introduction will help to understand why chiral differentiation of enantiomeric ions by chiral cryptands remained little explored, whereas a great deal of work has been done in this respect with "all-oxygen" crown ethers incorporating as the chiral sensor the binaphthyl unit, carbohydrates or tartaric acid building blocks [8]. It turns out that the complexation of primary alkylammonium cations by mean of nitrogen containing ligands is interfered with the transfer of a proton from the ammonium cation to a nitrogen atom of the ligand, especially in protic solvents. So far, very little is known about the association constants of aza-macrocycles with primary ammonium cations. On the other hand, the availability of chiral cryptands is more problematic than of chiral crown ethers.

Lehn and coworkers have performed a chiral recognition study using the macrotricyclic ligand (*54*) which incorporates a binaphthyl unit [48] (Scheme 16). Two modes of complexation may occur in solution for ligand *54*:

1) binding of primary ammonium cations via the formation of hydrogen bonds between the NH_3^+ group and one of the lateral binding sites;
2) cascade complexation where a metal cation is first bound to one of the macrocyclic rings, and this new created positively charged site now binds a molecular anion.

Scheme 16

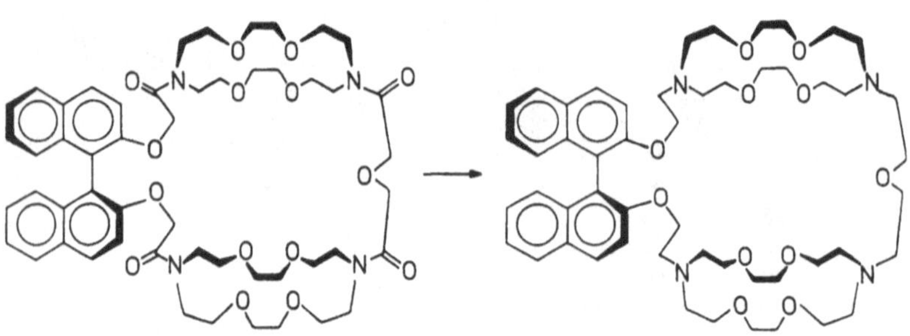

53 54

In order to avoid possible transprotonation, the tetra-amide *53* was used instead of the tetra-amine *54* for the complexation studies with primary ammonium cations, although the binding strength of a macrocyclic amide is lower compared with the corresponding macrocyclic amine. Diastereoisomeric complexes formed between a chiral ligand and optically active substrates differ in their association constants, reflecting therefore the chiral discrimination. The magnitude of the chiral discrimination can be evaluated conveniently (as an enantiomeric excess) by extraction methods. Partitioning experiments between water and $CDCl_3$ containing the tetra-amide *53* were carried out for racemic phenyl- and naphthylethylammonium hydrochloride as well as for racemic phenylalanine methyl ester hydrochloride. The stoichiometry and enantiomeric excess of the formed complexes were determined by ^1H NMR spectroscopy. It was found that ligand *53* forms complexes with 1:1 stoichiometry. Only for phenylethylammonium cation the chiral discrimination was distinct. An enantiomeric excess of ca. 15% was evaluated for the (—)-S-isomer.

Chiral discrimination studies with the tetra-amine *54* were also performed with respect to racemic molecular anions via the formation of cascade complexes. Regulation of the chiral discrimination may be achieved to some extent by the nature of the cation initially complexed. Efficient ion pairing with molecular anions is favoured in solvents of low polarity. In this respect, aqueous solutions of alkali metal salts of (±)-mandelic acid or (±)-α-hydroxy-1-naphthaleneacetic acid were extracted into a $CDCl_3$ phase where the ligand *54* is dissolved. In case of the mandelate anion, the anion: ligand ratios in the $CDCl_3$ layer were as follows: Na^+ (0.6:1), K^+ (1:1), Rb^+ (1.2:1), Cs^+ (0.6:1), NH_4^+ (1.2:1). With K^+, Rb^+ and Cs^+ mandelate the signal in ^1H NMR spectrum for the benzylic protons splits into two components arising from two diastereoisomeric complexes, whereas for Na^+ mandelate no splitting was observed. Evaluation of the enantiomeric excess by integration of the benzylic signals indicated that K^+ (—)-mandelate is extracted preferentially (e.e. 15%); in contrast, Cs^+ (+)-mandelate is extracted preferentially (e.e. ≈10%). No chiral discrimination was observed for Rb^+ and NH_4^+ mandelate. With α-hydroxy-1-naphthaleneacetic acid chiral discrimination did not occur, although splittings of the ^1H NMR signals for the methine protons in the diastereoisomeric complexes were present.

Though precise geometries of the ion pair inclusion complexes are unknown, it is clear that the cation size plays an important role in modifying the central cavity of the ligand *54* which allows the regulation of the chiral discrimination. Larger anions can not be included in the central cavity which is evident as a lack of chiral discrimination, however, less structured diastereoisomeric ion pair complexes were observed.

4.2 Complexing Properties of Chiral Cryptands Incorporating Carbohydrate Units

Two chiral [2.2.1]cryptands, one incorporating the methyl 4,6-O-benzylidene-α-D-mannopyranoside unit (*10*), the other the methyl 4,6-O-[(S)-phenylethylidene]-α-D-mannopyranoside unit (*11*), may exhibit the same two different modes of complexation as before: binding of primary alkylammonium cations via hydrogen bonds between the NH_3^+ group and the heteroatoms of the ligand or cascade complexation of ion pairs. Chart 2 shows the constitutions of the two cryptands *10* and *11* and is also indicative that the approach of guest molecules may occur from the more sterically hindered side.

So far, it was found that these two cryptands are able to dissolve a variety of primary alkylammonium tetraphenylborates in methylene chloride which confirmed that the inclusion complexes are formed instantly [49]. Furthermore, cryptands *10* and *11* dissolve potassium thiocyanate [49], hence, cascade binding may also be envisaged. These preliminary results encouraged more systematic complexation studies which are still in progress.

In conclusion, the design, the synthesis, and the complexation properties of chiral cryptands are still open areas for the creativity of chemists.

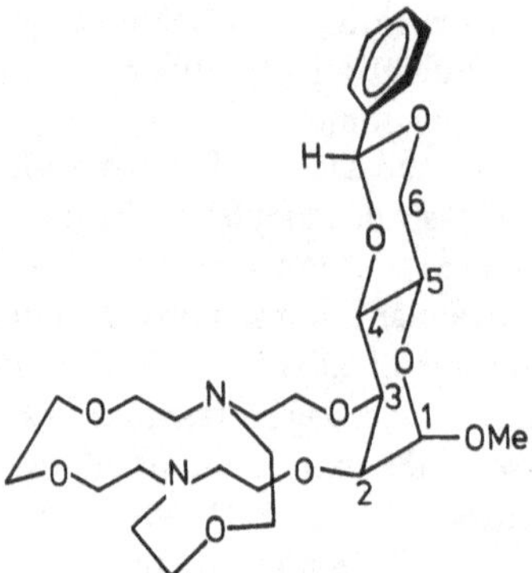

10

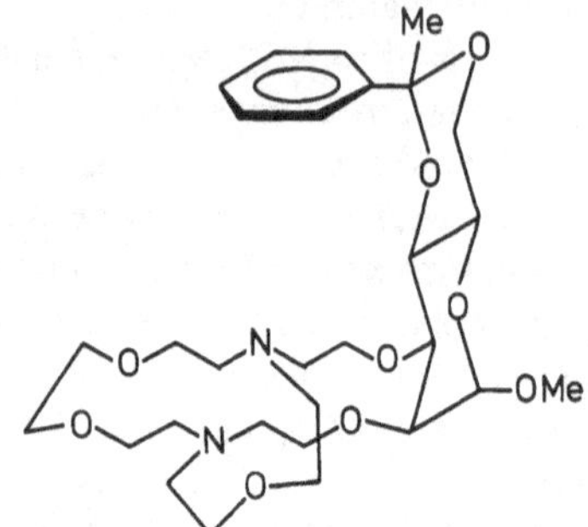

11

Chart 2

5 Summary

Undoubtly, cryptands remain the class of macrobicyclic and macropolycyclic receptors looking forward to a great future. The number of possible structures is practically unlimited and depends on a researcher imagination and creativity. Initially, the chemistry of cryptands had focused on the development of new synthetic strategies, either to improve the yields or to built up more complicated molecular architectures, like cylindrical, spherical or lateral macropolycycles, bi- or polynucleating, with hard and soft binding sites, etc.

Simultaneously, extensive investigations concerning the physico-chemical properties of cryptands have been conducted covering several sub-topics: their complexing properties, X-ray analysis, molecular dynamics of inclusion complexes, etc.

One important feature of cryptands is their enzyme analogy in certain instances, which could be very helpful to understand the natural enzyme activities. The next step in the chemistry of cryptands therefore will probably be the design, synthesis, and the study of the properties of more elaborated polyfunctionalyzed macropolycyclic structures allowing a deeper insight into the principles of enzyme reactivity. Actually in the synthetic respect, high-dilution methods enjoy a great popularity because of its generality. Nevertheless, as far as the introduction of sensitive groups into macropolycyclic architectures is concerned, the high-pressure method which offers milder reaction conditions than high-dilution techniques is among the most favoured alternatives, although the high-pressure method has one serious drawback, i.e. the availability of the apparative set-up. It is hoped that this difficulty will be overcome in a near future.

6 Acknowledgment

This work was supported by Grant 03.10 from the Polish Academy of Sciences.

7 References

1. Pedersen, C. J.: J. Am. Chem. Soc. *89*, 7017 (1967)
2. Gokel, G. W., Korzeniowski, S. H.: Macrocyclic Polyether Syntheses, Springer, Berlin, Heidelberg, New York, 1982
3. Hiraoka, M.: Crown Compounds. Their Characteristics and Applications. Studies in Organic Chemistry, Vol. 12, Kodansha, Tokyo, 1982
4. Melson, G. A.: Coordination Chemistry of Macrocyclic Compounds, Plenum Press, New York, 1979
5. Izatt, R. M., Christensen, J. J.: Synthetic Multidentate Macrocyclic Compounds, Academic Press, New York, San Francisco, London,1978;
Izatt, R. M., Christensen, J. J.: Progress in Macrocyclic Chemistry, Vol. 1 and 2, J. Wiley and Sons, New York, 1979
6. Vögtle, F., Weber, E.: Host Guest Complex Chemistry, Macrocycles, Springer, Berlin, Heidelberg, 1985
7. Bradshaw, J. S., Stott, P. E.: Tetrahedron *36*, 461 (1979);
Bradshaw, J. S., Baxter, L.: J. Heterocyclic Chem. *18*, 233 (1981);
Bradshaw, J. S., Jolley, S. T., Izatt, R. M.: ibid. *19*, 3 (1982)
8. Stoddart, J. F.: Chem. Soc. Rev. *8*, 85 (1979)
9. Lehn, J.-M.: Accts Chem. Res. *11*, 49 (1978)
10. Lehn, J.-M.: Pure Appl. Chem. *50*, 871 (1978)
Lehn, J.-M.: ibid. *52*, 2303 (1980);
Lehn, J.-M.: ibid. *52*, 2441 (1980)
11. Vögtle, F., Weber, E.: Angew. Chem. *91*, 813 (1979); Angew. Chem., Int. Ed. Engl. *18*, 753 (1979)
12. Vögtle, F., Weber, E., in: The chemistry of functional groups, Suppl. E, The chemistry of ethers, crown ethers, hydroxyl groups and their sulphur analogues, part 1 (ed.) Patai, S., p. 59, Wiley, London 1981
13. Gokel, G. W., Dishong, D. M., Schultz, R. A., Gatto, V. J.: Synthesis 997 (1982)
14. Host-Guest Complex Chemistry I–III: Topics in Current Chemistry (ed.) Vögtle, F., Weber, E., Springer, Berlin, Heidelberg, New York, 1981, 1982, and 1984
15. Cram, D. J.: Science *219*, 1177 (1983)
16. Tabushi, I.: Tetrahedron *40*, 269 (1984)
17. Weber, E.: Kontakte (Merck) *1984(1)*, 26
18. Atwood, J. L., Davies, J. E. D.: Journal of Inclusion Phenomena *1* (1983), Reidel, Dordrecht, Holland
19. Weber, E., Vögtle, F.: Inorg. Chim. Acta *45*, L65 (1980, Weber, E., Josef, H. P.: J. Ind. Phenom. *1*, 79 (1983)
20. Dietrich, B., Lehn, J.-M., Sauvage, J.-P.: Tetrahedron *29*, 1647 (1973)
21. Dietrich, B., Lehn, J.-M., Sauvage, J.-P.: Tetrahedron Lett. 2885 (1969)
22. Alberts, A. H., Annunziata, R., Lehn, J.-M.: J. Am. Chem. Soc. *99*, 8502 (1977)
23. Cheney, J., Lehn, J.-M., Sauvage, J.-P., Stubbs, M. E.: J. Chem. Soc., Chem. Commun. 1100 (1972)
24. Lehn, J.-M., Simon, J.: Helv. Chim. Acta *60*, 141 (1977)
25. Lehn, J.-M., Wu Cheng-tai, Plumere, P.: Acta Chim. Sinica 67 (1983)
26. Graf, E., Lehn, J.-M.: Helv. Chim. Acta *64*, 1040 (1981)
27. Canceill, J., Collet, A., Gabart, J., Kotzyba-Hibert, F., Lehn, J.-M.: Helv. Chim. Acta *65*, 1894 (1982)
28. Kulstad, S., Malmsten, L. A.: Tetrahedron Lett. 643 (1980)
29. Annunziata, R., Montanari, F., Quici, S., Vitali, M. T.: J. Chem. Soc., Chem. Commun. 777 (1981)
30. Newkome, G. R., Majestic, V. K., Fronczek, F. R.: Tetrahedron Lett. 3039 (1981)
31. Tse-Lok Ho: Synth. Commun. *3*, 99 (1973)
32. Newkome, G. R., Majestic, V. K., Fronczek, F. R., Atwood, J. L.: J. Am. Chem. Soc. *101*, 1047 (1979)
33. Landini, D., Maia, A., Montanari, F., Tundo, P.: J. Am. Chem. Soc. *101*, 2526 (1979)
34. Lukyanenko, N. G., Bogatsky, A. V., Kiruchenko, T. I., Scherbakov, S. V., Nazarova, N. Yu.: Synthesis 137 (1984)

35. Asano, T., leNoble, W. J.: Chem. Rev. *78*, 407 (1978)
36. Jenner, G.: Angew. Chem. Int. Ed. Engl. *14*, 137 (1975)
37. Jurczak, J., Tkacz, M., Majchrzak-Kuczynska, U.: Synthesis 920 (1983)
38. Jurczak, J.: in preparation
39. Jurczak, J.: Bull. Chem. Soc. Jpn *52*, 3438 (1979)
40. Pietraszkiewicz, M., Sałański, P., Jurczak, J.: submitted for publication
41. Pietraszkiewicz, M., Sałański, P., Jurczak, J.: submitted for publication
42. Pietraszkiewicz, M., Sałański, P., Jurczak, J.: J. Chem. Soc., Chem. Commun. 1184 (1983)
43. Pietraszkiewicz, M., Sałański, P., Jurczak, J.: Tetrahedron *40*, 2971 (1984)
44. Pietraszkiewicz, M., Jurczak, J.: J. Chem. Soc., Chem. Commun. 132 (1983)
45. Pietraszkiewicz, M., Jurczak, J.: Tetrahedron *40*, 2967 (1984)
46. Pietraszkiewicz, M., Jurczak, J.: submitted for publication
47. Pietraszkiewicz, M., Sałański, P., Ostaszewski, R., Jurczak, J.: submitted for publication
48. Lehn, J.-M., Simon, J., Modrapour, A.: Helv. Chim. Acta *61*, 2407 (1978)
49. Pietraszkiewicz, M.: unpublished results

Author Index Volumes 101–130

Contents of Vols. 50–100 see Vol. 100
Author and Subject Index Vols. 26–50 see Vol. 50

The volume numbers are printed in italics

ISBN 3-05-500195-8